Biology of Mammalian Germ Cell Mutagenesis

A Banbury Center Meeting

Banbury Report Series

Banbury Report 1: Assessing Chemical Mutagens
Banbury Report 2: Mammalian Cell Mutagenesis
Banbury Report 3: A Safe Cigarette?
Banbury Report 4: Cancer Incidence In Defined Populations
Banbury Report 5: Ethylene Dichloride: A Potential Health Risk?
Banbury Report 6: Product Labeling and Health Risks
Banbury Report 7: Gastrointestinal Cancer: Endogenous Factors
Banbury Report 8: Hormones and Breast Cancer
Banbury Report 9: Quantification of Occupational Cancer
Banbury Report 10: Patenting of Life Forms
Banbury Report 11: Environmental Factors in Human Growth and Development
Banbury Report 12: Nitrosamines and Human Cancer
Banbury Report 13: Indicators of Genotoxic Exposure
Banbury Report 14: Recombinant DNA Applications to Human Disease
Banbury Report 15: Biological Aspects of Alzheimer's Disease
Banbury Report 16: Genetic Variability in Responses to Chemical Exposure
Banbury Report 17: Coffee and Health
Banbury Report 18: Biological Mechanisms of Dioxin Action
Banbury Report 19: Risk Quantitation and Regulatory Policy
Banbury Report 20: Genetic Manipulation of the Early Mammalian Embryo
Banbury Report 21: Viral Etiology of Cervical Cancer
Banbury Report 22: Genetically Altered Viruses and the Environment
Banbury Report 23: Mechanisms in Tobacco Carcinogenesis
Banbury Report 24: Antibiotic Resistance Genes: Ecology, Transfer, and Expression
Banbury Report 25: Nongenotoxic Mechanisms in Carcinogenesis
Banbury Report 26: Developmental Toxicology: Mechanisms and Risk
Banbury Report 27: Molecular Neuropathology of Aging
Banbury Report 28: Mammalian Cell Mutagenesis
Banbury Report 29: Therapeutic Peptides and Proteins: Assessing the New Technologies
Banbury Report 30: Eukaryotic Transposable Elements as Mutagenic Agents
Banbury Report 31: Carcinogen Risk Assessment: New Directions in the Qualitative and Quantitative Aspects
Banbury Report 32: DNA Technology and Forensic Science
Banbury Report 33: Genetics and Biology of Alcoholism
Banbury Report 34: Biology of Mammalian Germ Cell Mutagenesis

Biology of Mammalian Germ Cell Mutagenesis

Edited by

JAMES W. ALLEN
Genetic Toxicology Division
U.S. Environmental Protection Agency

BRYN A. BRIDGES
MRC Cell Mutation Unit
University of Sussex

MARY F. LYON
MRC Radiobiology Unit

MONTROSE J. MOSES
Duke University Medical Center

LIANE B. RUSSELL
Biology Division
Oak Ridge National Laboratory

34
Banbury Report

COLD SPRING HARBOR LABORATORY PRESS
1990

Banbury Report 34: Biology of Mammalian Germ Cell Mutagenesis

Copyright 1990 by Cold Spring Harbor Laboratory Press
All rights reserved
Printed in the United States of America
Cover and book design by Emily Harste

Library of Congress Cataloging-in-Publication Data

Biology of mammalian germ-cell mutagenesis / edited by James W. Allen
... [et al.].
    p. cm.—(Banbury report; 34)
    Based on a conference held at Banbury Center, Cold Spring Harbor, N.Y., in Nov. 1989.
    Includes bibliographical references.
    ISBN 0-87969-234-0
    1. Mutation (Biology)—Congresses. 2. Germ cells—Congresses. 3. Mutagenicity testing—Congresses.   I. Allen, James W., 1943- .   II. Series.
    [DNLM: 1. Chromosome Aberrations—congresses. 2. Germ Cells—congresses.
3. Mutagenicity Tests—congresses. 4. Mutation—congresses. 5. Risk Factors—congresses.
W3 BA19 v. 34 / QH 460 B615 1989]
QH460.B56 1990   90-1570
599'.01592—dc20   CIP

The articles published in this book have not been peer-reviewed. They express their authors' views, which are not necessarily endorsed by the Banbury Center or Cold Spring Harbor Laboratory.

Authorization to photocopy items for internal or personal use, or the internal or personal use of specific clients, is granted by Cold Spring Harbor Laboratory Press for libraries and other users registered with the Copyright Clearance Center (CCC) Transactional Reporting Service, provided that the base fee of $1.00 per article is paid directly to CCC, 27 Congress St., Salem, MA 01970 [0-87969-234-0/90.$1.00 + .00]. This consent does not extend to other kinds of copying, such as copying for general distribution, for advertising or promotional purposes, for creating new collective works, or for resale.

All Cold Spring Harbor Laboratory Press publications may be ordered directly from Cold Spring Harbor Laboratory Press, Box 100, Cold Spring Harbor, New York 11724 (Phone: 1-800-843-4388) in New York State (516) 367-8325. FAX: (516) 367-8432.

Corporate Sponsors

Alafi Capital Company
American Cyanamid Company
Amersham International plc
AMGen Inc.
Applied Biosystems, Inc.
Becton Dickinson and Company
Beecham Pharmaceuticals
Boehringer Mannheim Corporation
Ciba-Geigy Corporation/Ciba-Geigy Limited
Diagnostic Products Corporation
E.I. du Pont de Nemours & Company
Eastman Kodak Company
Genentech, Inc.
Genetics Institute
Glaxo Research Laboratories
Hoffmann-La Roche Inc.
Johnson & Johnson
Life Technologies, Inc.
Eli Lilly and Company
Millipore Corporation
Monsanto Company
Oncogene Science, Inc.
Pall Corporation
Perkin-Elmer Cetus Instruments
Pfizer Inc.
Pharmacia Inc.
Schering-Plough Corporation
Tambrands Inc.
The Upjohn Company
The Wellcome Research Laboratories, Burroughs Wellcome Co.
Wyeth-Ayerst Research Laboratories

Special Program Support

Alfred P. Sloan Foundation

The meeting at the Banbury Center on which this book is based was initiated, developed, and financially supported jointly by the

UNITED STATES ENVIRONMENTAL PROTECTION AGENCY

and the

INTERNATIONAL COMMISSION FOR PROTECTION AGAINST ENVIRONMENTAL MUTAGENS AND CARCINOGENS

Banbury 34 constitutes ICPEMC Meeting Report 5.

This volume is dedicated to the memory of

Samuel A. Latt (1938–1988)

and

Sewall Wright (1889–1988)

Many colleagues from diverse areas of genetic research deeply miss their friendship
and leadership.
Their lives and works remain an inspiration.

Participants

Ilse-Dore Adler, GSF-Institut für Säugetiergenetik, Neuherberg, Federal Republic of Germany

James W. Allen, Genetic Toxicology Division, U.S. Environmental Protection Agency, Research Triangle Park, North Carolina

Norman Arnheim, Department of Molecular Biology, University of Southern California, Los Angeles

Jack B. Bishop, National Institute of Environmental Health Sciences, Research Triangle Park, North Carolina

Brigitte F. Brandriff, Biomedical Sciences Division, Lawrence Livermore National Laboratory, University of California, Livermore

Bryn A. Bridges, MRC Cell Mutation Unit, University of Sussex, Falmer, Brighton, United Kingdom

Bruce M. Cattanach, MRC Radiobiology Unit, Chilton, Didcot, Oxon, United Kingdom

Michael C. Cimino, U.S. Environmental Protection Agency, Washington, D.C.

Muriel T. Davisson, The Jackson Laboratory, Bar Harbor, Maine

Kerry L. Dearfield, U.S. Environmental Protection Agency, Washington, D.C.

Vicki L. Dellarco, Office of Health and Environmental Assessment, U.S. Environmental Protection Agency, Washington, D.C.

Dirk G. de Rooij, Department of Cell Biology, University of Utrecht Medical School, The Netherlands

Michael E. Dresser, Oklahoma Medical Research Foundation, Oklahoma City

Jack Favor, GSF-Institut für Säugetiergenetik, Neuherberg, Federal Republic of Germany

Irving B. Fritz, Banting and Best Department of Medical Research, University of Toronto, Ontario, Canada

Walderico M. Generoso, Biology Division, Oak Ridge National Laboratory, Tennessee

Mary Ann Handel, Department of Zoology, University of Tennessee, Knoxville

Norman B. Hecht, Department of Biology, Tufts University, Medford, Massachusetts

Bruce W. Kovacs, Women's Hospital, University of Southern California School of Medicine, Los Angeles

Susan E. Lewis, Center for Life Sciences and Toxicology, Research Triangle Institute, Research Triangle Park, North Carolina

Mary F. Lyon, MRC Radiobiology Unit, Chilton, Didcot, Oxon, United Kingdom

J. David McDonald, McArdle Laboratory for Cancer Research, University of Wisconsin, Madison

Harvey W. Mohrenweiser, Biomedical Sciences Division, Lawrence Livermore National Laboratory, Livermore, California

Montrose J. Moses, Department of Cell Biology, Duke University Medical Center, Durham, North Carolina

Josephine Peters, MRC Radiobiology Unit, Chilton, Didcot, Oxon, United Kingdom

Liane B. Russell, Biology Division, Oak Ridge National Laboratory, Tennessee

Lonnie D. Russell, Department of Physiology, Southern Illinois University, Carbondale

William L. Russell, Biology Division, Oak Ridge National Laboratory, Tennessee

Gary A. Sega, Biology Division, Oak Ridge National Laboratory, Tennessee

Paul B. Selby, Biology Division, Oak Ridge National Laboratory, Tennessee

Oliver Smithies, Pathology Department, University of North Carolina, Chapel Hill

Davor Solter, The Wistar Institute of Anatomy and Biology, Philadelphia, Pennsylvania

Lynn M. Wiley, Department of Obstetrics and Gynecology, University of California School of Medicine, Davis

Richard P. Woychik, Biology Division, Oak Ridge National Laboratory, Tennessee

Andrew J. Wyrobek, Lawrence Livermore National Laboratory, Biomedical Sciences Division, Livermore, California

Top: J. Allen; R. Woychik
Bottom: J. Favor; M. Davisson, L. Russell

Top: Conference Participants
Middle: N. Arnheim, O. Smithies; M. Lyon
Bottom: D. Solter, N. Hecht; S. Lewis, V. Dellarco

Preface

Progress in understanding the role of mutation in heritable disease has led to a revitalization of interest in germ-cell mutagenesis and risk assessment. Geneticists have long been concerned that exposures to mutagenic agents in the environment may be causing genetic damage transmissible to offspring, and a considerable body of experimental evidence exists to document the quantity and quality of heritable mutations induced by physical and chemical agents. Despite the technical and interpretive problems associated with the estimation of genetic risk to humans, much vital research in this area has been accomplished, and significant new advances are emerging. Views of unique barriers and targets impeding or mediating the induction of mutation in germ cells are maturing, and new molecular and cytological methods are converging with traditional experimentation to facilitate the detection of mutations and the assessment of their nature, transmission, and expression. These advances have come at a time when various U.S. Government agencies, European commissions, and international scientific organizations are giving new, or renewed, attention to the importance of heritable mutation as a public health concern. It seems timely to take account of current thinking and directions in the field. To this end, a conference aimed at communicating new research on the biology of mammalian germ-cell mutagenesis was held at the Banbury Center on November 12–15, 1989. Proceedings of the conference are summarized in this volume.

The meeting brought together researchers working in areas of reproductive biology, molecular and cellular mechanisms of mutagenesis, mutation expression, and risk assessment. The agenda was not intended to cover all relevant areas of investigation; rather, it was aimed at promoting exchange of information and discussion among experts from different fields who share a common interest in mechanistic aspects of germ-cell mutagenesis. Because most experimental work has involved exposure/analysis of rodent spermatogenic cells, and in some cases human sperm, the emphasis of the presentations was on male systems. Subjects of scientific discussion ranged from testicular physiology to the analysis of altered DNA sequences and addressed the modulation, characterization, and consequences of induced damage. Diverse gamete maturational stages and chromosome-, gene-, and DNA-level endpoints of analysis were considered. This breadth served to highlight many different germ-line properties affecting the induction and recovery of mutations, as well as variables affecting their nature and rate. Significant nonmutational genetic effects were described, and a series of talks provided an update on studies utilizing powerful new molecular methodologies for germ-cell mutation analysis. These sessions dealt with phenomena such as genomic imprinting; nonconventional pathways/expression of damage in the embryo; and new technological tools (e.g., PCR amplification, transgenic techniques) for investigating mutational pathways or endpoints. Discussions in these different areas set the stage for a final session on genetic risk estimation, which focused on the central problems and issues currently facing those responsible for assessing risks associated with germ-cell mutations.

This meeting would not have been possible without the support and cooperation of the U.S. Environmental Protection Agency; the International Commission for Protection Against Environmental Mutagens and Carcinogens; and the Banbury Center, Cold Spring Harbor Laboratory. I am most grateful to Dr. Jan Witkowski, Director of the Banbury Center, and to staff member Beatrice Toliver, for their careful administration of this conference. I was, indeed, fortunate to have Drs. Liane Russell, Montrose Moses, Mary Lyon, and Bryn Bridges as co-organizers of the meeting. The strength of the scientific program was achieved only through the concerted efforts of this committee. The support and helpful suggestions of many participants were also important to the success of the meeting. In this regard, I would like to especially thank Drs. Vicki Dellarco, Susan Lewis, William Russell, Jack Bishop, and Harvey Mohrenweiser. Many thanks are due to Nancy Ford, Managing Director of Publications; Patricia Barker, technical editor; Inez Sialiano, editorial assistant; and others in the Publications Department for their dedication to the rapid, careful production of this volume; and to Sharon Hammer for excellent transcription of the discussions. Katya Davey was a most gracious and attentive hostess at the Robertson House. Finally, the immense contributions of all the speakers should be acknowledged. It was their expertise that formed the basis for the meeting, and their enthusiastic interactions that made it a success.

William Russell, a pioneer in the field of mammalian germ-cell mutagenesis, has stated (this volume) "...germ-cell mutations can act from conception to old age to cause major disorders in every bodily structure and function...relative inattention to genetic risk [as compared to cancer risk] may be out of balance." It seems that many now share this view, as a growing number of geneticists are turning their attention to the special problems of germ-cell mutagenesis. This volume offers fresh perspectives of many important questions and challenges in the field, and is presented with the hope that it will prove to be a valuable resource for further work.

Note to the Reader

In many instances, comments and discussion pertained more to the conference presentation than to the written chapter. An editorial decision was made to keep such comments in the text.

James W. Allen

Contents

Participants, ix

Preface, xiii

SECTION 1: GERM-LINE PROPERTIES AFFECTING INDUCTION AND RECOVERY

Barriers to Entry of Substances into Seminiferous Tubules: Compatibility of Morphological and Physiological Evidence / Lonnie D. Russell — 3

Cell-Cell Interactions in the Testis: A Guide for the Perplexed / Irving Bamdas Fritz — 19

Correlation between Proliferative Activity of Mouse Spermatogonial Stem Cells and Their Sensitivity for Cell Killing and Induction of Reciprocal Translocations by Irradiation / Dirk G. de Rooij, Yvonne van der Meer, Ans M.M. van Pelt, and Paul P.W. van Buul — 35

Use of Heritable Mutations in the Analysis of Meiosis / Mary Ann Handel — 51

Regulation of Testicular Postmeiotic Genes / Norman B. Hecht — 67

Molecular Targets, DNA Breakage, and DNA Repair: Their Roles in Mutation Induction in Mammalian Germ Cells / Gary A. Sega — 79

Detecting Specific-locus Mutations in Human Sperm / Andrew J. Wyrobek, Marge Currie, Jackie L. Stilwell, Rod Balhorn, and Larry H. Stanker — 93

SECTION 2: ABERRANT CHROMOSOME STRUCTURE / BEHAVIOR

Clastogenic Effects of Acrylamide in Different Germ-cell Stages of Male Mice / Ilse-Dore Adler — 115

The Synaptonemal Complex as an Indicator of Induced Chromosome Damage / Montrose J. Moses, Patricia Poorman-Allen, James H. Tepperberg, James B. Gibson, Lorraine C. Backer, and James W. Allen ... 133

Synaptonemal Complex Analysis of Mutagen Effects on Meiotic Chromosome Structure and Behavior / James W. Allen, Patricia Poorman-Allen, Lorraine C. Backer, Barbara Westbrook-Collins, and Montrose J. Moses ... 155

Meiosis and Experimentation in the Yeast *Saccharomyces cerevisiae* / Michael E. Dresser ... 171

Human Sperm Cytogenetics and the One-cell Zygote / Brigitte F. Brandriff and Laurie A. Gordon ... 183

Chromosome Aberrations Associated with Induced Mutations: Effect on Mapping New Mutations / Muriel T. Davisson and Susan E. Lewis ... 195

SECTION 3: VARIABLES AFFECTING THE RATE AND NATURE OF MUTATIONS

Factors Affecting Mutation Induction by X-rays in the Spermatogonial Stem Cells of Mice of Strain 101/H / Bruce M. Cattanach, Carol Rasberry, and Colin Beechey ... 209

Toward an Understanding of the Mechanisms of Mutation Induction by Ethylnitrosourea in Germ Cells of the Mouse / Jack Favor ... 221

Mosaic Mutants Induced by Ethylnitrosourea in Late Germ Cells of Mice / Susan E. Lewis, Lois B. Barnett, and Raymond A. Popp ... 237

Analysis of Electrophoretically Detected Mutations Induced in Mouse Germ Cells by Ethylnitrosourea / Josephine Peters, Janet Jones, Simon T. Ball, and John B. Clegg ... 247

Investigating Inborn Errors of Phenylalanine Metabolism by Efficient Mutagenesis of the Mouse Germ Line / J. David McDonald, Alexandra Shedlovsky, and William F. Dove ... 259

Factors Affecting the Nature of Induced Mutations /
Liane B. Russell, William L. Russell, Eugene M. Rinchik, and
Patricia R. Hunsicker   271

SECTION 4: NONMUTATIONAL GENETIC EFFECTS ON EARLY DEVELOPMENT

Effect of Imprinting on Analysis of Genetic Mutations /
Julie DeLoia and Davor Solter   293

What Is the Radiosensitive Target of Mammalian Gametes and Embryos at Low Doses of Radiation? / Lynn M. Wiley   299

Developmental Anomalies: Mutational Consequence of Mouse Zygote Exposure / Walderico M. Generoso, Joe C. Rutledge, and John Aronson   311

Manipulation of the Mouse Genome by Homologous Recombination / Oliver Smithies and Beverly H. Koller   321

SECTION 5: UTILIZATION OF DNA TECHNIQUES IN THE DETECTION OF GERM-LINE MUTATIONS

Molecular Approaches to the Estimation of Germinal Gene Mutation Rates / Harvey W. Mohrenweiser, Brian A. Perry, and Stephen A. Judd   335

Quantitation and Characterization of Human Germinal Mutations at Hypervariable Loci / Bruce W. Kovacs, Bejan Shahbahrami, and David E. Comings   351

Analysis of DNA Sequences in Individual Human Sperm Using PCR / Norman Arnheim   363

Insertional Mutagenesis in Transgenic Mice / Richard P. Woychik, Barbara R. Beatty, William L. McKinney, Debra K. Andreadis, Anne J. Chang, and P. Eugene Barker   377

SECTION 6: GENETIC RISK ESTIMATION

Problems and Possibilities in Genetic Risk Estimation /
William L. Russell   385

Quantification of Germ-cell Risk Associated with the Induction of Heritable Translocations / Vicki L. Dellarco and Lorenz R. Rhomberg   397

Use of Germ-cell Mutagenicity Testing by the U.S. Environmental Protection Agency Office of Toxic Substances /
Michael C. Cimino and Angela E. Auletta   413

Mammalian Heritable Effects Research in the National Toxicology Program / Jack B. Bishop and Michael D. Shelby   425

The Importance of the Direct Method of Genetic Risk Estimation and Ways to Improve It / Paul B. Selby   437

Aspects of Germ Cells Relevant to Mutagenic Risk Evaluations: Some Concluding Remarks / Bryn A. Bridges   451

Author Index   455

Subject Index   457

Germ-line Properties Affecting Induction and Recovery

Barriers to Entry of Substances into Seminiferous Tubules: Compatibility of Morphological and Physiological Evidence

LONNIE D. RUSSELL
Laboratory of Structural Biology
Department of Physiology
Southern Illinois University
School of Medicine
Carbondale, Illinois 62901

OVERVIEW

Substances that damage the genome of male gametes must gain access to germ cells by a complex pathway to exert their effect(s). How substances reach the germ-cell population is often not known. It is sometimes difficult to reconcile morphological and physiological data pertaining to the barriers to entry of substances. In this paper, I review the morphological evidence for the presence of separate compartments within the testis and how the morphological evidence corresponds to physiological data. Areas that have not been previously explored or where our information on the topic is lacking are discussed. The postnatal development of a Sertoli cell barrier and the development of a barrier during maturation of germ cells are described.

Compartments within the Testis and the Seminiferous Tubule

Vascular Compartment

From the perspective of a substance in the vascular system "willing" and able to gain entrance into the seminiferous tubule, there are several morphological barriers to surmount before the majority of germinal elements can be reached (Fig. 1). First, since the capillaries are not fenestrated as they are at numerous other sites in the body, the substance must either pass through the capillary wall (e.g., transcytosis) or between capillary endothelial cells. Tight junctions that would restrict passage of substances from the endothelium are apparently absent between endothelial cells (Fawcett et al. 1970). Furthermore, there are few micropinocytotic vesicles that would act in transcytosis. Although movement through the intercellular clefts is apparently the means by which substances exit the vascular system, there is virtually no morphological evidence to document to what extent this occurs in the testis. The role of the basal lamina outside the endothelial cell in restricting passage of materials is not known.

Banbury Report 34: Biology of Mammalian Germ Cell Mutagenesis
Copyright 1990. Cold Spring Harbor Laboratory Press. 0-87969-234-0/90.$1.00 + .00

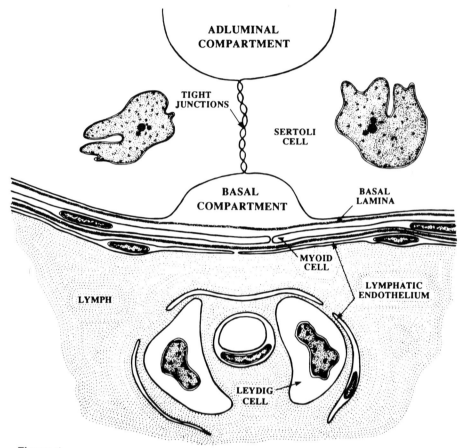

Figure 1
Simplified diagram of the potential barriers to penetration of substances and the compartments of the rat testis. Beginning with the vascular compartment, the various compartments are indicated: capillary, interstitial, lymphatic, lymphatic endothelium, basal lamina, myoid cell, basal lamina, basal compartment of the seminiferous tubule, Sertoli-Sertoli tight junctions, adluminal compartment. (Modified, with permission, from Russell et al. 1990.)

Intravascularly injected electron-opaque substances readily traverse the capillary wall. There have been few physiological investigations of what is and what is not capable of moving out of the vascular system into the interstitial space, but those that have been performed indicate that the capillary presents little restriction to the movement of substances (Setchell 1974). Some high-molecular-weight substances such as luteinizing hormone may traverse the capillary wall in a few minutes or less.

The best terminology to describe any restriction to movement of substances is the vasculature-testis barrier. This cumbersome term must be applied, since the only other appropriate term, blood-testis barrier, has, unfortunately, come to mean something quite different.

Interstitial Compartment

Having exited the vascular system, the substance is found within the interstitial area of the testis. It percolates or is transported among the perivascular Leydig cells and other cells of the interstitium, such as macrophages. The concentrations of substances in the interstitial and lymphatic compartments are assumed to be similar, although there have been no studies directed at answering this question.

Lymphatic Compartment

The relationship of the interstitial cells to the lymphatic system is highly variable (Fawcett et al. 1973). In some species, such as the ram, monkey, and human, the lymphatics are formed by discrete vessels bounded by endothelial cells. In these species, substances need not enter the lymphatic system to gain entrance to the seminiferous tubule. In others species, such as the mouse, rat, and guinea pig, the lymphatics are incompletely circumscribed by endothelial cells and, as such, are best termed peritubular lymphatics. A sinusoidal space is present between the intertubular cells and the seminiferous tubules. In these species, the intertubular cells show a discontinuous endothelial cell lining in which there appears to be open communication between the spaces between Leydig cells and the lymphatic system. Thus, substances readily pass, or are carried, into the lymphatic system on their way to the tubule. To what degree the lymphatic system drainage lowers the concentration of a substance before the substance has an opportunity to enter the tubule is not known.

The Boundary Tissue or Limiting Membrane

The outer boundary of the seminiferous tubule limiting membrane is composed of flattened, polygonal, connective tissue elements (Fig. 2). In those species where the lymphatics are described as being "peritubular," the term given to this outermost connective tissue cell layer is lymphatic endothelium of the seminiferous tubule. Endothelial cells appear to present no barrier to substances destined for the seminiferous tubule, since they often overlap each other and contain no apparent junctions that would exclude passage of materials. Their intercellular spaces are irregular and darken upon introduction of electron-opaque tracers, indicating free flow of electron-dense tracers. Furthermore, micropinocytotic vesicles are frequently seen (Fig. 3) that darken upon administration of electron-dense tracers (Fig. 4).

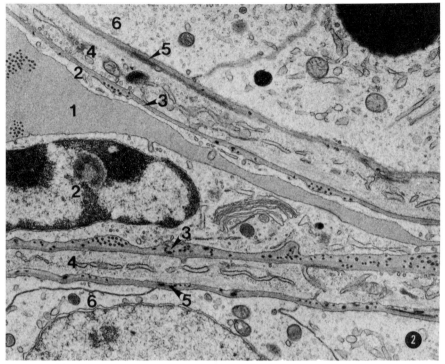

Figure 2
The limiting membranes or boundary tissues of two adjacent tubules and their relationship on one side to the lymphatic space and on the other side to the cells of the base of the seminiferous tubule. Layers are numbered; lymphatic space (1), lymphatic endothelium of the seminiferous tubule (2), basement membrane (3), myoid cell (4), basement membrane (5), cells forming the base of the seminiferous tubule (6). Micropinocytotic vesicles are indicated by arrows.

A basal lamina and a continuous basement membrane are present between the endothelial cell and the more deeply placed myoid cell. Although a basal lamina, in general, may function as a selective filter in some instances and a barrier to the penetration of cellular elements, there is virtually no information that the particular basal lamina situated in the tubular wall will restrict the passage of substances.

The next morphological obstacle in the path of substances headed in the direction of the seminiferous tubule is the myoid cell, a flattened, polygonal cell that lines the tubule. This cell is a bit thicker, but shaped like the peritubular endothelial cell. Most species have only one layer of myoid cells that surround each tubule, but others contain several layers. In the first electron microscopy investigations of the morphological restrictions to entry of molecules into the seminiferous tubules

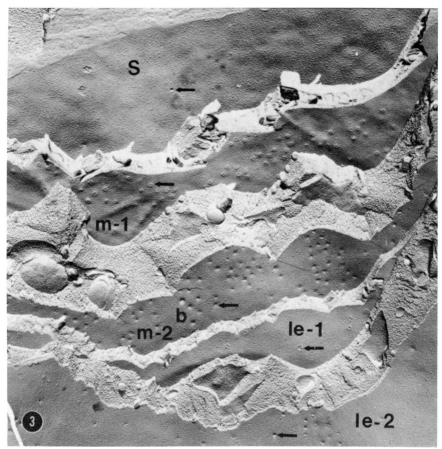

Figure 3
Freeze-fracture micrograph showing the relative abundance of transport vesicles (isolated arrows) on membranes of the lymphatic endothelium (two membrane faces seen; le-1, le-2), myoid cell (two membrane faces seen; m-1, m-2), and Sertoli cell (S).

(Fawcett et al. 1969; Dym and Fawcett 1970), the rat myoid cell showed partial permeability to tracers. Partial permeability meant that some of the higher molecular weight tracers did not pass the myoid cell barrier, whereas other lower molecular weight tracers were seen on the distal side of the myoid cell. It also meant that the lower molecular weight tracers did not pass through the myoid cell at all sites. The myoid cell was nonetheless considered a significant barrier. In a subsequent report where the monkey was studied (Dym 1973), it was found that myoid cells overlapped and presented no restriction to the passage of lanthanum. There seem to

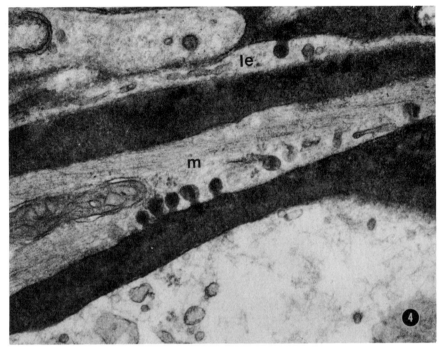

Figure 4
Micropinocytotic vesicles containing lanthanum in a lymphatic endothelial cell (le) and a myoid cell (m). Lanthanum is seen on both sides of the myoid cell.

be tight junctions between myoid cells in some regions (Dym and Fawcett 1970); however, the majority of areas examined where myoid cells join display a wide intercellular space that apparently presents little or no restriction to passage of materials via the intercellular route (Fig. 5). Thus, although there is some evidence for restriction of molecules at the level of the myoid cells (cited above), most investigators currently believe that the myoid cell presents little or no restriction to molecules entering the seminiferous tubule.

What is often not considered in the literature is the possible transfer of substances across, rather than between, the myoid cells. Nonspecific transport of materials via micropinocytosis is apparently common, since tracer substances may be captured in the process of moving through the cell (Fig. 4). There have been no reports of receptor-mediated transport of substances across the myoid cell.

A basement membrane is present between the myoid cells and the cells at the basal aspect of the seminiferous tubule. There is no information about its role in regulating the passage of substances into the tubule.

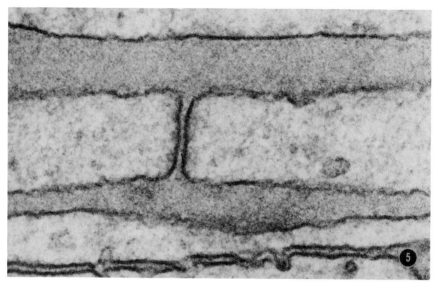

Figure 5
Junction of two myoid cells. The intercellular space measures about 20 nm throughout. No tight junctions are seen.

Basal Compartment of the Seminiferous Tubule

The substance, having gained entrance to the base of the seminiferous tubule, now has direct access to Sertoli cells and relatively immature germ cells. Approximately 4.84% of the total surface area of the Sertoli cell (rat) is directly exposed to the substance (Weber et al. 1983). All spermatogonia and young spermatocytes are exposed to substances (Russell 1978). These germ cells are said to reside in the basal compartment of the seminiferous tubule (Dym and Fawcett 1970), although in no way does the term compartment necessarily imply that these cells are sequestered from any of the aforementioned compartments. The basal aspects of the Sertoli cell and spermatogonia and young spermatocytes are generally considered, correctly or incorrectly, to have ready access to substances in the interstitial spaces.

The Body of the Sertoli Cell

If substances are to proceed farther into the tubule, they may either pass between adjacent Sertoli cells or pass through the body of Sertoli cells. There are few studies on the movement of molecules by uptake into the Sertoli cell from its basal surface. The basal aspects of Sertoli cells have long been known to possess bristle-coated pits as one of their surface specializations (Brökelmann 1963). These specializations

(Fig. 6) are infrequently seen (Fig. 3) and presumably function in the receptor-mediated uptake of substances in the tubule. An autoradiographic study by Morales and Clermont (1986) suggests that transferrin enters the tubule by receptor-mediated endocytosis. In addition, very small and irregularly shaped, uncoated vesicles are also present, but neither variety of vesicle is commonly seen. It is possible that these vesicles are related to nonspecific entry of substances, but virtually nothing is known about this method of uptake of substances into the Sertoli cell. Although little is known about the passage of substance through the body of the Sertoli cell, the cell should be regarded as a potential barrier to the passage of molecules into the tubule.

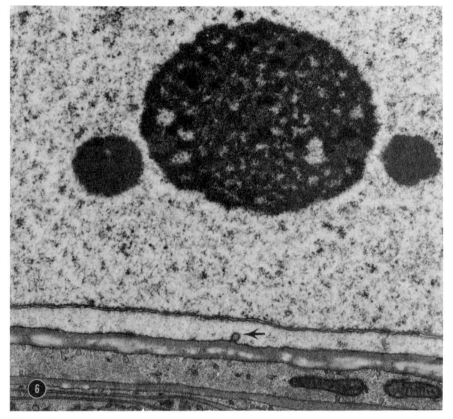

Figure 6
A single bristle-coated pit (arrow) is seen on the basal aspect of a Sertoli cell.

Sertoli-Sertoli Tight Junctions Forming the Sertoli Cell Barrier

Movement of substances by the paracellular route is greatly restricted by the occluding junctions that are present at the basolateral aspect of the Sertoli cell (Fig. 7; Dym and Fawcett 1970). Sertoli cells joined in this manner ring the entire tubule

Figure 7
The junctional region of interdigitating Sertoli cells is displayed. Freeze-fracture micrograph showing the rows of membrane particles forming the tight junctions of the Sertoli cell barrier. Over 50 tight junctional rows are seen.

(Ross 1970), ensuring that this paracellular barrier is not leaky at any one site.

The paracellular barrier has been referred to as the "blood-testis" barrier. Unfortunately, this term is a catchy one and has been used frequently over a long period. It is not a barrier between the blood and the seminiferous tubule. In fact, the term often leads to confusion since inappropriate parallels are often made between it and the blood-brain barrier. For example, the blood-brain barrier is largely at the level of vessel tight junctions. The barrier in the testis, however, is a barrier between one compartment of the seminiferous tubule and another. Since the Sertoli cell is solely responsible for its formation, it is most aptly termed the Sertoli cell barrier (Russell and Peterson 1985). In practice, the term Sertoli cell barrier should be used to refer not only to the restrictions imposed along the paracellular route, but also to those imposed by the cell itself.

Electron-dense tracers have been used to demonstrate the paracellular barrier formed by Sertoli cells (Dym and Fawcett 1970; Fawcett et al. 1970). Tracers pass between adjacent Sertoli cells and between Sertoli cells and basal compartment cells until they encounter Sertoli-Sertoli occluding junctions where the intercellular tracer is no longer seen. These junctions are conspicuous using electron microscopy, since they are flanked by prominent bundles of actin filaments and by cisternae of endoplasmic reticulum. Freeze-fracture micrography demonstrates from 30 to over 100 rows of occluding junctions at this site (Fig. 7) (Gilula et al. 1976; Russell and Peterson 1985), one of the most extensive tight junctional complexes in the body.

No electron-dense tracer has ever been shown to penetrate the Sertoli-Sertoli tight junctional complexes in a normal mammalian testis except in the mink, a species in which the testes go through cyclic periods of activity and inactivity (Pelletier 1988). As judged by the penetration of electron-dense tracers, the germ cells located along the body of the Sertoli cell and along its lateral and apical surfaces are said to reside in the adluminal compartment. In the rat, the adluminal surface area of the Sertoli cell represents about 91% of the total Sertoli cell surface area. Except for spermatogonia and spermatocytes, all other germ cells are assumed to be present in the adluminal compartment. It is assumed, but not proven, that the intercellular spaces around adluminal compartment cells are of the same composition as the fluid in the tubular lumen.

It should be noted that the elongate spermatids that lie in the deep crypts of the Sertoli cell in most mammalian species may have particular secretory products directed toward them that are not present on the adluminal surface of the cell. However, there is no evidence that the environment around these cells is any different from that around other adluminal compartment germ cells.

Physiological experiments have shown that the compositions of fluids in the blood or lymph and in the seminiferous tubular lumen differ considerably. It is not the aim of this paper to review these differences, since they are amply reviewed elsewhere (see Setchell and Brooks 1988). It is often assumed that differences in lymph and seminiferous tubule fluid are also reflective of the concentrations of the same substances in the basal compartment and adluminal compartment, respectively, although this assumption has never been tested.

Physiological Implications of Tracer Studies Used to Demonstrate Compartments of the Testis

Tracers almost by their very nature are nonphysiological substances that pass along a paracellular route. They demonstrate open pathways and indicate the potential for a substance to move from one compartment of the testis to another. Tracers as a group possess a variety of molecular weights, but molecular weight alone is not the determining factor for movement of substances across the Sertoli cell barrier. Setchell and Brooks (1988) have indicated that lipid solubility is more important than molecular size. Substances that are lipid-soluble are more likely to pass through the Sertoli cell. Hydrophilic molecules are more apt to pass via the paracellular route. To be considered also are substances that enter the tubular lumen more readily than would be expected, suggesting that transport and facilitated means of movement are operative.

Entry of substances into tubules by the paracellular route is not necessarily an all-or-nothing event. Labeled insulin, for example, a substance that does not enter cells, will equilibrate in a few hours to about 5% of its concentration in serum (Weber et al. 1988). Thus, tracers do not necessarily predict the extent of entry of substances into seminiferous tubules. The best way to do this is empirically by determining the concentration of the particular substance in lymph and in seminiferous tubule after a period of time in which the substance has been exposed to the seminiferous tubules. Okamura et al. (1975) have shown that selective drugs and chemicals move into the seminiferous tubules. Water-soluble compounds entered seminiferous tubules according to particle size. A diameter of 3.6 Å was considered the maximum size for ready access of water-soluble molecules. Lipid-soluble compounds entered according to their partition coefficient. There was a high correlation between lipid solubility and the ease with which substances entered the tubule (Lee 1981).

The literature contains little information about the concentration of substances within many of the various compartments of the testis, since it is difficult to measure substances in certain locales. Assuming that a particular substance is in the blood coursing through the testis, a graph could be plotted of its concentration in the blood, in the testicular lymph, and in the seminiferous tubule fluid. The techniques for collection of each of these fluids are available. Such a graph would allow one to determine, in a rough way, the barriers to substances at various levels and might allow an approximate prediction of the level of a substance reaching germinal cells at any given blood level.

Development of a Sertoli Cell Barrier during Puberty

Few studies are available that pinpoint Sertoli cell barrier development. In the rat, the barrier develops either by the end of postnatal day 17 (Russell et al. 1989) or day 19 (Vitale et al. 1973). In the rabbit, the barrier develops between the ninth and tenth postnatal week (Sun and Gondos 1986). These studies and others (not cited) show that a barrier to tracers and, in one instance, rapid penetration of fluids (Russell et al. 1989) develop coincident with tight junction formation between Sertoli cells. Caution should be exercised when interpreting these data, since Setchell has shown that the functional barrier gradually matures with time (for review, see Setchell and Brooks 1988) and in the rat may not attain full maturity until after the thirtieth or nearer the fortieth postnatal day.

Barrier Development during the Cyclic Maturation of Germ Cells

Once spermatogenesis is fully established, cells must be translocated from one compartment of the testis to another without a breach in the barrier. Intuitively, one assumes that a significant breach in the barrier would in some way cause leakage of the tubule along its length and be detrimental to the spermatogenic process. It appears that cells move across the barrier approximately at the time that they enter meiotic prophase (rat: leptotene phase of meiosis; Russell 1977, 1978) or shortly after they enter meiosis (monkey: leptotene or zygotene phase of meiosis; Dym and Cavicchia 1977). They do so by entering an intermediate compartment of the testis (Russell 1977) that acts as a transit chamber believed to cause minimal leakage of the barrier during movement of spermatocytes. Christensen et al. (1985) have shown that as cells move into the intermediate compartment, they carry with them albumin, which soon disappears before, or as, cells enter the adluminal compartment. Other than the work of Christensen et al. (1985), there is no evidence that substances gain entry to the adluminal compartment in this manner, or if they do, that they have any appreciable effect on the development of cells.

Disruption of the Sertoli Cell Barrier by Various Substances

When considering the entry of various substances into the seminiferous tubules, it is important to consider also the effect of the substance on the Sertoli cell barrier itself. The substances that have an effect on the barrier have been reviewed (Russell and Peterson 1985), but the methodologies used to show disruption of the barrier are not always sound, bringing into question the findings of many of the earlier studies. Furthermore, morphological techniques may not be in agreement with physiological techniques. Nevertheless, the Sertoli cell barrier is generally regarded as being highly resistant to a variety of substances and treatments. Efferent duct ligation (Neaves 1973), *cis*-platinum (Pogash et al. 1989), and cytochalasin (Weber et al.

1988) have been clearly shown to disrupt the Sertoli cell barrier. In such instances, it is assumed that the general effect of the loss of the barrier is superimposed on the effect of the substance or treatment administered.

Note Added in Proof

O'Brien et al. 1989 (*Endocrinology* **125**: 2973) describe receptor-mediated endocytosis of mannose 6-phosphate receptors in Sertoli cells and germ cells. They describe other recent evidence for Sertoli cell receptor-mediated endocytosis.

REFERENCES

Brökelmann, I. 1963. Fine structure of germ cells and Sertoli cells during the cycle of the seminiferous epithelium in the rat. *Z. Zellforsch.* **59**: 820.

Christensen, A.K., T.E. Komorowski, B. Wilson, S.-F. Ma, and R.W. Stevens III. 1985. The distribution of serum albumin in the rat testis, studied by electron microscope immunocytochemistry on ultrathin frozen sections. *Endocrinology* **116**: 1983.

Dym, M. 1973. The fine structure of the monkey *(Macaca)* Sertoli cell and its role in maintaining the blood-testis barrier. *Anat. Rec.* **175**: 639.

Dym, M. and J.C. Cavicchia. 1977. Further observations on the blood-testis barrier in monkeys. *Biol. Reprod.* **17**: 390.

Dym, M. and D.W. Fawcett. 1970. The blood-testis barrier in the rat and the physiological compartmentation of the seminiferous epithelium. *Biol. Reprod.* **3**: 308.

Fawcett, D.W., P.M. Heidger, Jr., and L.V. Leak. 1969. Lymph vascular system of the interstitial system of the testis as revealed by electron microscopy. *J. Reprod. Fertil.* **19**: 109.

Fawcett, D.W., L.V. Leak, and P.M. Heidger, Jr. 1970. Electron microscopic observations on the structural components of the blood-testis barrier. *J. Reprod. Fertil.* (suppl. 10): 105.

Fawcett, D.W., W.B. Neaves, and M.N. Flores. 1973. Comparative observations on intertubular lymphatics and the organization of the interstitial tissue of the mammalian testis. *Biol. Reprod.* **9**: 500.

Gilula, N.B., D.W. Fawcett, and A. Aoki. 1976. The Sertoli cell occluding junctions and gap junctions in mature and developing mammalian testes. *Dev. Biol.* **50**: 142.

Lee, I.P. 1981. Effect of drugs and chemicals on male reproduction. *Colloq. INSERM* **103**: 335.

Morales, C. and Y. Clermont. 1986. Receptor-mediated endocytosis of transferrin by Sertoli cells of the rat. *Biol. Reprod.* **35**: 393.

Neaves, W.B. 1973. Permeability of Sertoli cell tight junctions to lanthanum after ligation of ductus deferens and ductuli efferentes. *J. Cell Biol.* **59**: 559.

Okamura, K., I.P. Lee, and R.L. Dixon. 1975. Permeability of selected drugs and chemicals across the blood-testis barrier of the rat. *J. Pharmacol. Exp. Ther.* **194**: 89.

Pelletier, R.-M. 1988. Cyclic modifications of Sertoli cell junctional complexes in a seasonal breeder: The mink *(Mustela vison)*. *Am. J. Anat.* **182**: 130.

Pogash, L.M., Y. Lee, S. Gould, W. Giglio, M. Meyenhofer, and H.F.S. Huang. 1989.

Characterization of *cis*-platinum-induced Sertoli cell dysfunction in rodents. *Toxicol. Appl. Pharmacol.* **98**: 350.

Ross, M.H. 1970. The Sertoli cell and the blood-testicular barrier: An electronmicroscopic study. In *Morphological aspects of andrology* (ed. A.F. Holstein and E. Horstmann), p. 83. Grosse, Berlin.

Russell, L.D. 1977. Movement of spermatocytes from the basal to the adluminal compartment of the rat testis. *Am. J. Anat.* **148**: 313.

———. 1978. The blood-testis barrier and its formation relative to spermatocyte maturation in the adult rat: A lanthanum tracer study. *Anat. Rec.* **190**: 99.

Russell, L.D. and R.N. Peterson. 1985. Sertoli cell junctions: Morphological and functional correlates. *Int. Rev. Cytol.* **94**: 177.

Russell, L.D., A. Bartke, and J.C. Goh. 1989. Postnatal development of the Sertoli cell barrier, tubular lumen, and cytoskeleton of Sertoli and myoid cells in the rat, and their relationship to tubular fluid secretion and flow. *Am. J. Anat.* **184**: 179.

Russell, L.D., E.D. Clegg, R.A. Ettlin, and A.P. Sinha Hikim. 1990. *Histological evaluation of the testis*. (Published by senior author). (In press.)

Setchell, B.P. 1974. The entry of substances into the seminiferous tubules. In *Male fertility and sterility* (ed. R.E. Mancini and L. Martini), p. 35. Academic Press, New York.

Setchell, B.P. and D.E. Brooks. 1988. Anatomy, vasculature, innervation, and fluids of the male reproductive tract. In *The physiology of reproduction* (ed. E. Knobil and I. Neill), vol. 1, p. 753, Raven Press, New York.

Sun, E.L. and B. Gondos. 1986. Formation of the blood-testis barrier in the rabbit. *Cell Tissue Res.* **243**: 575.

Vitale, R., D.W. Fawcett, and M. Dym. 1973. The normal development of the blood-testis barrier and the effects of clomiphene and estrogen treatment. *Anat. Rec.* **176**: 333.

Weber, J.E., L.D. Russell, V. Wong, and R.N. Peterson. 1983. Three-dimensional reconstruction of a rat stage V Sertoli cell: II. Morphometry of Sertoli-Sertoli and Sertoli-germ-cell relationships. *Am. J. Anat.* **167**: 163.

Weber, J.E., T.T. Turner, K.S.K. Tung, and L.D. Russell. 1988. Effects of cytochalasin D on the integrity of the Sertoli cell (blood-testis) barrier. *Am. J. Anat.* **182**: 130.

COMMENTS

Hecht: Lonnie, why has the Sertoli cell barrier evolved?

Lonnie Russell: Why? That question needs a 10-minute answer. One of the older postulates, that antigens are developed on germ cells above the barrier which could be recognized by the system as foreign and that the barrier is there to isolate these antigens from the system, is beginning to crumble with some of Ken Tung's work. There are antigens on preleptotene spermatocytes before these germ cells move up. The question is: Why doesn't the system recognize them? It may be because we have an immunoprotective system there, such as the high steroid levels. My feeling is that we need to provide a special environment for these germ cells in the adluminal compartment. The Sertoli cell has the opportunity to provide products directly to the germ cell that are not necessarily bloodborne products. For example, if you do a hypophysectomy, the cells degenerate at one specific time during the long process of spermatogenesis, and they are primarily cells that are in the compartment above the barrier. A lot of evidence suggests that if you have primary Sertoli cell damage, the cells most susceptible are those above the barrier. These features point to disruption of the environment of the Sertoli cell.

Hecht: That doesn't say why the barrier has to be there. That just says there's a special relationship between those germ cells and the Sertoli cell. You could maintain that relationship without the barrier.

Lonnie Russell: How? If you have material moving fast across adluminal germ cells from the blood, without maintaining a concentration gradient that the Sertoli cell could maintain, the movement would be difficult.

Hecht: I'm thinking of localized regions in the membrane that are going to be sequestering particular components that would simply interact with the germ cells, and if it wouldn't make any difference whether it was in the whole cell or just in particular regions.

Handel: Don't you think the soluble environment is important, too? I always think of the barrier's existence as one of the main reasons we haven't been able to successfully achieve spermatogenesis in vitro.

Lonnie Russell: I think so, too. In addition, there is an obvious reason why the barrier is there: Sperm cannot move down the lumina of the seminiferous tubule without a lumen being formed, and a formation of lumen is like blowing up a balloon; otherwise, the fluid would leak, by pressure, back against the Sertoli cell. So, the fluid in the seminiferous tubule is there; it does not move backward; it is the vehicle for the transport of the sperm. You have asked an important question. There is no current answer.

Moses: Let me go over the edge of teleological speculation and suggest that, perhaps, the processes that are going on in the adluminal compartment are very sensitive processes and need to be protected in the special environment.

Lonnie Russell: I would agree; again, the evidence is lacking.

Selby: Are the spermatogonial stem cells inside or outside the Sertoli cell barrier?

Lonnie Russell: Everything is outside the Sertoli cell barrier until about the beginning of meiosis. We find that, as cells undergo transition in the rat from preleptotene to leptotene, they go on the other side of the barrier.

Cell-Cell Interactions in the Testis: A Guide for the Perplexed

IRVING BAMDAS FRITZ
Banting and Best Department of Medical Research
University of Toronto, Ontario, M5G 1L6, Canada

OVERVIEW

In providing a guide to the literature on cell-cell interactions in the testis, we initially survey the sorts of cells in the testis and the types of cell interactions that take place. The role of gonadal somatic cells in the control of spermatogenesis is emphasized, and systemic hormones are shown to act directly only on somatic cells. Evidence is reviewed that supports the conclusion that germ-cell development is dependent on the local chemical microenvironment generated by surrounding somatic cells in the unique cytoarchitecture within the seminiferous tubule. Chemical agents that impair spermatogenesis when administered in vivo may do so not only by direct actions on germ cells, but also by disturbing the function of Sertoli cells or other gonadal somatic cells. Chemical agents injected by usual routes can readily gain access to germinal cells in the basal compartment. However, such agents will not have access to germinal cells in the adluminal compartment until the chemical penetrates the Sertoli cell barrier. Somatic cells in the gonad (primarily but not exclusively Sertoli cells) influence the development of germinal cells in a variety of ways other than by direct cell interactions. Sertoli cells secrete specific components (growth factors, nutrients, cofactors, and modulators) into the adluminal compartment; Sertoli cells generate and maintain a barrier that sustains this chemical microenvironment and contributes to some degree to the immune privilege enjoyed by germ cells; and Sertoli cells are intimately involved in the restructuring processes required for the synchronized movement of clones of germinal cells within the seminiferous tubule during spermatogenesis.

INTRODUCTION AND PERSPECTIVES

The testis, surrounded by a capsule (the tunica albuginea), contains seminiferous tubules separated from each other by an interstitium consisting primarily of Leydig cells, blood vessels, and lymphatic vessels (for excellent reviews of the range of anatomical arrangements within testes from different mammals, see Setchell 1970, 1978; de Kretser and Kerr 1988). Blood vessels do not penetrate the walls of the seminiferous tubule. Cells within the seminiferous tubule interact directly with each other, and they interact with cells in the interstitium via chemical messengers. Similarly, cells within the interstitium are influenced by paracrine agents secreted by cells in the seminiferous tubule. Cells within each of the two major compartments are regulated by pituitary hormones, with the most obvious effects being the control by luteinizing hormone (LH) of rates of steroidogenesis by Leydig cells (Christensen 1975; Sharpe 1979; Hall 1988) and the stimulation by follicle stimulating hormone (FSH) of many Sertoli cell functions (for review, see Fritz 1978). Classic hormones, especially androgens, are essential to permit spermatogenesis to proceed efficiently. It is now generally accepted, however, that systemic hormones do not act directly on germinal cells. Instead, the controls of mammalian spermatogenesis by FSH, LH, and androgens are thought to be mediated by direct actions on testicular somatic cells, which then generate conditions within the seminiferous tubule required for germinal cell development. In this scheme, the optimal expression of programs in germinal cells is dependent on the cytoarchitectural arrangements and the chemical microenvironment provided by somatic cells adjacent to the different classes of developing germinal cells. We have previously reviewed much of the evidence in support of the postulates outlined (Fig. 1) (Fritz 1978, 1985; Fritz et al. 1986).

The nature of cell interactions in the testis is complex. Direct contacts among cells in the seminiferous tubule include several sorts of junctional complexes and ectoplasmic specializations involving somatic cell interactions with each other and with germinal cells (for reviews, see Russell and Peterson 1985; de Kretser and Kerr 1988). Germinal cells are in intimate contact with daughter cells via intercellular bridges, which remain because cytokinesis is incomplete (Fawcett 1975; de Kretser and Kerr 1988). Specific integral plasma membrane proteins ("integrins") have been characterized that bind to extracellular matrix (ECM) proteins, such as laminin in the basement membrane (Hynes 1987). Sertoli cells have been shown to contain receptors for laminin (P.S. Tung et al. 1987a). Interactions of Sertoli cells with each other result in the formation of a layer of polarized Sertoli cells. These cells are anchored to a basal lamella deposited by cooperative interactions between Sertoli cells and peritubular myoid cells. Sertoli cells extend all the way to the lumen, making contacts of various sorts with each class of germinal cell during development. Tight junctional complexes form among adjacent Sertoli cells during gonadal maturation and generate a circumferential barrier that divides the seminiferous tubule into basal and adluminal compartments having a highly ordered structure (for reviews, see Fawcett 1975; L.D. Russell, this volume).

In addition to direct contacts, gonadal cells communicate by means of soluble messengers, with testicular cells responding to a variety of growth factors and modulating factors secreted by neighboring cells. One or more cell types in the testis have been shown to secrete the following growth factors: transforming growth factor type alpha (TGF-α) and type beta (TGF-β); insulin-like growth factor, type 1, synonymous with somatomedin C (IGF); basic fibroblast growth factor (bFGF); interleukin-1 (IL-I); B-nerve growth factor (β-NGF); seminiferous growth factor (SGF); and inhibin (for review, see Bellvé and Zheng 1989). The synthesis of para-

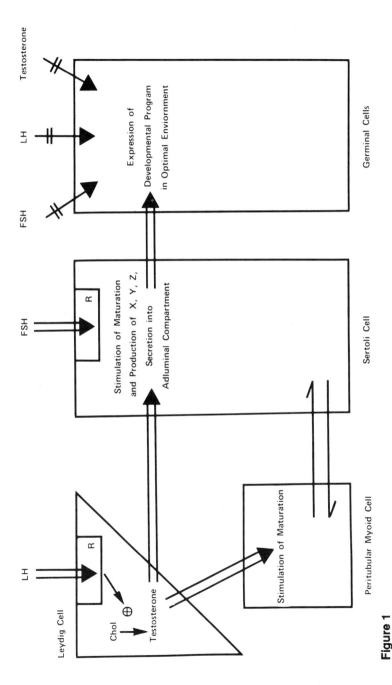

Figure 1
Summary of cellular sites of controls of spermatogenesis by luteinizing hormone (LH), follicle stimulating hormone (FSH), and testosterone. For details, see the text and previous reviews (Fritz 1978, 1985; Fritz et al. 1986).

crine factors by somatic cells may be regulated by systemic hormones, as indicated in the case of peritubular myoid cells that secrete P-Mod-S under androgenic modulation (Skinner and Fritz 1985) (for reviews, see Fritz 1978; Mather 1984; Stefanini et al. 1985; Fritz and Tung 1986; Fritz et al. 1986; Skinner 1987).

The unique cytoarchitectural arrangements in the testis, combined with the complex cell interactions outlined, are required to permit spermatogonia to develop into successively more differentiated classes of germinal cells and eventually into spermatozoa. This development of germ cells occurs synchronously with specific cell association patterns (Clermont 1972; Schulze and Rehder 1984; de Kretser and Kerr 1988). It has not been possible to duplicate these processes in vitro. Only the barest of beginnings, in fact, has been made in understanding the molecular and cellular basis for mechanisms governing complex interactions among testis cells during spermatogenesis. The maintenance in vivo of intercellular bridges between developing germinal cells, closely enveloped by Sertoli cell cytoplasmic extensions, has not been convincingly duplicated in vitro in mammalian germ cells undergoing meiosis. The remarkable cell association patterns among classes of germinal cells and Sertoli cells, shown to exist in vivo at different stages of the cycle of the seminiferous epithelium (Clermont 1972; Parvinen 1982; Schulze and Rehder 1984), add to the difficulties of duplicating all aspects of spermatogenesis in vitro.

It is apparent that too large a body of material exists on these topics to permit an adequate review within these pages. Fortunately, however, much is readily accessible in recent reviews and articles. It therefore seems reasonable to try to present a guide to the literature pertaining to the nature of cell interactions in the testis and to offer a summary of major points. In evaluating this summary, however, the reader should be aware that many of the data have been obtained from investigations on the properties of cells in culture under specific sets of conditions. Consequently, the reader is cautioned that information from these in vitro experiments indicates only the *potentials* of cell behavior described. The physiological relevance, in all too many cases, remains to be proven. Correspondence between in vitro and in vivo observations has been demonstrated in a few situations, and this cause for celebration will be underlined wherever possible.

Types of Cells in the Testis

By simply examining a list of various cells in the testis (Table 1), the reader can quickly gain an appreciation of the scope of problems involved in considering the nature of cell interactions in this organ. Nearly everything known about the cytology of each class of cell in the testis, except possibly that of rete testis epithelial cells, has recently been extensively reviewed (de Kretser and Kerr 1988). Consequently, we refer the reader to this well-illustrated treatise in good conscience, without attempting to summarize encyclopedic information covered. The general organization of the cells within the seminiferous tubule is shown in a

Table 1
List of Types of Cells in the Testis

I. Somatic cells

    A. Tunica albuginea
        1. fibroblasts
        2. smooth muscle cells
        3. blood vessels (vascular endothelial cells)

    B. Interstitium
        1. Leydig cells
        2. blood vessels (and pericytes)
        3. lymphatic vessels (lymphatic endothelial cells)
        4. white blood cells (macrophages, leukocytes, mast cells)
        5. fibroblasts

    C. Seminiferous tubule
        1. lymphatic endothelial cells
        2. peritubular myoid cells
        3. Sertoli cells

    D. Rete testis (see text)
        1. Rete testis epithelial cells
        2. Rete testis duct cells
        3. fibroblasts

II. Classes of germinal cells

    A. Spermatogonia (stem cells [A_s]; types A1, A2...; intermediate; and B)

    B. Primary spermatocytes (preleptotene, leptotene, zygotene, pachytene, and diplotene)

    C. Secondary spermatocytes

    D. Round spermatids (multiple classes, $t_1...t_7$)

    E. Elongated spermatids (multiple classes, $t_7...t_{19}$)

scanning electron micrograph (Fig. 2), simply to refresh the memories of those who may wish to have a picture before them of the overall structure of the testis while considering topics under discussion. A series of scanning electron micrographs, showing relationships between Sertoli cells and germinal cells in normal and in irradiated rat testes, is being prepared (K. Burdzy and I.B. Fritz) for publication in a forthcoming issue of the *Journal of Electron Microscopy Technique.*

Note that cells included in Table 1 have been listed in an order generally corresponding to their location, proceeding from the outer portion of the testis inward toward the wall and then to the lumen of the seminiferous tubules. The anastomosing network of ducts of the rete testis may appear to be an anomaly, since

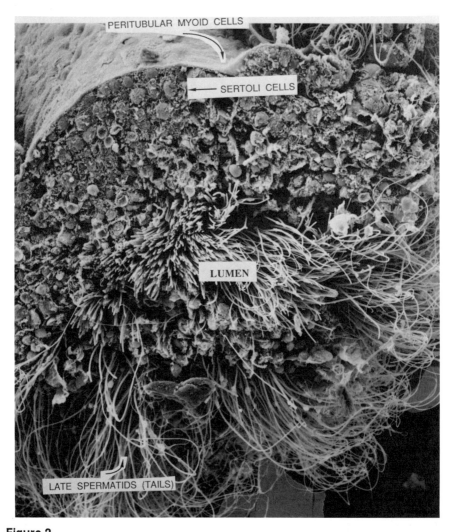

Figure 2
Scanning electron micrograph, showing a freeze-fractured cross-section of an adult rat seminiferous tubule at stage VIII of the cycle of the seminiferous epithelium. Sperm tails are evident in the lumen, and groups of spermatogonia are present just below the tubule wall, adjacent to Sertoli cells. The tubule wall consists of peritubular myoid cells covered by lymphatic endothelial cells and connective tissue cells. For SEM experimental details, see Ailenberg and Fritz 1989.

these ducts often lie close to the tunica. Rete testis epithelial cells themselves, however, first appear in the transitional zone of the seminiferous tubule (called the tubuli recti) devoid of germinal cells and are located in subsequent distal regions, leading eventually into rete ducts, which in turn merge to form efferent ducts emptying into the epididymis. The nature of the rete cells alters, changing from "squamous-type" cells to "prismatic-type" cells (Bustos-Obregon and Holstein 1976). The anatomy of the rete testis is, in fact, quite complex, consisting of several zones (Roosen-Runge and Holstein 1978; also, see Fig. 2 in de Kretser and Kerr 1988). The structure of Sertoli cells located at the transitional zone of seminiferous tubules is not identical with the structure of Sertoli cells in proximal regions of seminiferous tubules. At distal regions, rete testis epithelial cells of different classes replace Sertoli cells as the major somatic cells within the tubule (Dym 1974; Bustos-Obregon and Holstein 1976).

This digression to consider the rete was inserted to emphasize that various types of rete testis cells exist that have properties clearly different from those of Sertoli cells and that Sertoli cells themselves are not homogeneous. From a functional vantage point, the different arrangements may be of considerable importance in governing the exchange of components between the adluminal compartment and the basal compartment. For example, in autoimmune orchitis, initial lesions are usually detected in the rete (K.S.K. Tung et al. 1987). From this and related types of information, several authors have suggested that the rete testis epithelial cell barrier may not be as tight as that of the Sertoli cell barrier or that immune tolerance may not be as great in the more distal regions as in the proximal portions of seminiferous tubules.

Properties of ram rete testis epithelial cells in culture have been compared with those of ram Sertoli cells, and some of the differences have been partially characterized (P.S. Tung et al. 1987b). For example, rete testis epithelial cells from adult testes retain the capacity to proliferate in culture, but Sertoli cells do not. In addition, rete testis epithelial cells contain true desmosomal proteins, and intermediary filaments in these cells are rich in cytokeratins. In contrast, Sertoli cells isolated from adult testes do not have detectable desmosomal proteins, and the intermediary filaments are rich in vimentin, without detectable cytokeratins (P.S. Tung et al. 1987b).

Classes of Cells Involved in Interactions in the Testis

Among the many possible permutations and combinations of interactions possible among the 50 or more distinguishable types of cells listed in Table 1, the interactions about which most information is available are those in which Sertoli cells are implicated. We consider briefly the kinds of interactions that Sertoli cells have with each other, with peritubular myoid cells, with Leydig cells, and with different classes of germinal cells, respectively (Fig. 3). In this summary of inter-relations depicted schematically, the dash between letters signifies modulation of the activities of the second cell type shown, represented by the letter appearing last.

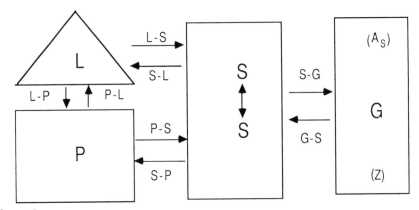

Figure 3
Types of cell interactions in the testis. Cells depicted schematically are Leydig cells (L), peritubular myoid cells (P), Sertoli cells (S), and germ cells (G), of which the stem cells (A_s) and spermatozoa (Z) represent various stages of spermatogenesis. Dashes between letters denote modulation by the first cell of the activities of the second cell type listed, indicated by arrows. For details, see the text and previous reviews (Fritz 1978, 1985; Fritz et al. 1986).

Thus, "P-S" means that peritubular myoid cells contain factors that modulate the activities of Sertoli cells. In this instance, P-S includes "P-Mod-S," a protein secreted by peritubular myoid cells under androgenic regulation that enhances the secretion of transferrin and androgen-binding protein by Sertoli cells (Skinner and Fritz 1985, 1986). P-Mod-S has been isolated and partially characterized (Skinner et al. 1988).

Reviews describing information about various sets of cell interactions in the testis are annotated in Table 2. These reviews provide evidence documenting that Sertoli cells secrete proteins that modulate steroidogenesis by Leydig cells in the presence or absence of LH (see reviews, part 3 in Table 2) and that they secrete other proteins that modulate the activities of peritubular myoid cells or that bind to different classes of germinal cells. In addition, Sertoli cells secrete a variety of factors (growth factors, steroids, inositol, metabolites of glucose, etc.) that influence the activities of neighboring cells (see references in Table 2, parts 2 and 4). Reciprocally, peritubular cells secrete growth factors and proteins that modulate several known activities of Sertoli cells: aromatase (Verhoeven and Cailleau 1988); secretion of androgen-binding proteins and transferrin (Skinner and Fritz 1985, 1986); net plasminogen activator activity (Fritz et al. 1989); and other Sertoli cell functions, independent of the presence of FSH. For example, TGF-β increases the levels of mRNA for plasminogen activator inhibitor in peritubular cells and decreases the levels of mRNA for plasminogen activator in Sertoli cells (Fritz et al. 1989; Nargolwalla et al. 1990). TGF-β also increases contractility of peritubular

Table 2
Annotation of Reviews on Different Types of Cell Interactions in the Testis

Cells involved	Reference	Orientation and emphasis
1. Sertoli-Sertoli	Russell and Peterson 1985	Junctional complexes, morphological and ultrastructural
2. Peritubular-Sertoli	Fritz and Tung 1986; Fritz et al. 1986; Verhoeven and Calleau 1988	Overview; metabolic-cooperativity; tubulogenesis; Androgen responsiveness of Sertoli cells, influenced by peritubular cells
	Fritz et al. 1989	Interactions in controls of net proteinase activities in the tubule
3. Sertoli-Leydig	Avallet et al. 1989	Control by Sertoli cell products of biochemical functions of Leydig cells
4. Sertoli-germ cell	Dighe et al. 1986	Biochemistry: uptake of ABP and transferrin by germ cells
Sertoli-germ cell	Griswold et al. 1988, 1989	Influences on germ-cell development
Sertoli-germ cell	Russell 1980; de Kretser and Kerr 1988 (pp. 866–892)	Morphological and ultrastructural; general overviews
Pachytene spermatocyte-Sertoli cell	Djakiew and Dym 1988	Influences of pachytene spermatocyte products on proteins secreted by Sertoli cells
Sertoli-germ cell	Wright et al. 1988	"Cyclic protein 2"
Sertoli-germ cell	Grootegoed et al. 1988	Energy metabolism
5. Multiple cell-cell interactions in the testis	Fritz 1978; Sharpe 1983, 1984; Mather 1984; Fritz 1985; Stefanini et al. 1985; Parvinen et al. 1986; Skinner 1987;	General reviews on the nature of cell-cell interactions involved in spermatogenesis
	Bellvé and Zheng 1989	Growth factors in testis

cells (Ailenberg et al. 1990). Androgens synthesized and secreted by Leydig cells have direct actions on peritubular cells by increasing P-Mod-S secretion. Androgens thereby indirectly act on Sertoli cells, mediated by effects of P-Mod-S. Androgens also have direct effects on Sertoli cells, mediated by binding to specific androgen receptors in Sertoli cells (see references in Table 2, part 2, and Skinner and Fritz

1985). Finally, germ cells secrete proteins that influence Sertoli cell functions (Parvinen et al. 1986; Djakiew and Dym 1988; Griswold et al. 1988; Fritz et al. 1989).

Possible Relevance of Somatic Cell/Germ Cell Interactions in the Testis to Problems Pertaining to the Sensitivity of Germ Cells to Mutagenic Agents

Toxic agents can cause disruption of spermatogenesis by directly altering germ-cell metabolism or function. They can also do so, however, by interfering with somatic cell/germinal cell relations. As examples, consider that busulfan presumably acts directly on spermatogonia; glycol ether solvents on spermatocytes; and ethyl methane sulfonate on spermatids. In contrast, several drugs (such as phthalate esters, dinitrobenzene, and hexanedione) appear to exert their toxic effects on the testis by acting directly on Sertoli cells. Still others (such as ethane-1,2-dimethane sulfonate and acetaldehyde) elicit a cessation of spermatogenesis by destroying Leydig cells and thereby preventing steroidogenesis (for reviews, see Edwards et al. 1988; Foster 1988). It is possible that some drugs act by blocking responses of testis cells to paracrine agents secreted by neighboring cells. These data demonstrate the need to determine precisely the cellular target of each chemical found to perturb spermatogenesis.

If the complex restructuring in the seminiferous tubule that normally occurs during spermatogenesis at discrete stages of the cycle were to be disrupted, sperm production would cease. I believe this would happen, for example, if there were disturbances of net proteinase activity at specific stages. I base this belief on data indicating that proteinases secreted by Sertoli cells at stages VII and VIII are implicated in tissue restructuring events, and that local proteolytic activity can be modulated by proteinase inhibitors secreted by neighboring peritubular myoid cells (for review, see Fritz et al. 1989). The control of net proteinase activity at a specific locus represents a more subtle form of cell-cell interaction than the direct effects of a paracrine factor on the activities of target cells. It is also different from the metabolic cooperativity between Sertoli cells and peritubular myoid cells that takes place during the organized deposition of ECM components to form a basement membrane at sites of contact between the two cell types (for review, see Fritz and Tung 1986). It is possible that diseases of cell-cell interactions exist, in which such cooperativity may be impaired. Could any of the mutagenic agents act at such loci?

If a chemical agent acts directly on spermatocytes or spermatids, the compound clearly must gain access to the germ cell in question. Does the chemical cross the barrier by being transported through the Sertoli cell? Or does the chemical agent enter spermatogonia or preleptotene spermatocytes in the basal compartment and then exert a delayed response on these germ cells as they develop into more advanced stages and become translocated into the adluminal compartment? At another level of organization, some chemical agents could perturb the metabolic activities of more than one cell type. For example, if retinoic acid metabolism were impaired, spermatogenesis would cease, with few if any germ cells remaining more advanced than preleptotene spermatocytes (Griswold et al. 1988, 1989). From these considerations, it follows that an appreciation of the nature of cell-cell interactions in the testis is required to evaluate the cellular sites of action of drugs that disturb spermatogenesis and also to investigate mechanisms of action of such agents.

Fortunately, the availability of increasingly sophisticated cell culture techniques, in which the phenotype of the cell can be better maintained by employing suitable ECM substrata and culture conditions, makes it possible to determine which cells are likely to be influenced by chemical agents of interest and to investigate biochemical mechanisms involved. The direct correspondence observed between the in vivo and in vitro modulation of mRNA levels of transferrin in Sertoli cells by retinoic acid and by FSH (Griswold et al. 1988, 1989) is encouraging and offers hope that it will become increasingly possible to evaluate the physiological significance and relevance of results obtained from well-controlled experiments on the properties of cells in culture.

ACKNOWLEDGMENTS

It is a pleasure to express appreciation to the Canadian Medical Research Council, which has provided uninterrupted grant support for the past 22 years for researches in my laboratory related to topics reviewed. I thank my colleagues, especially Pierre Tung, Menachem Ailenberg, and Krystyna Burdzy, for stimulating and rewarding interactions during the past few years. Without their work, there would be little to report from our group.

REFERENCES

Ailenberg, M. and I.B. Fritz. 1989. Influences of follicle stimulating hormone, proteases and anti-proteases on permeability of the barrier generated by Sertoli cells in a two-chambered assembly. *Endocrinology* **124**: 1399.

Ailenberg, M., P.S. Tung, and I.B. Fritz. 1990. TGF-β elicits shape changes and increases contractility of testicular peritubular cells. *Biol. Reprod.* **126**: (in press).

Avallet, O., M.H. Perrard-Sapori, M. Vigier, P.G. Chatelain, and J.M. Saez. 1989. Regulation of Leydig cell functions by Sertoli cells. In *Development and function of the reproductive organs* (ed. N. Josso), vol. 3, p. 187. Serono Symposium Review no. 21, Rome.

Bellvé, A.R. and W. Zheng. 1989. Growth factors as autocrine and paracrine modulators of male gonadal functions. *J. Reprod. Fertil.* **85**: 771.

Bustos-Obregon, E. and A.F. Holstein. 1976. The rete testis in man: Ultrastructural aspects. *Cell Tissue Res.* **175**: 1.

Christensen, A.K. 1975. Leydig cells. In *Handbook of physiology: Male reproductive system* (ed. D.W. Hamilton and R.-O. Greep), vol. 5, sect. 7, p.57. Williams and Wilkins, Baltimore.

Clermont, Y. 1972. Kinetics of spermatogenesis in mammals. Seminiferous epithelial cycle and spermatogonial renewal. *Physiol. Rev.* **52:** 198.

de Kretser, D.M. and J.B. Kerr. 1988. Cytology of the testis. In *Physiology of reproduction* (ed. E. Knobil et al.), p. 837. Raven Press, New York.

Dighe, R.R., M. Meistrich, and A. Steinberger. 1986. Sertoli-germ cell interaction: Uptake of Sertoli cell secretory proteins by germ cells during spermatogenesis. In *Development and function of the reproductive organs* (ed. A. Eskkol et al.), p. 161. Ares-Serono Symposia Review no. 11, Rome.

Djakiew, D. and M. Dym. 1988. Pachytene spermatocyte proteins influence Sertoli cell function. *Biol. Reprod.* **39:** 1193.

Dym, M. 1974. The fine structure of monkey Sertoli cells in the transitional zone at the junction of the seminiferous tubules with the tubuli recti. *Am. J. Anat.* **140:** 1.

Edwards, G., B.W. Fox, H. Jackson, and J.D. Morris. 1988. Leydig cell cytotoxicity of putative male antifertility compounds related to ethane 1, 2-dimethane sulphonate. *Serono Symp. Publ. Raven Press* **50:** 185.

Fawcett, D. 1975. The ultrastructure and functions of the Sertoli cell. In *Handbook of physiology: Male reproductive systems* (ed. D.W. Hamilton and R.-O. Greep), vol. 5, sect.7, p. 21. Williams and Wilkins, Baltimore.

Foster, P.M.D. 1988. The potential uses and misuses of compounds affecting spermatogenesis. *Symp. Publ. Raven Press* **50:** 231.

Fritz, I.B. 1978. Sites of actions of androgen and follicle stimulating hormone on cells of the seminiferous tubule. In *Biochemical actions of hormones* (ed. G. Litwack), vol. 5, p. 249. Academic Press, New York.

―――. 1985. Past, present and future of molecular and cellular endocrinology of the testis. *Colloq. INSERM* **123:** 15.

Fritz, I.B. and P.S. Tung. 1986. Role of interactions between peritubular cells and Sertoli cells in mammalian testicular functions. In *Gametogenesis and the early embryo* (ed. J. Gall), p. 151. A.R. Liss, New York.

Fritz, I.B., M.K. Skinner, and P.S. Tung. 1986. The nature of somatic cell interactions in the seminiferous tubule. In *Development and function of the reproductive organs* (ed. A. Eshkol et al.), p. 185. Ares-Serono Symposia Review, no. 11, Rome.

Fritz, I.B., M., Lacroix, F. Smith, A. Hettle, C. Nargolwalla, P. Tung, and M. Ailenberg. 1989. The functional roles of plasminogen activator in the testis. In *Development and function of the reproductive organs* (ed. N. Josso), vol. 3, p. 133. Serono Symposia Review no. 21, Rome.

Griswold, M.D., C. Morales, and S.R. Sylvester. 1988. Molecular biology of the Sertoli cell. *Oxford Rev. Reprod. Biol.* **10:** 124.

Griswold, M.D., L. Heckert, A.F. Karl, L. Law, V. Roberts, J.E. Siiteri, B. Stallard, J. Tsuruta, and S.R. Sylvester. 1989. Sertoli cell secretory proteins in normal and synchronized testes. In *Development and functions of the reproductive organs* (ed. N. Josso), vol. 3, p. 207. Serono Symposium Review no. 21, Rome.

Grootegoed, A., P.J. den Boer, and P. Mackenbach. 1988. Sertoli cell-germ cell communication. *Ann. N.Y. Acad. Sci.* **564:** 232.

Hall, P.F. 1988. Testicular steroid synthesis: Organization and regulation. In *Physiology of reproduction* (ed. E. Knobil et al.), p. 975. Raven Press, New York.

Hynes, R.O. 1987. Integrins: A family of cell surface receptors. *Cell* **48**: 549.

Mather, J. 1984. Intra-testicular regulation: Evidence for autocrine and paracrine control of testicular function. In *Mammalian cell culture* (ed. J. Mather), p. 167. Plenum Press, New York.

Nargolwalla, C., D. McCabe, and I.B. Fritz. 1990. Modulation of levels of messenger RNA for tissue-type plasminogen activator in rat Sertoli cells, and levels of messenger RNA for plasminogen activator inhibitor in testis peritubular cells. *Mol. Cell. Endocrinol.* (in press).

Parvinen, M. 1982. Regulation of the seminiferous epithelium. *Endocr. Rev.* **3**: 404.

Parvinen, M., K.K. Vehko, and J. Toppari. 1986. Cell interactions during the seminiferous epithelial cycle. *Int. Rev. Cytol.* **104**: 115.

Roosen-Runge, E.C. and A.F. Holstein. 1978. The human rete testis. *Cell Tissue Res.* **89**: 409.

Russell, L.D. 1980. Sertoli-germ cell interrelations: A review. *Gamete Res.* 3: 179.

Russell, L.D. and R.N. Peterson. 1985. Sertoli cell junctions: Morphological and functional correlates. *Int. Rev. Cytol.* **94**: 177.

Schulze, W. and W. Rehder. 1984. Organization and morphogenesis of the human seminiferous epithelium. *Cell Tissue Res.* **237**: 395.

Setchell, B.P. 1970. Testicular blood supply, lymphatic drainage and secretion of fluids. In *The testis* (ed. A.D. Johnson et al.), vol. 1, p. 101. Academic Press, New York.

———. 1978. *The mammalian testis*. Paul Elek, London.

Sharpe, R.M. 1979. The hormonal regulation of the Leydig cell. *Oxf. Rev. Reprod. Biol.* **1**: 241.

———. 1983. Local control of testicular function. *Quart. J. Exp. Physiol.* **68**: 265.

———. 1984. Intratesticular factors controlling testicular function. *Biol. Reprod.* **30**: 29.

Skinner, M.K. 1987. Cell-cell interactions in the testis. *Ann. N.Y. Acad. Sci.* **513**: 158.

Skinner, M.K. and I.B. Fritz. 1985. Testicular peritubular cells secrete a protein under androgen control that modulates Sertoli cell functions. *Proc. Natl. Acad. Sci.* **82**: 114.

———. 1986. Identification of a non-mitogenic paracrine factor involved in interactions between testicular peritubular cells and Sertoli cells. *Mol. Cell. Endocrinol.* **44**: 85.

Skinner, M.K., P.M. Fetterolf, and C.T. Anthony. 1988. Purification of a paracrine factor, P-Mod-S, produced by testicular peritubular cells that modulates Sertoli cell function. *J. Biol. Chem.* **263**: 2884.

Stefanini, M., M. Conti, R. Geremia, and E. Ziparo. 1985. Regulatory mechanisms of mammalian spermatogenesis. In *Biology of fertilization* (ed. C.B. Metz and A. Monroy), vol. 2, p. 59. Academic Press, New York.

Tung, K.S.K., T.D. Yule, C.A. Mahi-Brown, and M.B. Listrom. 1987. Distribution of histopathology and I_a positive cells in actively induced and adoptively transferred experimental autoimmune orchitis. *J. Immunol.* **138**: 752.

Tung, P.S., A.H.C. Choi, and I.B. Fritz. 1987a. Topography and behavior of Sertoli cells in sparse culture during the transitional remodeling phase. *Anat. Rec.* **220**: 11.

Tung, P.S., J. Rosenior, and I.B. Fritz. 1987b. Isolation of culture of ram rete testis epithelial cells: Structural and biochemical characteristics. *Biol. Reprod.* **36**: 1297.

Verhoeven, G. and J. Cailleau. 1988. The androgen responsiveness of Sertoli cells depends on a complex interplay between androgens, FSH and peritubular cell factors. *Serono Symp. Publ. Raven Press* **50**: 297.

Wright, W.W., S.D. Sabludoff, M. Erickson-Lawrence, and A.W. Karzai. 1988. Germ cell-Sertoli cell interactions. Studies of cyclic protein-2 in the seminiferous tubule. *Ann. N.Y. Acad. Sci.* **564**: 173.

COMMENTS

Wiley: What was the dose of radiation that you used?

Fritz: That was fairly low. It was a cesium-137 source. It was about 125 rads total body radiation to the pregnant rat.

Wiley: How can you differentiate between direct effect on the Sertoli cells or that radiation effect mediated by the germ cell?

Fritz: You can't. The whole point of that morphology was to show that the Sertoli cell was developing in what appears to be a normal fashion in the absence of germ cells. All we can say is, from the morphology you saw, they form junctional contacts; they form what looks to be a barrier. You can isolate the cells from such irradiated animals and study them in culture and find many of the characteristics of Sertoli cells obtained from a non-irradiated testis. I'm not saying there is no damage to them at all. I'm saying that's what the morphology looks like. I think that, in the absence of germ cells, one thing for sure is clearly different, namely, the course of formation of the protease. So, I think that there are subtle changes on the Sertoli cells. I think it's a way to get at what germ cells might be doing to the Sertoli.

Wiley: You still can't really differentiate between the ones that reflect a direct effect of the radiation on the Sertoli cell itself?

Handel: No, except that we have made use of a genetically germ-cell-free model also, to look at Sertoli cell development. We find similar aspects of apparently normal early development Sertoli cells.

Moses: Is there any evidence for transfer of materials from Sertoli cell to spermatocyte in the adluminal compartment?

Fritz: The best example I have found so far has been the iron transferrin story. It has been done by a number of labs. I think the best work has been done by Griswold and his colleague Morales. The story there is that the Sertoli cell makes a unique transferrin from the same gene product as liver transferrin, but the glycosylation pattern is different. The plasma transferrin carrying iron comes to that transferrin in the Sertoli cell, which picks up the iron and then translocates it by secreting it into the adluminal compartment. There is a receptor for the transferrin in the spermatocyte that literally takes up the iron. The only part that hasn't been shown experimentally is the requirement of that iron in the spermatocyte. All the other parts of the sequence have been shown.

Moses: Is there anything unusual about the two membranes facing each other morphologically that could correlate to the possible transfer of material?

Lonnie Russell: Well, that was my question to Irving. We showed gap junctions between Sertoli cells and germ cells not quite 10 years ago, but the ability to experimentally manipulate gap junctions has not proceeded well, even in the cell biology field, where you can culture cells from all kinds of places. Certainly, gap junctions are mediators of many cell activities, especially in embryonic development. I was going to ask Irving if he thought that they were having an effect in the testis, and if so, if he knew anything about it.

Fritz: Well, you answered your own question. This is experimentally very difficult to come by. I think that the gap junctions between Sertoli cells and germ cells pale into insignificance when the intercellular bridges between germ cells are considered. There are enormous intercellular bridges between each of the germ cells in a given syncytium, so that anything that happens in one germ cell is easily transferred to all the others. This begs the question: Could the Sertoli cell transfer to a germ cell, via gap junctions, something from Sertoli cells? I don't know. I have seen no hard evidence.

You probably are aware of some other work that Gilula did some years ago that is quite lovely. One could imagine doing this if one had appropriate in vitro germ-cell development. He took a cardiac cell and cocultured it with a granulosa cell in close contact. When the contact was sufficient, they formed gap junctions with each other. Cardiac cells have no receptor to FSH, but the granulosa cells do. After the gap junctions were formed, when he added FSH, he got an increased contractility of the heart muscle. In this case, he was able to show functional coupling of gap junctions. If we had a marker of germ cell secretion that we knew was dependent on the Sertoli cell and could find a marker that was independent of Sertoli cells, and then if we find after gap junction formation an increased activity of the germ cell when adding FSH directly, that might be very interesting. We could even take an extraneous cell that might form a gap junction and again test that. Such experiments have not been done to my knowledge. In fact, it is very hard to prove what a germ cell product is that is independent of everything else.

Hecht: Can you induce the protease by coculture of the previously irradiated Sertoli cells with, let's say, germ cells?

Fritz: I have not done that experiment. What you can do is increase the activity. The slide I showed indicates that there is a synergism between residual body and FSH or anything that is being taken up by phagocytosis and FSH on Sertoli cells. Therefore, if you elicit increased phagocytosis in Sertoli cells, you can increase protease activity. During spermiation the residual bodies that were released by spermatids may be a trigger to Sertoli cells to increase protease activity. In other words, if there is a constant level of hormone across the entire length of the seminiferous tubule, why is a Sertoli cell at stage 8 more sensitive than a Sertoli cell at any other stage? The possibility exists that residual body uptake during spermiation may be involved. Experimentally, that fits.

Griswold has a beautiful model; again, this is with Morales. You take vitamin A-deficient animals and wait until everything has regressed; the most advanced germ cell is apparently the preleptotene spermatocyte. Then, you give a walloping dose of retinoic acid to reinitiate and maintain spermatogenesis. Now, instead of having a single tubule show all the stages of the cycle, you have in essence elicited a "grasshopper" type of spermatogenesis—that is, everything goes in phase: The preleptotene goes to leptotene to the zygotene and so on, and spermatogenesis remains synchronized for at least two cycles of the seminiferous epithelium. That gives you a long time—it's two whole waves—it gives you about 60 days to play. So you take out a testis, and all the testis is at stage 2 or stage 3 or stage 4, depending upon the time after initiating vitamin A treatment. When you do this, you discover that the transferrin mRNA is going up in a certain way. In intact tubules, the Sertoli cell transferrin mRNA is highest at stages 13 and 14. In Sertoli cells isolated from seminiferous tubules by enzymatic fractionation, there is no difference in levels of transferrin mRNA detected in Sertoli cells at different stages of the cycle. When he adds back germ cells, differences are restored. It therefore appears that the germ-cell association pattern influences levels of transferrin mRNA in Sertoli cells. This is a beautiful demonstration—the best one that has been done, to my knowledge.

Correlation between Proliferative Activity of Mouse Spermatogonial Stem Cells and Their Sensitivity for Cell Killing and Induction of Reciprocal Translocations by Irradiation

DIRK G. DE ROOIJ,[1] YVONNE VAN DER MEER,[1]
ANS M.M. VAN PELT,[1] AND PAUL P.W. VAN BUUL[2]
[1]Department of Cell Biology, University of Utrecht Medical School
3508 TD Utrecht, The Netherlands
[2]Department of Radiation Genetics and Chemical Mutagenesis
University of Leiden, and J.A. Cohen Institute
Interuniversity Institute of Radiopathology and Radiation Protection
2333 AL Leiden, The Netherlands

OVERVIEW

The proliferative activity of the A_s spermatogonia, the spermatogonial stem cells, varies during the cycle of the seminiferous epithelium. In this study, the sensitivity of stem cells during the epithelial cycle for cell killing by X-rays was determined in CBA mice.

It was found that from the middle of stage I to stage III, when the stem cells are actively proliferating, they have a D_0 value of approximately 1.7 Gy. In stages VI and VII, where these cells are quiescent, the D_0 was about 1.0 Gy, and in stages IX–XII, in which the quiescent stem cells gradually start to proliferate again, the D_0 was about 2.3 Gy. This relation between proliferative activity and cell killing is closely similar to that found in Cpb-N mice after fission neutron irradiation. Hence, spermatogonial stem cells are highly sensitive to the cell killing effect of irradiation when they are quiescent; they have an intermediate radiosensitivity when they are actively proliferating; and they pass through a very radioresistant phase when they are triggered out of quiescence.

To study the correlation between radiosensitivity for cell killing and the induction of reciprocal translocations, the seminiferous epithelium of Cpb-N mice was synchronized by the vitamin A deficiency/vitamin A replacement procedure. Groups of these mice were irradiated when their seminiferous epithelium was in stages I–II, VI–VII, or X–XII. The number of translocations induced in the testes that were in stages VI–VII at the time of irradiation, in which the stem cells are quiescent and sensitive for cell killing, tended to be higher than in the testes that were in other stages. These experiments will now be continued with higher doses.

INTRODUCTION

Spermatogenesis

Huckins (1971a), Oakberg (1971), de Rooij (1973), Oud and de Rooij (1977), and Lok et al. (1982) have postulated that the A_s spermatogonia are the stem cells of spermatogenesis in the rat, mouse, Chinese hamster, and the ram (for review, see de Rooij 1983). The daughter cells of an A_s spermatogonium either become two new stem cells or stay together as A_{pr} spermatogonia, interconnected by an intercellular bridge. The A_{pr} spermatogonia will divide further to form chains of four, eight, or sixteen A_{al} spermatogonia. The A_s, A_{pr}, and A_{al} spermatogonia together are called undifferentiated spermatogonia.

The proliferative activity of the undifferentiated spermatogonia follows a cyclic pattern (Huckins 1971b; Lok and de Rooij 1983). During a particular period of the epithelial cycle, the undifferentiated spermatogonia are mostly quiescent and their number is low, with only A_s, A_{pr}, and a few A_{al} spermatogonia being present. Subsequently, these cells go through a period of active proliferation, during which increasing numbers of A_{al} spermatogonia are formed, while the total number of A_s and A_{pr} spermatogonia remains about the same. The period of active proliferation ends when most of the cells become arrested in G_1 phase. After a successive period of relative quiescence, almost all of the A_{al} spermatogonia differentiate into the first generation of the differentiating-type spermatogonia, in most animals called A_1 spermatogonia. After that, the differentiating spermatogonia go through a series of six divisions, ultimately giving rise to spermatocytes. When the differentiating spermatogonia have started their series of divisions, the remaining undifferentiated spermatogonia resume proliferation to produce new A_1 spermatogonia for the following epithelial cycle.

The stem cells, the A_s spermatogonia, follow the general pattern of the proliferative activity of the undifferentiated spermatogonia, except that they continue to proliferate somewhat longer in time than the A_{pr} and A_{al} spermatogonia (Huckins 1971b; Lok and de Rooij 1983).

Properties of the Spermatogonial Stem Cell Population

Huckins (1971c, 1978) has postulated the existence of so-called "long-cycling" stem cells. These cells were supposed to be a minority group of cells among the A_s spermatogonia and to be the real stem cells. The remaining, fast-cycling A_s spermatogonia would be destined to differentiate. Furthermore, in rats and mice, experiments were done in which labeling with [^{3}H]thymidine was combined with irradiation (Oakberg and Huckins 1976; Huckins 1978; Huckins and Oakberg 1978; Oakberg et al. 1986). The results were interpreted to mean that the long-cycling stem cells are more resistant to irradiation. However, the existence of long-cycling and radioresistant stem cells still is a matter of debate. Lok et al. (1984) concluded that the concept of long-cycling stem cells is just one of a number of possibilities to explain the [^{3}H]thymidine labeling pattern observed. It is also questionable whether or not there exists a special type of radioresistant stem cells. van Beek et al. (1986a) have determined the dose-response relationships of the stem cells for fission neutron irradiation in various stages of the epithelial cycle. Their results seem to exclude the existence of a separate type of radioresistant stem cells, and these authors concluded that it is more likely that the radiosensitivity of the total population of stem cells varies with their proliferative behavior.

With respect to the induction of genetic damage in spermatogonial stem cells, most authors agree that there is a correlation between cell killing and induction of mutations or reciprocal translocations (Russell 1956; Oftedal 1968; Gerber and Léonard 1971; de Ruiter-Bootsma et al. 1977; Leenhouts and Chadwick 1981; Cattanach and Kirk 1987; van Buul and Goudzwaard 1990). However, some contrasting reports have also been published (Oakberg 1978; van Buul and de Boer 1982). For a review on this subject, see van Buul (1983).

We now have further tested the correlation between the proliferative activity of the stem cells and their radiosensitivity with respect to cell killing, using a different strain of mice and a different type of radiation than in the previous study (van Beek et al. 1986a). The sensitivity of spermatogonial stem cells for the cell-killing effect of X-rays in each stage of the cycle of the seminiferous epithelium in CBA mice was determined, using a modification of the technique described by van Beek et al. (1984, 1986a,b). Furthermore, we started an investigation into the relation between cell killing and translocation induction by X-rays in Cpb-N mice, the strain used by van Beek et al. (1986a). This was done by irradiating mice in which the seminiferous epithelium was synchronized by vitamin A replacement after they had been made vitamin-A-deficient (A.M.M. van Pelt and D.G. de Rooij, in prep.).

MATERIALS AND METHODS

Dose-Response Curves

Male CBA/p mice, 19–25 weeks old, received doses of 0.5–8 Gy whole body irradiation with a Philips Müller X-ray tube, operating at 300 kVp and 5 mA (HVL 2.1 mm Cu; dose rate 0.3 Gy/min). The animals were sacrificed at 10 days after irradiation. The testes were fixed in Bouin's fluid and embedded in glycol methacrylate (Technovit 7100; Kulzer & Co., Wehrheim, Germany). Sections (5 μm) were stained with the periodic acid-Schiff (PAS) reaction and hematoxylin. In these sections, the numbers of undifferentiated A spermatogonia per 1000 Sertoli cells, in each stage of the cycle of the seminiferous epithelium, were determined by way of cell counts after doses of 0.5, 1, 2, 3, 4, 5, 6, and 8 Gy. Dose-response curves were constructed for each epithelial stage, and preliminary D_0 values were calculated by way of regression analysis.

Translocation Induction Studies

Synchronization of the Seminiferous Epithelium

Breeding pairs of Cpb-N mice were fed a vitamin-A-deficient diet (Teklad Trucking, Madison, Wisconsin) for at least 4 weeks. The male young born after these 4 weeks received the same diet for about 15–17 weeks, and then they became vitamin-A-deficient. At this time, each animal received an intraperitoneal injection of 0.5 mg of retinol-acetate (Sigma, St. Louis, Missouri), and further on received a normal vitamin-A-containing diet. The retinol-acetate was dissolved in 25 µl ethanol and mixed with 75 µl sesame oil. At intervals of 43, 46, and 48 days after vitamin A replacement, groups of 7–8 mice received 3 Gy whole body irradiation. The irradiation was given with an Orthovolt X-ray machine (Philips RT 250), operating at 250 kV and 20 mA, resulting in an HVL of 2 mm Cu. The mice were sacrificed between 189 and 215 days after irradiation.

Determination of Reciprocal Translocations

Meiotic chromosome preparations were made according to the technique described by Wessels-Kaalen (1984). Analysis for the presence of multivalent configurations at diakinesis-metaphase I, indicative for reciprocal translocations, was done on coded slides, C-banded with a modified BSG method of Sumner (1972).

RESULTS

Radiosensitivity of Spermatogonial Stem Cells

In the sections of the testes of each mouse, sacrificed 10 days after irradiation, the numbers of undifferentiated spermatogonia were counted in tubular cross-sections until 5000–7000 Sertoli cells were encountered. Separate scores were made for each epithelial stage. For each stage of the cycle of the seminiferous epithelium, a dose-response curve was made, in which the number of spermatogonia per 1000 Sertoli cells present after the graded doses of X-rays was taken as a measure of the number of surviving stem cells. The D_0 values obtained from these curves are plotted in Figure 1. Using a value of 196.7 hours for the duration of the epithelial cycle in the CBA/p mouse (A.L. Bootsma, unpubl.), the corresponding epithelial stage at the moment of irradiation was calculated for each epithelial stage present at day 10. In Figure 1, the stages at the time of irradiation are indicated. The radiosensitivity of the stem cells varied between the epithelial stages with a factor of about two. Three levels of sensitivity could be discerned. In stages mid-I–III, the D_0 varied from 1.7 to 1.8 Gy. In stages VI–VII, D_0 values of 1.0 and 1.1 Gy were found, and in stages IX–XII, the D_0 ranged from 2.2 to 2.3 Gy.

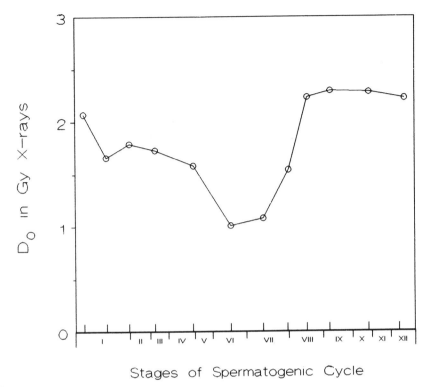

Figure 1
D_0 values obtained from dose-response curves for the cell-killing effect of X-rays on spermatogonial stem cells in each stage of the cycle of the seminiferous epithelium, in CBA mice. The effect of the irradiation was ascertained by counts of undifferentiated spermatogonia at 10 days after irradiation in all epithelial stages. It was then calculated to which stage at the time of irradiation each stage at day 10 corresponds. At the x-axis the stages at the time of irradiation are indicated.

Induction of Reciprocal Translocations in Spermatogonial Stem Cells

After vitamin A replacement in the vitamin-A-deficient Cpb-N mice, a synchronized reinitiation of spermatogenesis was induced. After 43 days, the great majority of the tubular cross-sections showed stages VI or VII of the cycle of the seminiferous epithelium, with only a few showing stages V or VIII (Fig. 2). After 46 and 48 days, most tubular cross-sections showed stages IX–XII and stages I–II, respectively. Assuming a duration of the epithelial cycle of exactly 8.6 days for the Cpb-N mouse, it was calculated at what time these testes would show stage XII, i.e., meiotic divisions, between 180 and 215 days after vitamin A replacement. How-

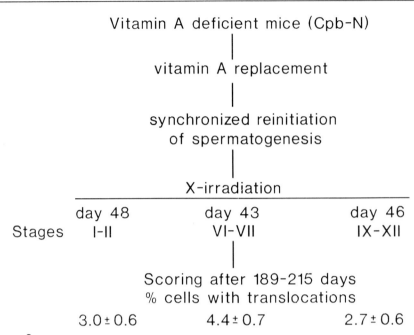

Figure 2
Design and most important results of the experiment ascertaining the translocation induction in mouse spermatogonial stem cells in various stages of the cycle of the seminiferous epithelium, following 3 Gy of X-rays.

ever, it was not checked whether or not the testes were still synchronized at that time. In any case, enough cells in diakinesis-metaphase I were obtained to enable the determination of translocation induction. Some of the testes had become completely atrophied (Table 1).

The lowest testis weights were obtained for the stage VI/VII mice, but the differences between the three groups were not significant (Table 1). Similarly, the percentage of degenerated testes was highest for the stage VI/VII mice. In addition, most translocations were found in the group of mice of which the seminiferous epithelium was in stages VI and VII at the time of irradiation. However, the differences between the percentages of cells with translocations in the three groups were not significant. The translocation data were statistically analyzed with the Kruskal-Wallis test and analysis of variance, and the p values were found to be 0.12 and 0.13, respectively.

DISCUSSION

In the mouse, it is not possible to study the proliferative activity of the A_s spermatogonia by analyzing the [^{3}H]thymidine incorporation in autoradiographs of tubular

Table 1
Induction of Translocations in Stem Spermatogonia of Cpb-N Mice by 3 Gy of X-rays

Irradiation in stage	No. of mice	No. of cells/ translocations	% Translocations (± S.E.M.)	Testis weight mg ± S.E.M.	% Degenerated testis
I–II	8	1300/39	3.0 ± 0.6	68 ± 1.4	19
VI–VII	7	1000/44	4.4 ± 0.7	60 ± 3.6	29
IX–XII	7	1140/31	2.7 ± 0.6	65 ± 1.7	14

whole mounts like in the rat (Huckins 1971b) and the Chinese hamster (Lok et al. 1983). However, with the help of the cell counts performed in the mouse (e.g., van Beek et al. 1984) and the knowledge obtained from the Chinese hamster and the rat, the most likely pattern of the proliferative activity of the A_s spermatogonia during the epithelial cycle can be deduced. Hence, mouse spermatogonial stem cells are very likely quiescent in stages VI and VII. In subsequent stages IX–XII, the A_s spermatogonial population is stimulated to proliferate, since the numbers of undifferentiated spermatogonia start to increase again during these stages, and in the Chinese hamster, the labeling index of the A_s spermatogonia doubles in stage X (Lok and de Rooij 1983). However, since the growth fraction of the A_s spermatogonia in the Chinese hamster was about 60%, apparently these cells are not all activated at the same time. Consequently, in stages IX–XII there will be a mixture of A_s spermatogonia that still are quiescent, cells that are stimulated to proliferate and will start their first cell cycle, and cells that are already proliferating. In subsequent stages I–III, virtually all cells will be proliferating, and finally, the proliferative activity becomes more and more inhibited during stages III–VI. In stages III–VI, probably a mixture of proliferating and quiescent A_s spermatogonia exists. During the cycle of the seminiferous epithelium, the number of A_s spermatogonia does not vary to a great extent (Table 2).

Assuming that the A_s spermatogonia are the spermatogonial stem cells, the important question arises whether or not these cells form a homogeneous population of cells. Huckins (1971c, 1978) has postulated the existence of so-called long-cycling stem cells that were supposed to have a higher self-renewal probability than the fast-cycling cells and to be the actual stem cells. Lok et al. (1984) performed a quantitative autoradiographic study in the Chinese hamster and concluded that firm evidence for the concept of long-cycling stem cells is lacking and that it is just one of a number of possibilities to explain the [^{3}H]thymidine labeling pattern observed (for review, see de Rooij 1988).

The next question deals with a possible correlation between the proliferative activity of the stem cell population and its sensitivity for cell killing by irradiation. van Beek et al. (1986a) have found that in stages IX–XII of the epithelial cycle, part

Table 2
Properties of Spermatogonial Stem Cells during the Epithelial Cycle

Epithelial stage	I	II	III	IV	V	VI	VII	VIII	IX	X	XI	XII
Proliferative activity[a]		high		decreasing			low		stimulation			
Cpb-N mice Number of stem cells[b]		14	13		18		19	19			17	
D_0 1 MeV fission neutrons (Gy)[c]		0.55 ± 0.02					0.22 ± 0.01				0.76 ± 0.02	
% stem cells[c,d]		≈90					≈100				≈33	
% Cells with translocations after 3 Gy X-rays		3.0 ± 0.6					4.4 ± 0.7				2.7 ± 0.6	
CBA mice D_0 X-rays (range; Gy)		1.66–1.79					1.01–1.08				2.22–2.29	

[a] Deduced from data of Huckins (1971b) and de Rooij (1983).
[b] Numbers of A_s spermatogonia per 1000 Sertoli cells. van Beek et al. (1984).
[c] Data from van Beek et al. (1986a).
[d] Percentage of stem cells present in these stages having the indicated D_0 value.

of the stem cells are very resistant to the cell-killing effect of 1 MeV fission neutrons. In stages I–III, virtually all stem cells have an intermediate radiosensitivity, and in stage VII, all stem cells are very radiosensitive (Table 2). It was concluded that quiescent stem cells are the most sensitive, that proliferating stem cells have an intermediate radiosensitivity, and that during the transition from quiescence to active proliferation the stem cells go through a highly radioresistant phase. In accordance with this, Bootsma and Davids (1988) found an intermediate D_0 value of 0.43 Gy for fission neutron irradiation for CBA mouse spermatogonial stem cells in S phase.

The present results confirm and extend those of van Beek et al. (1986a). Using another strain of mice and X-irradiation instead of fission neutrons, exactly the same pattern of sensitivity for cell killing was found (Table 2). For the first time, D_0 values were determined for the cell killing effect of X-rays on the spermatogonial stem cell population in each stage of the cycle of the seminiferous epithelium. From the middle of stage I to stage III, where the stem cells are actively proliferating, a D_0 of approximately 1.7 Gy was found. In stages VI and VII, during which these cells are quiescent, the D_0 was about 1.0 Gy, and in stages IX–XII, where the quiescent stem cells gradually start to proliferate again, the D_0 was about 2.3 Gy. The D_0 of 2.3 Gy of the most resistant stem cells is close to the 2.42 Gy value found previously in the CBA mouse with the repopulation index method, which only shows the effects of irradiation on the most resistant stem cells (de Ruiter-Bootsma et al. 1977).

Oakberg and Huckins (1976), Huckins (1978), Huckins and Oakberg (1978), and Oakberg et al. (1986) have postulated the existence of a subclass of A_s spermatogonia that are long cycling, radioresistant, and uniquely capable of self-renewal. In this respect, two important conclusions can be drawn from the studies of van Beek et al. (1986a). First, the most resistant stem cells were only found in stages IX–XII, where they comprised about one third of the number of A_s spermatogonia (Table 2). However, in stages I–III and in stage VII, nearly all stem cells have an intermediate radiosensitivity or are highly sensitive. These results exclude the existence of a subpopulation of A_s spermatogonia being a special type of radioresistant stem cells. Second, the studies of van Beek et al. (1986a) showed that also highly sensitive stem cells and stem cells with an intermediate radiosensitivity, being virtually all stem cells in stages I–VII, are able to form viable repopulating colonies. Apparently, nearly all A_s spermatogonia in stages I–VII are capable of self-renewal. This is at variance with the suggestion of Huckins (1971c, 1978) that only a small number of the A_s spermatogonia have the capacity of self-renewal.

In view of the ongoing debate on the correlation between the radiosensitivity of spermatogonial stem cells for cell killing and their sensitivity for mutation or translocation induction (van Buul 1983), we have made a first attempt to determine this correlation in a more direct way. To do this, the seminiferous epithelium of Cpb-N mice was synchronized with the vitamin A deficiency/ vitamin A replacement method (Fig. 2). In this way, mice could be irradiated when, for example, nearly the whole testis was in stages VI and VII of the epithelial cycle. Consequently, in such mice virtually all spermatogonial stem cells were quiescent and hence highly sensitive to cell killing at the time of irradiation. Indeed, more translocations, lower testes weights, and more degenerated testes were found in these mice compared to those in which the seminiferous epithelium was in other stages at the time of irradiation.

Leenhouts and Chadwick (1981) published a theoretical model that fitted most of the data on the induction of translocations by ionizing radiation in mouse spermatogonial stem cells. One of the most important parameters in this model was the ratio between cell killing and translocation induction, the so-called "p/c" ratio. According to their calculations, the p/c ratio should be 13, meaning that the probability that a basic lesion in the DNA leads to cell killing is 13 times higher than the probability that it leads to translocation formation. Indeed, van Buul and

Goudzwaard (1990), using combined hydroxyurea and X-ray treatments, could obtain experimental evidence that for mouse spermatogonial stem cells the p/c ratio was about 10. In the present experiments, the differences in cell killing between the most sensitive and most resistant epithelial stages are a factor of 2–3, in fact leading to a 20–30% change in translocation induction. However, the differences were not significant. These studies will be continued using higher doses of irradiation.

ACKNOWLEDGMENTS

The technical assistance of H.L. Roepers-Gajadien and C.M.J. Seelen is gratefully acknowledged. Parts of this work were supported by the Netherlands Cancer Foundation, grant UUKC-86-5, and the association of Euratom and the University of Leiden, contract BIO-E-406-81 NL.

REFERENCES

Bootsma, A.L. and J.A.G. Davids. 1988. The cell cycle of spermatogonial colony forming stem cells in the CBA mouse after neutron irradiation. *Cell Tissue Kinet.* **21:** 105.

Cattanach, B.M. and M.J. Kirk. 1987. Enhanced spermatogonial stem cell killing and reduced translocation yield from X-irradiated 101/H mice. *Mutat. Res.* **176:** 69.

de Rooij, D.G. 1973. Spermatogonial stem cell renewal in the mouse. I. Normal situation. *Cell Tissue Kinet.* **6:** 281.

———. 1983. Proliferation and differentiation of undifferentiated spermatogonia in the mammalian testis. In *Stem cells. Their identification and characterization* (ed. C.S. Potten), p. 89. Churchill Livingstone, Edinburgh.

———. 1988. Regulation of the proliferation of spermatogonial stem cells. *J. Cell Sci.* Suppl. **10:** 181.

de Ruiter-Bootsma, A.L., M.F. Kramer, D.G. de Rooij, and J.A.G. Davids. 1977. Survival of spermatogonial stem cells in the mouse after split-dose irradiation with fission neutrons of 1-MeV mean energy or 300-kV X rays. *Radiat. Res.* **71:** 579.

Gerber, G.B. and A. Léonard. 1971. Influence of selection—Non uniform cell population and repair on dose-effect curves of genetic effects. *Mutat. Res.* **12:** 175.

Huckins, C. 1971a. The spermatogonial stem cell population in adult rats. I. Their morphology, proliferation and maturation. *Anat. Rec.* **169:** 533.

———. 1971b. The spermatogonial stem cell population in adult rats. II. A radioautographic analysis of their cell cycle properties. *Cell Tissue Kinet.* **4:** 313.

———. 1971c. The spermatogonial stem cell population in adult rats. III. Evidence for a long-cycling population. *Cell Tissue Kinet.* **4:** 335.

———. 1978. Behavior of stem cell spermatogonia in the adult rat irradiated testis. *Biol. Reprod.* **19:** 747.

Huckins, C. and E.F. Oakberg. 1978. Morphological and quantitative analysis of spermatogonia in mouse testes using whole mounted seminiferous tubules. II. The irradiated testes. *Anat. Rec.* **192:** 529.

Leenhouts, H.P. and K.H. Chadwick. 1981. An analytical approach to the induction of

translocations in spermatogonia of the mouse. *Mutat. Res.* **82**: 305.
Lok, D. and D.G. de Rooij. 1983. Spermatogonial multiplication in the Chinese hamster. III. Labelling indices of undifferentiated spermatogonia throughout the cycle of the seminiferous epithelium. *Cell Tissue Kinet.* **16**: 31.
Lok, D., M.T. Jansen, and D.G. de Rooij. 1983. Spermatogonial multiplication in the Chinese hamster. II. Cell cycle properties of undifferentiated spermatogonia. *Cell Tissue Kinet.* **16**: 19.
――――. 1984. Spermatogonial multiplication in the Chinese hamster. IV. Search for long cycling stem cells. *Cell Tissue Kinet.* **17**: 135.
Lok, D., D. Weenk, and D.G. de Rooij. 1982. Morphology, proliferation and differentiation of undifferentiated spermatogonia in the Chinese hamster and the ram. *Anat. Rec.* **203**: 83.
Oakberg, E.F. 1971. Spermatogonial stem-cell renewal in the mouse. *Anat. Rec.* **169**: 515.
――――. 1978. Differential spermatogonial stem-cell survival and mutation frequency. *Mutat. Res.* **50**: 327.
Oakberg, E.F. and C. Huckins. 1976. Spermatogonial stem cell renewal in the mouse as revealed by ^{3}H-thymidine labeling and irradiation. In *Stem cells of renewing cell populations* (eds. A.B. Cairnie et al.), p. 287. Academic Press, New York.
Oakberg, E.F., D.G. Gosslee, C. Huckins, and C.C. Cummings. 1986. Do spermatogonial stem cells have a circadian rhythm? *Cell Tissue Kinet.* **19**: 367.
Oftedal, P. 1968. A theoretical study of mutant yield and cell killing after treatment of heterogeneous cell populations. *Hereditas* **60**: 177.
Oud, J.L. and D.G. de Rooij. 1977. Spermatogenesis in the Chinese hamster. *Anat. Rec.* **187**: 113.
Russell, W.L. 1956. Lack of linearity between mutation rate and dose for X-ray induced mutations in mice. *Genetics* **41**: 658.
Sumner, A.T. 1972. A simple technique for demonstrating centromeric heterochromatin. *Exp. Cell Res.* **75**: 304.
van Beek, M.E.A.B., J.A.G. Davids, and D.G. de Rooij. 1986a. Variation in the sensitivity of the mouse spermatogonial stem cell population to fission neutron irradiation during the cycle of the seminiferous epithelium. *Radiat. Res.* **108**: 282.
van Beek, M.E.A.B., J.A.G. Davids, H.J.G. van de Kant, and D.G. de Rooij. 1984. Response to fission neutron irradiation of spermatogonial stem cells in different stages of the cycle of the seminiferous epithelium. *Radiat. Res.* **97**: 556.
――――. 1986b. Non-random distribution of mouse spermatogonial stem cells surviving fission neutron irradiation. *Radiat. Res.* **107**: 11.
van Buul, P.P.W. 1983. Inductions of chromosome aberrations in stem cell spermatogonia of mammals. In *Progress and topics in cytogenetics* (ed. T. Ishihara and M.S. Sasaki), vol. 4, p. 369. A.R. Liss, New York.
van Buul, P.P.W. and P. de Boer. 1982. The induction by X-rays of chromosomal aberrations in somatic and germ cells of mice with different karyotypes. *Mutat. Res.* **92**: 229.
van Buul, P.P.W. and J.H. Goudzwaard. 1990. The relation between induced reciprocal translocations and cell killing of mouse spermatogonial stem cells after combined treatments with hydroxyurea and X-rays. *Mutat. Res.* (in press).
Wessels-Kaalen, M.C.A. 1984. A new procedure for meiotic preparations for mammalian testes. *Mamm. Chromosome Newslett.* **25**: 66

COMMENTS

Generoso: Is the fluctuation of the proportion of resistant cells with cell proliferation a function of the proportion of long-cycling and short-cycling stem cells?

de Rooij: No, I don't think it is. There is no conclusive evidence for the existence of long-cycling and fast-cycling stem cells. We have tried to find long-cycling stem cells. We have done a study in the Chinese hamster, but we couldn't find conclusive evidence that they really exist. There are other explanations for that.

Generoso: You are not in agreement with Oakberg and Huckins, then, as to the existence of the long-cycling versus the fast-cycling spermatogonia in the mouse and the rat?

de Rooij: I am not. I think all data still fit with a model in which all stem cells are alike and divide two to three times each cycle on the average.

Generoso: You mentioned A_s as being in a quiescent stage. What do you mean by that?

de Rooij: They are inhibited from proliferating at a certain stage of the epithelium. They are excited again from stage IX onward, then they slowly begin to proliferate.

Adler: Isn't that long cycling?

de Rooij: Well, they proliferate; there is a cell cycle. In the Chinese hamster, the cell cycle was 90 hours. It's a long cycle, of course, but I do not think that it is very long cycling.

Adler: Nothing like 6 to 8 days?

de Rooij: No. Well, perhaps some of them, but just by coincidence. I don't think there is a separate class of stem cells that are long cycling.

Bridges: There is a separate population that always goes slowly and another population that always goes fast?

de Rooij: Yes.

Bridges: You're saying it's the same population; sometimes they go fast, sometimes they go slow?

de Rooij: Yes.

Generoso: Do you also mean that there is a sort of reserve A_s?

de Rooij: No, not reserve cells.

Lonnie Russell: If a cell of a clone of differentiating spermatogonia degenerates, does the rest of that clone degenerate as a result of that one cell, or can they pinch off and leave both viable and degenerating cells?

de Rooij: It has been postulated by Huckins that when one cell dies, the whole clone dies.

Lonnie Russell: What do you think?

de Rooij: We didn't do any work on that.

Lonnie Russell: So the effect could be magnified depending on when the degeneration occurs?

de Rooij: Yes.

Lonnie Russell: In differentiating spermatogonia, Clermont says they look different in whole mouse A_1, A_2, A_3, and A_4. Do they look different to you?

de Rooij: No. They are very much alike. When I did not have the help of the other cells, I could not distinguish between A_2, A_3, A_4, although perhaps with A_4 there is some difference.

Lonnie Russell: A paper will be coming out that describes their differences a little more. I would really like to see something dramatic so we could get a handle on these cells and be able to pinpoint and know what they are. It is my hope that they would be different.

de Rooij: Well, they can be distinguished from the stage of cycle they are in.

Handel: But that's only when you have all the cells in the cycle. It becomes very difficult when you don't have them all.

de Rooij: Well, I couldn't distinguish them.

Fritz: What is your impression of Moens's work who addressed this issue some years ago when he did serial sections? His conclusion was to say, as you indicated earlier, that once you have a clone of gonia start, either they all die, the entire clone, or they survive, but you don't get some daughter cells being destroyed. That was Moens's conclusion.

Lonnie Russell: His conclusion couldn't have been right, because if they all survived you would have 4000 cells in a clone.

Fritz: Why 4000? It's 2^8.

Lonnie Russell: You have 16 A_1 cells together, as Huckins would calculate, so if you do the kinetics there, you have over 4000.

Fritz: I beg your pardon. I misstated. He said once you had a particular cell committed to go, at that stage, starting from A_1, then it all went or nothing went, so it was all or none.

Lonnie Russell: The answer is still the same. You would have over 4000 cells.

de Rooij: Yes, but maybe some part of the clone can get loose, it may break, it may disconnect. It would be interesting to give perhaps a low dose of radiation and look at the clones to see if now and then one cell dies or that whole clones die.

Hecht: Do you think that the differences in the radiosensitivity are due to the cells themselves or to the environment that the cells are in, i.e., the stage of the cycle?

de Rooij: It's very difficult to speculate on that. There can be other factors from outside that could influence radiosensitivity of cells.

Bridges: Getting back to this work of Oakberg and Huckins: My memory may be bad on this, but weren't they postulating rather more dramatic differences in radiosensitivity to explain their results than those that you measure, or are they compatible?

de Rooij: I think the differences are quite dramatic. For fission neutrons we found a D_0 value of 0.2 Gy for the most sensitive stem cells and of 0.7–0.8 Gy for the most resistant cells, so it's a factor of 3.

Bridges: For gamma rays it's only about a twofold difference?

de Rooij: Yes, somewhat more than 2.

Liane Russell: Do you think that every time an A_s divides it's going to make one A_s and one A_{pr}, or are there times when one A_s just makes two A_ss; and, if so, could these two types of events happen in different parts of the cycle?

de Rooij: When the two daughter cells stay together, there is no room for an A_s.

Liane Russell: No. What I mean is the A_s has to replace itself, so one A_s divides and makes another A_s.

Handel: She's asking if the A_s ever divides to form two A_ss.

de Rooij: Yes, about half of the divisions of the A_s will produce new A_ss.

Liane Russell: Do you think that happens in a different part of the cycle from that which makes an A_{pr}?

de Rooij: We have looked at that but we couldn't find it. It is difficult because you can have false parts that are A_s that are not connected by a bridge that still need time to migrate away from each other. There is some decrease in the number of the A_ss depending on the stages. Maybe there is more differentiation at late stages. It may also be that they need time to migrate away.

Lonnie Russell: When does spermatogenesis begin? If you had to clinically examine what you have just said, does it begin in stage 8 or does it begin when A divides to form A pairs?

de Rooij: I think the formation of spermatozoa begins with the formation of the pair.

Lonnie Russell: So that occurs all during the cycle and there is no real answer to the question, then?

de Rooij: The formation of pairs will be especially starting in about stage X of the cycle and go on up to the inhibition at about stage III or stage IV. During that time, the pairs will be formed.

Lonnie Russell: So we still have a vague answer.

Handel: There's no easy answer.

Use of Heritable Mutations in the Analysis of Meiosis

MARY ANN HANDEL
Department of Zoology
University of Tennessee
Knoxville, Tennessee 37996

OVERVIEW

Analysis of the effects of heritable mutations on spermatogenesis carries the dual promise of unraveling mechanisms of spermatogenesis and clarifying the basis of germ-cell stage specificity of mutation induction. Both autosomal recessive mutations and chromosomal aberrations can impair spermatogenesis in the mouse, interrupting the process at any of a number of stages in differentiation (Handel 1987). Here, the focus is on genetic aberrations that affect male germ-cell meiotic progress more severely than they affect female germ-cell meiosis. There are interesting sex differences in the process of meiosis. These include timing and coordination of the onset of meiosis (fetal and synchronous among female germ cells, and postpubertal and not synchronous among male germ cells) as well as remarkable differences in the behavior of the X chromosome, which is transcriptionally active at meiotic prophase in female germ cells and transcriptionally inactivated during meiosis in male germ cells. Mutations affecting male meiosis include an autosomal recessive mutation, skeletal fusions with sterility (*sks*), and reciprocal translocations between the X chromosome and an autosome. Although the mutant *sks* allele impairs completion of meiotic metaphases, it is unlikely that it is a "meiosis gene" (i.e., one expressed uniquely during meiosis), since its effects are pleiotropic. Reciprocal X-autosome translocations in mammals are invariably associated with male sterility. Arrest of germ-cell development occurs most frequently in meiotic prophase, and function of the X chromosome may be impaired. It is probable that the sexual dimorphism in the behavior of the X chromosome is reflective of fundamental processes that differ in the meiosis of male and female mammalian germ cells.

INTRODUCTION

Although the genetic events during meiosis are pivotal and obligatory for creating genetic variability and for gamete survival, we know relatively little about how these processes are controlled and executed in mammalian germ cells. Here, the focus is on those genetic aberrations in the mouse that interrupt spermatogenesis at meiosis. The term meiosis is used to refer to the entire process of the lengthy and involved meiotic prophase as well as the two meiotic divisions that produce the haploid gametes.

Despite the critical importance of meiotic prophase in genetic mechanisms of recombination, little is known about the biochemical and morphological events of meiosis, and even less about the controls over its initiation, regulation, and progress. Considerable differences exist between meiosis in female and male mammals; this sexual dimorphism will likely provide basic information about fundamental mechanisms of meiosis, as well as about the origin of parental genomic imprinting (see DeLoia and Solter, this volume). One sexually dimorphic feature of meiosis is the timing of the onset of meiotic prophase (for review, see Zamboni 1986). Onset of meiosis in female mammals occurs once and is fetal and synchronous in all germ cells; in males, however, meiotic prophase does not occur until puberty, is repeated throughout reproductive life, and occurs in subsets or cohorts of germ cells, rather than synchronously in the entire germ-cell population. Meiosis is discontinuous in females, with arrest occurring in all oocytes at the end of meiotic prophase. Resumption and completion of meiosis occurs in small groups of germ cells at the time of ovulation. There is considerable information about hormonal and other physiological controls of meiotic maturation of oocytes (Eppig 1987). Meiosis in males, on the other hand, appears to be continuous in that completion of the meiotic metaphases seems to be automatic and obligatory for the long-term survival of the germ cell. Little is known of the physiological controls, but it seems likely that they will be demonstrated to be the same common metaphase-promoting factors found to operate in oocytes at meiotic maturation and, indeed, in all mitotic cells (Murray and Kirschner 1989).

Another dramatic difference between male and female meiosis is the behavior and transcriptional activity of the sex chromosomes. In oocytes, both X chromosomes are transcriptionally active, whereas in spermatocytes there appears to be inactivation of the X chromosome. The inactivity of the X chromosome during spermatogenesis is manifest in its sequestration in the nucleus in a region of differentially condensed chromatin, the XY body or sex vesicle (Fig. 1). This structure, described in careful detail by Solari (1974; 1989), is adjacent to the nuclear envelope and is composed of the partially paired sex chromosomes. The nucleolus is apposed to the XY body on its nucleoplasmic surface and penetrates into the sex chromatin (Fig. 1). It is not known if there is functional significance of this intimate relationship or if it represents spurious heterochromatic associations. In this context, it is interesting that the nucleolar organizer is involved in mediating X-Y pairing in *Drosophila* (McKee et al. 1988; McKee and Karpen 1990). Transcriptional inactivation of the X chromosome is inferred from the failure of the XY body to incorporate tritiated uridine (for review, see Handel 1987). There is conflicting evidence, all indirect, as to whether the X chromosome is inactive in spermatogonia (Tiepolo et al. 1967; Tres and Solari 1968; Kofman-Alfaro and Chandley 1971; de Boer and Branje 1979); this issue clearly needs to be resolved with more direct approaches. It is important to bear in mind that neither the occurrence nor the timing of X-chromosome inactivation during spermatogenesis has been rigorously demonstrated at the

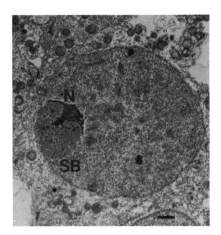

Figure 1
Electron micrograph of the nucleus of a pachytene spermatocyte. Note that the XY, or sex, body (SB) is penetrated by the nucleolus (N). Bar, 1 µm.

molecular level with probes for X-encoded gene products. It is not known why there might be reactivation of the previously inactive X chromosome at meiotic prophase in female germ cells and inactivation of the X chromosome, apparently at the same period of meiosis, in male germ cells.

This summary of some of the sexually dimorphic features of meiosis illustrates that there are many unanswered questions about the mechanisms underlying meiosis. Focusing on these sex differences may lead to better understanding about the processes of meiosis, about recombination, about sex differences sometimes observed in recombination frequencies, and about the differences that exist between paternally and maternally derived genomes. Thus, of particular interest to us are those mutations and genetic aberrations in the mouse that interfere more severely with male meiosis than they do with female meiosis. These include an autosomal recessive mutation, *sks*, and chromosomal translocations involving the X chromosome.

RESULTS

Skeletal fusions with sterility, or *sks*, is a mutation originally identified by Muriel Davisson at The Jackson Laboratory that affects male meiosis more severely than it does female meiosis (Handel et al. 1988). Affected individuals of both sexes are characterized by fusions of vertebrae and of ribs, so that homozygotes have shorter-than-normal bodies and tail kinks. The gene has been mapped to Chr 4, 16.6 cM distal to the brown locus.

Within the testes, the effects of the mutant gene are quite dramatic; rarely is there any spermatogenic development beyond the pachytene or metaphase I (MI) stages of meiosis (Fig. 2). We find that cells accumulate at all stages of meiotic prophase, but especially during mid-to-late pachytene and diplotene to MI. Thus, the meiotic progress of germ cells expressing this mutation is inhibited, with some cells failing at earlier stages of prophase than others. There are no morphologically normal cells developing past the pachytene or MI stage; however, occasional abnormal and apparently degenerating postmeiotic cells can be observed. Pachytene cells are characterized by development of an XY body, representing the paired and inactive sex chromosomes; so at least some aspects of meiotic prophase are executed more or less normally. However, many pachytene cells are characterized by misshapen nuclei and irregularly condensed chromatin, suggestive of aberrant or degenerative changes. These abnormalities are reflected in the analysis of meiotic pairing as visualized in microspreads of synaptonemal complexes (SCs), performed by Muriel Davisson. Although normal SC morphology is seen in control cells, the SCs of mutant meiotic cells exhibit frequent disruption in pairing, with abnormal morphology, including thin, extended, and attenuated SCs. However, it is difficult to determine if these SC abnormalities are specific effects of the mutation or simply reflective of meiotic delay or degeneration of meiotic cells, since some cells with normal SC morphology are found in all mutant mice.

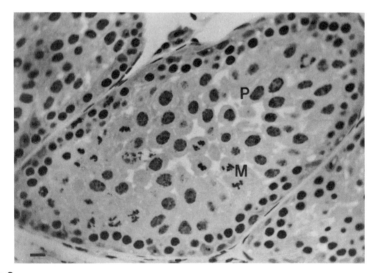

Figure 2
Cross-section of a seminiferous tubule of *sks/sks* mouse. Pachytene (P) and metaphase I (M) spermatocytes fill the center of the tubule, but no cells beyond MI are seen. Bar, 10 µm.

Despite the almost total inability of spermatocytes to complete meiosis, oocytes can mature, i.e., can complete meiotic divisions, in vivo or in vitro, at a frequency equivalent to that of littermate controls (Handel et al. 1988). However, these oocytes could not be fertilized successfully in vitro. Thus, the impairment of the progress of meiotic prophase and metaphase caused by mutant *sks* action is more severe in males than in females. Since it is not marked by inhibition of meiosis, it is not clear if the infertility of females is identical or related to meiotic impairment in males.

Pleiotropy of effects of this gene, with respect to both germ cells and skeleton formation, frustrates analysis of the primary action of the normal allele of the *sks* gene. It is not known if the mutant *sks* is expressed autonomously in the germ cells or if it is acting in the gonadal environment.

A different category of heritable defects that affect meiosis in the male are the X-autosome chromosome translocations. A striking feature of all X-autosome translocations that involve separation of the X chromosome to two centromeres is male sterility, characterized by meiotic arrest of germ-cell differentiation. Although these same translocations may be associated with some reduction in reproductive life span of females, they do not cause female sterility or total arrest of meiosis in oocytes. Thus, they may affect male meiosis in an instructive way. Possibly X-chromosome inactivation is involved, since this is unique to male meiosis. Handel and Park (1988) asked the question of whether a failure in X-chromosome inactivation might be associated with the sterility and arrest of meiotic progress of spermatocytes of mice bearing either of two X-autosome translocations, T(X;16)16H (T16H) and T(X;11)38H (T38H) (Searle et al. 1983).

Examination of histological sections of testes reveals that spermatogenesis is impaired at meiotic prophase, with inhibition primarily at the pachytene stage in T16H and prior to the onset of pachytene, at the leptotene or zygotene stages in T38H (Fig. 3). At least in T16H, the severity of impairment was age-related, being more severe in older males. We next sought to determine if the XY body, the morphological manifestation of X-chromosome inactivation, developed in these spermatocytes. Spermatocytes of T16H mice that progress to early to mid-pachytene stages formed an XY body (Fig. 4a), confirming earlier observations of Solari (1971). XY bodies are rarely observed in nuclei of spermatocytes of T38H mice, even in cells judged by the presence of SC to have progressed to pachytene (Fig. 4b). Whenever we observe an XY body, we find no incorporation of tritiated uridine, determined by autoradiography after an in vitro exposure to uridine (Fig. 5). Given this autoradiographic evidence, can we therefore conclude that there is no failure in X-chromosome inactivation in these translocation-sterile males? This may not be valid because, in addition to those spermatocytes with an XY body, we observe many spermatocytes that do not have an XY body. This suggests the possibility that, in many cells, failure of X-chromosome inactivation might prevent meiotic progress through the events of zygotene to pachytene stages. To test this,

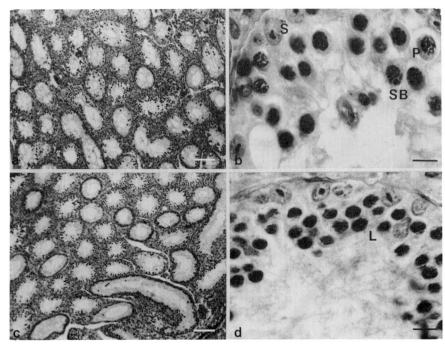

Figure 3
(*a* and *b*) Cross-sections of seminiferous tubules of a T16H mouse illustrating inhibition of spermatogenesis at the pachytene stage. Note the Sertoli cell (S), the XY, or sex, body (SB), and the pachytene spermatocyte (P). (*c* and *d*) Cross-sections of seminiferous tubules of a T38H mouse illustrating spermatogenic inhibition prior to the pachytene stage. Note the leptotene spermatocyte (L). (*a* and *c*) Bar, 100 μm. (*b* and *d*) Bar, 10 μm.

what is needed is a way to detect the X chromosome before it organizes a morphologically distinct XY body. We are presently assessing the issue of X-chromosome inactivation in these mutants using techniques to localize the X chromosome in the nucleus and to correlate its location with transcriptional activity.

DISCUSSION

Meiosis is a highly specialized differentiative event unique to germ cells. Meiotic events of recombination and segregation are pivotal for creating genetic variability and the genotype of the following generation. These events may also be requisite for the further normal development and function of male gametes, although this idea has not been tested empirically. DNA-repair processes during meiosis are likely to be of importance in establishing germ-cell stage sensitivities to induction of muta-

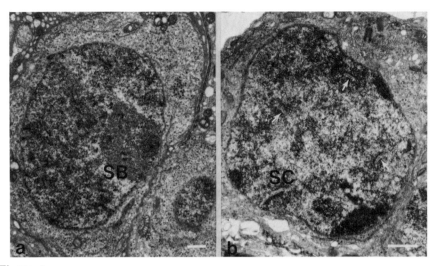

Figure 4
(a) Electron micrograph of a nucleus of a pachytene spermatocyte from a T16H mouse showing the XY, or sex, body (SB). Bar, 1 μm. (b) Electron micrograph of a spermatocyte nucleus from a T38H mouse. No XY, or sex, body is present, although profiles of synaptonemal complex (SC) are clearly evident (arrows). Bar, 1 μm.

tions. These various functions and processes of meiosis predict that gametocytes should express genes encoding proteins and enzymes for assembly of the SC (see Dresser, this volume) and for mediating recombination. Much progress has been made in identifying genes whose expression is unique to meiosis in yeast (Giroux 1988; Engebrecht and Roeder 1989); some progress has been made in identifying such genes in *Drosophila* (Baker et al. 1976); however, very little is known of meiosis-specific gene expression in mammalian spermatocytes. Promising approaches include the identification of genes encoding antigenic determinants of the mammalian SC (Moens et al. 1987; Heyting et al. 1988) and the construction and screening of germ-cell stage-specific cDNA libraries.

The focus here has been on the sexual dimorphism of meiosis and consequently on the use of genetic aberrations that affect male meiosis more severely than female meiosis. Analysis of such mutations might be expected to yield insight to differences in male and female meiosis and ultimately to lead to the identification of genes and gene products involved in establishing these differences.

sks is the first mutation identified in the mouse that impairs the process of meiosis. It is of considerable interest, therefore, that *sks* affects male meiosis more dramatically than female meiosis. However, the mutant *sks* allele does not interfere with either pairing of the sex chromosomes or condensation of sex chromatin into

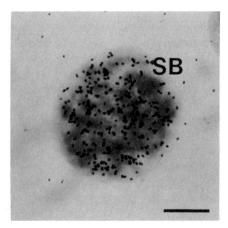

Figure 5
Nucleus of a pachytene spermatocyte of a T16H mouse after incubation in vitro with tritiated uridine. Note absence of grains over the XY, or sex, body (SB). Bar, 10 μm.

the XY body, the two morphological features most reflective of the sex differences of meiosis. It is possible that mutant *sks* does not really affect meiosis in a sex-specific manner as much as it does in a sex-sensitive manner. Thus, male gametocytes may be more sensitive to the effects of the mutant gene and arrest earlier than do female gametocytes. Oocytes, even though not inhibited in the completion of meiosis, may be sensitive in a way that is reflected in their inability to be fertilized and develop normally. Critical to understanding the effect of mutant *sks* will be determination of whether *sks* acts autonomously in the germ cell or acts in the gonadal environment. This could be demonstrated by recombining normal and mutant germ and somatic cells in chimeras (Handel et al. 1987), or, ultimately, by identification and in situ localization of the gene transcript and protein product.

As mentioned previously, pleiotropy of action of the *sks* genes complicates analysis but may eventually prove instructive (as has been the case for genes in the sex-determination pathway of *Drosophila* [Cline 1989]). Certainly *sks* is not a meiosis gene, that is, one expressed uniquely during meiosis. Nonetheless, the product of the normal *sks* gene appears to be essential, either directly or indirectly, for spermatocytes to complete meiosis. This mutation could be useful in differential or subtractive library screening strategies for the identification of gene transcripts essential for meiosis.

Although we might expect meiosis-specific gene expression during spermatogenesis, it is likely also that there are some aspects of meiosis that are chromosomal rather than genic in control. One of these is the behavior of the sex chromosomes, perhaps reflected in their transcriptional activity, which shows remarkable sexual dimorphism. Mitotically proliferating oogonia are characterized, as are the somatic cells of female fetuses, by a single active X chromosome. As oogonia differentiate to primary oocytes and enter meiotic prophase, the inactive X chromosome is reactivated so that progress through meiotic prophase to arrest at diplotene is marked by two active X chromosomes (Andina 1978; Gartler et al. 1980; Monk and McLaren 1981). In marked contrast, the single X chromosome of primary spermatocytes appears to be inactive, and it appears that the time of inactivation is similar to the time of X-chromosome reactivation in female germ cells, namely, as gametocytes enter meiotic prophase. The nature and extent of these events at the molecular level are not well defined.

Male-limited sterility, with meiotic arrest of germ-cell development, is a feature of the X-autosome translocations described above. We have demonstrated that at least in some gametocytes, both condensation of the sex-chromosome chromatin and transcriptional inactivation of the sex chromosomes occur. However, this analysis could not exclude the possibility that this process of inactivation and condensation to form an XY body was impaired in the majority of spermatocytes. Further study, involving more sensitive techniques for the cytological localization of the X chromosome, will be necessary to resolve this issue.

Although it might be supposed that these translocation aberrations affect the process of X-chromosome inactivation in male germ cells, it is also possible that sterility in males is caused by inactivation of genes critical for spermatogenesis that are on the autosomal region translocated to the X chromosome. If this were the case, we would have to suppose that there are a great many genes, throughout the genome, critical for spermatogenic passage through meiosis, since the sterility and meiotic impairment is a feature of all reciprocal X-autosome translocations thus far investigated, regardless of autosomal breakpoints. There are only two translocations involving the X chromosome and autosomal chromosomes of the mouse that do not cause male-limited sterility; both of these are special cases and not typical reciprocal translocations. One case is a Robertsonian fusion, Rb(X.2)2Ad, between the X chromosome and Chr 2 (Adler et al. 1989). The second is the unbalanced form of an insertion of a portion of Chr 7 into the X chromosome, Is(In7;X)1Ct. Interestingly, male mice bearing the balanced form of the insertional translocation (that is, with the insertion in the X, a normal Chr 7, and a Chr 7 deleted for the region of the insertion) are sterile, whereas mice with the unbalanced condition (with the insertion in X, two normal Chrs 7, and thus segmentally trisomic for the insertion) are fertile! Among these various chromosome aberrations—on the one hand, reciprocal translocations involving the X chromosome and the balanced form of Is1Ct, and the unbalanced form of Is1Ct and Rb2Ad on the other—the feature common to the sterile forms is the possibility that complete pairing of the autosomal segments may not be achieved, and the feature common to the fertile forms is the opportunity for complete pairing of all autosomal segments outside of the XY body, leaving the only unpaired autosomal segments (i.e., the insertion in the X) within the XY body, where being unpaired may be "allowable" for the progress of spermatogenesis.

We consider the possibility that inactivation of sex chromosomes is related, either causally or consequentially, to chromosome pairing and recombination. In this context, it is useful to return to the concept of the sexual dimorphism in the behavior of the sex chromosomes, and to consider the following facts. During female meiosis, both X chromosomes are transcriptionally active, the sex chromosomes are fully and homologously paired, and there is recombination and chiasmata formation along the length of the sex chromosomes. However, during male meiosis, the X (and Y) chromosomes appear to be transcriptionally inactive and exhibit a differential condensation of chromatin compared to the autosomes, and there is only spatially restricted and limited pairing and recombination between the sex chromosomes. Thus, the inactivation of the X chromosome during spermatogenesis may be a reflection of fundamental differences between male and female meiosis with respect to pairing and recombination of the sex chromosomes.

It is our hope and contention that continued investigation of these heritable mutations that affect male meiosis more severely than they affect female meiosis will provide resolution of these issues involving the function of X-chromosome inactivation and enhance our understanding of the sexually dimorphic facets of meiosis.

ACKNOWLEDGMENTS

Work in my laboratory reported here has been supported in part by a grant, HD-16978, from the National Institutes of Health. Collaborative work at The Jackson Laboratory on *sks* mutants was supported in part by BSR-84-18828 from the National Science Foundation. I am indebted to Cynthia Park for technical assistance.

REFERENCES

Adler, I.-D., R. Johannisson, and H. Winking. 1989. The influence of the Robertsonian translocation Rb(X.2)2Ad on anaphase I non-disjunction in male laboratory mice. *Genet. Res.* **53**: 77.

Andina, R.J. 1978. A study of X chromosome regulation during oogenesis in the mouse. *Exp. Cell Res.* **111**: 211.

Baker, B.S., A.T.C. Carpenter, M.S. Esposito, R.E. Esposito, and L. Sandler. 1976. The genetic control of meiosis. *Annu. Rev. Genet.* **10**: 53.

de Boer, P. and H.E.B. Branje. 1979. Association of the extra chromosome of tertiary trisomic male mice with the sex chromosome during first meiotic prophase, and its significance for impairment of spermatogenesis. *Chromosoma* **73**: 369.

Cline, T.W. 1989. The affairs of *daughterless* and the promiscuity of developmental regulators. *Cell* **59**: 231.

Engebrecht, J.A. and G.S. Roeder. 1989. Yeast *mer1* mutants display reduced levels of meiotic recombination. *Genetics* **121**: 237.

Eppig, J.J. 1987. Factors controlling mammalian oocyte maturation. In *The primate ovary* (ed. R. Stouffer), p. 77. Plenum Press, New York.

Gartler, S.M., M. Rivest, and R. Cole. 1980. Cytological evidence for an inactive X chromosome in murine oogonia. *Cytogenet. Cell Genet.* **28**: 203.

Giroux, C.N. 1988. Chromosome synapsis and meiotic recombination. In *Genetic recombination* (ed. R. Kucherlapati and G.R. Smith), p. 465. American Society for Microbiology, Washington, D.C.

Handel, M.A. 1987. Genetic control of spermatogenesis in mice. In *Results and problems in cell differentiation, spermatogenesis: Genetic aspects* (ed. W. Hennig), vol. 15, p. 1. Springer-Verlag, Berlin.

Handel, M.A. and C. Park. 1988. Spermatogenesis and sex-vesicle structure in translocation-sterile mice. *Biol. Reprod.* (suppl. 1) **36**: 81. (Abstr.)

Handel, M.A., P.W. Lane, A.C. Schroeder, and M.T. Davisson. 1988. New mutation causing sterility in the mouse. *Gamete Res.* **21**: 409.

Handel, M.A., L.L. Washburn, M.P. Rosenbery, and E.M. Eicher. 1987. Male sterility caused by p^{6H} and qk mutations is not corrected in chimeric mice. *J. Exp. Zool.* **243**: 81.

Heyting, C., R.J. Dettmers, A.J.J. Dietrich, E.J.W. Redeker, and A.C.G. Vink. 1988. Two major components of synaptonemal complexes are specific for meiotic prophase nuclei. *Chromosoma* **96**: 325.

Kofman-Alfaro, S. and A.C. Chandley. 1971. Meiosis in the male mouse. An autoradiographic investigation. *Chromosoma* **31**: 404.

McKee, B.D. and G.H. Karpen. 1990. *Drosophila* ribosomal RNA genes function as an X-Y pairing site during male meiosis. *Cell* **61**: 61.

McKee, B.D., G.H. Karpen, and C.D. Laird. 1988. Stimulation of X-Y meiotic pairing in *Drosophila melanogaster* by a transformed rDNA gene. *Genome* (suppl. 1) **30**: 124. (Abstr.)

Moens, P.B., C. Heyting, A.J.J. Dietrich, W. van Raamsdonk, and Q. Chen. 1987. Synaptonemal complex antigen location and conservation. *J. Cell Biol.* **105**: 93.

Monk, M. and A. McLaren. 1981. X-chromosomal activity in fetal germ cells of the mouse. *J. Embryol. Exp. Morphol.* **63**: 75.

Murray, A.W. and M.W. Kirschner. 1989. Dominoes and clocks: The union of two views of the cell cycle. *Science* **246**: 614.

Searle, A.G., C.V. Beechey, E.P. Evans, and M. Kirk. 1983. Two new X-autosome translocations in the mouse. *Cytogenet. Cell Genet.* **35**: 279.

Solari, A.J. 1971. The behavior of chromosomal axes in Searle's X-autosome translocation. *Chromosoma* **34**: 99.

———. 1974. The behavior of the XY pair in mammals. *Int. Rev. Cytol.* **38**: 273.

———. 1989. Sex chromosome pairing and fertility in the heterogametic sex of mammals and birds. In *Fertility and chromosome pairing: Recent studies in plants and animals* (ed. C. B. Gillies), p. 77. CRC Press, Boca Raton, Florida.

Tiepolo, L., M. Fraccaro, M. Hulten, J. Lindsten, A. Mannini, and P.M.L. Ling. 1967. Timing of sex chromosome replication in somatic and germ-line cells of the mouse and the rat. *Cytogenetics* **6**: 51.

Tres, L.L. and A.J. Solari. 1968. The ultrastructure of the nuclei and the behavior of the sex chromosomes of human spermatogonia. *Z. Zellforsch. Mikrosk. Anat. Abt. Histochem.* **91:** 75.

Zamboni, L. 1986. Meiosis as a sexual dimorphic character of germinal cell differentiation. *Tokai J. Exp. Clin. Med.* **11:** 377.

COMMENTS

Lonnie Russell: Could these abnormalities be Sertoli cell defects that are subsequently expressed in germ cells?

Handel: I think that it is quite possible. I would like to think that in the case of the translocations we are affecting a fundamental property in the germ cells. As Lee Russell has shown, you certainly cannot rescue the translocation-carrying cells by recombining them with normal cells in a chimeric testis. You would surmise that you might be able to do that if it were a defect in the Sertoli cell. I think we almost certainly get into a vicious circle here of failure of cells to progress to later stages, therefore Sertoli cells are not receiving the right cross talk from differentiating germ cells, and they are failing to provide the right environment. That may be one explanation for the age-related severity of the effects in these mutations, in that they are less severe in the younger animals than in the older animals.

Fritz: Some years ago, in collaboration with Mary Lyon, we examined barrier function looking at the behavior in response to hypertonic salts in vitro using the technique that Lonnie (Russell) showed briefly. This was in vitro, using the procedure of Aoki and Gilula, where you study the barrier by simply immersing the whole testis in hypertonic lithium chloride solution. With the translocations that were studied, there was no impairment in barrier function, whereas there was impairment in barrier function in testis of *Sxr* mice.

Lonnie Russell: The barrier is probably one of the most resistant things in Sertoli function. You can have all kinds of genetic abnormalities. We have looked at *Tfm*, the H^{re} rat, we've looked at the barrier there, and it persists despite obvious information that there are some Sertoli cell problems itself. In one of your slides, huge Sertoli cell vacuoles were present. It's hard to get those vacuolizations unless something has gone wrong with the Sertoli cell; whether it's primary or secondary, who knows?

Handel: That is, I think, the big problem. I would be almost willing to guarantee that these are not normally functioning Sertoli cells because they're not getting the cross talk from differentiating spermatids. It's a chicken-and-the-egg problem.

Lonnie Russell: Often, when you have differentiating spermatids gone, you will not get vacuolization in the Sertoli cells.

Generoso: The male sterile/female fertile situation that you see in the X-autosome translocations is also very common among autosome translocations. That raises a question as to whether it is in fact meiotic genes that are a factor or whether it is a secondary factor. Are there real meiotic mutants in the mouse similar to the ones we know in *Drosophila*?

Handel: None that I know of.

Smithies: Mary Ann, you mentioned that you could rescue the translocations in chimeras. Can you rescue the *sks*?

Handel: That hasn't been attempted.

Moses: Just as a point of information, *sks* is completely sterile in the males?

Handel: Yes.

Moses: And it has reduced or complete fertility in the female?

Handel: It's completely sterile in the female.

Wiley: Do those undergo gamete fusion? You said they ovulated.

Handel: That question was not addressed specifically in the experiments. The experiments were designed to look at polar body formation and cleavage.

Wiley: What happens in vivo if you just let the animals mate and then you flush. If you take up the oocyte masses after natural mating, do you see secondary polar body formation?

Handel: That has not been done.

Wiley: That way you could rule out experimental intervention to see if they can just naturally fertilize.

Handel: Right. But if they do fertilize, they don't develop.

Wiley: So you don't know about gamete fusion?

Handel: No, we have no direct evidence for gamete fusion, but we don't have critical evidence against it either.

Lonnie Russell: In normal rats, if one quantitates the number of cells going through meiosis that degenerate, some animals are extremely impressive, that in meiosis 2, they enter metaphase early and degenerate in metaphase 2 early, at 20 times the rate in other animals.

Handel: Is that a strain difference?

Lonnie Russell: No, it's not.

Handel: Just inter-animal within the strain?

Lonnie Russell: Inter-animal differences. This represents a major loss during spermatogenesis, at meiosis. Is there any way to approach that to see why such loss occurs?

Handel: You cannot predict in advance which cells are going to degenerate. It's quite possible to do cytogenetic analyses, and that might be informative, but they would only be prospective and correlational studies.

Hecht: You speak of X-chromosome inactivation occurring; yet, in the testis, mRNAs are notoriously stable and stored, so that would basically invalidate that whole hypothesis.

Handel: It invalidates a hypothesis that I didn't state, which is that X-linked products aren't expressed.

Hecht: Right.

Handel: But I didn't say that; we know that is not the case.

Hecht: Have you looked in round spermatids for some of the X-linked genes to ask whether the message is there?

Handel: Yes. Shannon in my laboratory is looking at HPRT activity and now beginning to look at *Hprt* transcripts. The results are preliminary at this point, but there is certainly enzyme activity in the later cells.

Moses: In *sks*, have you looked at the synaptonemal complex during meiosis in females?

Davisson: No. We haven't looked at it because we haven't had a way to look at known females. We haven't had enough mice to kill a whole litter in the embryonic stage and try to tell which was which.

Bishop: I recall that chromosomes in meiosis condense with some pattern in terms of sequential number. If you have translocations, whether X-autosome or autosome-autosome translocations, is it possible, or do you know of anybody that has looked at the possibility, that those exchanged types mess up the condensation pattern such that meiosis is shut down, at least for some cells? I think, in fact, the X-inactivation comes after a certain amount of condensation, so there is some order to that.

Handel: I think that's a valid point. It has not been addressed critically. What Solari did was serial sectioning of T16H carriers and found that the autosomal segment is not included within the XY body. That's the only relevant information I know.

Regulation of Testicular Postmeiotic Genes

NORMAN B. HECHT
Department of Biology
Tufts University
Medford, Massachusetts 02155

OVERVIEW

To explore mechanisms that regulate gene expression in the mammalian testis, the expression of the protamines, the predominant proteins of the sperm nucleus, have been studied as a model system. Data describing their regulation at the chromosomal, transcriptional, translational, and posttranslational levels are presented.

INTRODUCTION

As in most differentiating tissues, gene expression is temporally regulated during spermatogenesis. The regulation of numerous testicular enzymes and structural proteins including testis-specific isozyme variants, unique DNA-binding proteins, testicular variants of structural proteins, and proto-oncogenes has begun to be understood (for review, see Goldberg 1977; Bellvé 1979; Bellvé and O'Brien 1983; Dixon et al. 1985; Mezquita 1985; Hecht 1986, 1989a,b, 1990; Poccia 1986; Handel 1987; Willison and Ashworth 1987; Eddy 1988; Griswold et al. 1988; Olds-Clarke 1988; Balhorn 1989; Kistler 1989; Meistrich 1989; Wolgemuth et al. 1989). In this paper, special attention will be focused on spermiogenesis, the postmeiotic phase of spermatogenesis, a time when a large number of genes have been demonstrated to be first expressed in the testis. Among this group of postmeiotically expressed genes are three DNA-binding proteins—transition protein 1 and protamines 1 and 2.

The chromatin composition of male germ cell nuclei changes dramatically as the nucleosomes of the premeiotic, meiotic, and early postmeiotic cell types are replaced by the nucleoprotamine complex of the spermatozoon. Concomitant with these changes, transcription terminates during mid-spermiogenesis. Starting with the chromosomal events of meiosis, a group of testis-specific and testis-enriched histones supplement and replace the somatic histones of the earlier stage germ cells (Meistrich et al. 1981). As the differentiating gamete enters spermiogenesis, several novel basic proteins, the transition proteins, replace the majority of histones. The transition proteins are in turn replaced toward the end of spermatogenesis by the protamines, the primary nuclear proteins of the spermatozoon.

We have isolated cDNA probes for mouse transition protein 1 (mTP1) and the protamines 1 and 2 by differentially screening a mouse testis library with radiolabeled cDNAs prepared against poly(A)$^+$ RNA from the meiotic pachytene spermatocytes and the postmeiotic round spermatids (Kleene et al. 1983). The predicted amino acid sequences from the cDNAs indicate that mTP1 is a 54-amino-acid protein rich in lysine and arginine (Kleene et al. 1988), whereas the mouse protamines 1 and 2 (mP1 and mP2) are arginine-rich proteins of 50 and 63 amino acids, respectively (Kleene et al. 1985; Yelick et al. 1987). All three proteins are initially transcribed during spermiogenesis and are translationally regulated (Kleene et al. 1984; Yelick et al. 1989). mTP1 mRNAs are detected approximately 3 days before transition protein 1 is synthesized, and the protamine 1 and 2 mRNAs are stored in the cytoplasm of the differentiating spermatids about 1 week before translation. The gene for mTP1 is located on chromosome 1 (Heidaran et al. 1989), and the two mouse protamines are encoded by single-copy genes tightly linked approximately 7 map units from the centromere on the proximal part of chromosome 16 (Reeves et al. 1990). The synthesis of mouse protamines 1 and 2 is coordinated, ultimately serving to compact the nuclear DNA of the spermatozoon in the ratio of 1:2 (mP1:mP2) (Balhorn et al. 1984).

To date, protamine 1 has been found in the spermatozoa of all mammalian species examined. The protamine 1 proteins generally contain a central basic core, a highly conserved amino terminus, and a highly variable carboxyl terminus. Protamine 2 has only been found in the spermatozoa of the mouse, human, stallion, hamster, and certain primates (for review, see Balhorn 1989; Hecht 1989a,b). Although like protamine 1 it contains over 50% arginine, the nucleotide sequences of mP1 and mP2 are sufficiently different that their cDNAs do not cross-hybridize. Moreover, in the mouse, protamine 2 is synthesized as a precursor protein of 106 amino acids that is proteolytically cleaved stepwise in the nucleus of the elongating spermatid into the 63-amino-acid protein found in spermatozoa (Yelick et al. 1987).

We have been using mTP1 and the protamines as markers to investigate the temporal regulation of gene expression in the mammalian testis. In the following sections I shall describe (1) our studies analyzing methylation changes of the mTP1, mP1, and mP2 genes; (2) an in vitro transcription system developed to define essential *cis*-acting elements of mP2; and (3) translational regulatory elements operating during spermiogenesis.

RESULTS AND DISCUSSION

Methylation Changes of the mTP1 and Protamine Genes

The methylation state of cytosine in the dinucleotide sequence 5'-CpG-3' has been correlated with transcriptional activity of genes, changes in chromatin structure, and in the imprinting of genomic DNA. Since the expression of mTP1, mP1, and mP2 has been shown to be temporally restricted to spermiogenesis, we set out to determine whether there are methylation changes in the three genes encoding these proteins during germ-cell development (Trasler et al. 1990). To detect methylation changes in the differentiating germ cells, DNA was prepared from isolated populations of germ cells ranging from type A spermatogonia to spermatozoa. The methylation state of specific CpG sites in the coding and flanking regions of the mTP1, mP1, and mP2 genes was monitored by digestion with methylation-sensitive and methylation-insensitive restriction enzymes. At the sites examined, mTP1 was less methylated in testis DNA than in control somatic spleen DNA. In contrast, mP1 and mP2 were more methylated in testis DNA. Analyses of prepubertal testis DNA from 6 days after birth to sexual maturity revealed that the mTP1 gene became progressively demethylated, whereas the mP1 and mP2 genes became progressively more methylated. These changes were confirmed when DNA from isolated testicular cells was examined. The demethylation of sites in the mTP1 gene and methylation of sites in the mP1 and mP2 genes occurred during meiotic prophase. These de novo methylation and demethylation events are not coupled with DNA synthesis, since the last replication of DNA occurs in the preleptotene spermatocyte. These studies, although only examining a subset of CpG sites for three postmeiotic genes, demonstrate a switching of methylation patterns during male germ-cell development. The changes in methylation during meiotic prophase also argue for methylase/demethylase activities that function independently of DNA replication in the mammalian testis.

Transcriptional Regulation

Sequence analysis of genomic clones for mP1 and mP2 has identified several similar sequences present in the 5'-flanking regions of both protamine genes and a conserved region in the 3'-untranslated region adjacent to the poly(A) addition site (Johnson et al. 1988). Using DNA fragments containing these 5'-flanking sequences to direct expression of reporter genes in transgenic mice, cell type expression has been obtained with a 465-bp mP1 sequence (Peschon et al. 1987; Behringer et al. 1988) or an 859-bp mP2 sequence (Stewart et al. 1988). Usually a more detailed analysis of these regulatory regions would be conducted by transfection analysis. However, a permanent spermatogenic germ-cell line is not available, and it is not possible to maintain primary cultures of dissociated testicular germ cells for more than a few days. Therefore, to dissect the *cis*-acting 5' sequences of the protamine genes, we have developed a testicular in vitro transcription system (Bunick et al. 1990b). In addition to being able to detect sites needed for the initiation of transcription, an in vitro transcription system also provides a means to identify specific factors required for promoter recognition.

A transcriptionally active in vitro system was established from the nuclei of testes from prepubertal (16 days old) or sexually mature mice. Extracts made from the prepubertal testis transcribe optimally at 30°C, similar to somatic in vitro transcription systems. In contrast, the temperature optimum for extracts prepared from the sexually mature testis was 20°C. The different temperature optima of the prepubertal and sexually mature testis extracts parallel the temperature sensitivities seen in vivo for the differentiating male germ cells. Since 16-day mouse testes contain germ cells up to and including the meiotic stages, we believe the reduced temperature optimum seen with the adult testis in vitro extract is due to components provided by postmeiotic germ cells. Whether a novel temperature-sensitive factor is synthesized in round spermatids or a modification of a pre-existing factor occurs is not known.

Deletion analysis has revealed both positive and negative regulatory regions in the mP2 gene. Positive promotion lies within the region −170 to −82 from the start of transcription, whereas sequences further upstream (from −360 to −170) appear to repress in vitro transcription in a heterologous HeLa cell system. This suggests that the −360 to −170 region binds factor(s) that act as repressors.

Comparison of the mouse and rat protamine 2 genes has proven to be informative in helping to understand protamine gene expression (Fig. 1). Unlike mouse spermatozoa, which contain protamines 1 and 2, rat spermatozoa only contain protamine 2 (Calvin 1976). Although a protamine 2 gene with greater than 90% homology to mP2 at both the nucleotide and amino acids levels is present in the rat (Tanhauser and Hecht 1989), they are not expressed equally. Comparison of protamine 1 and 2 mRNA levels in mouse and rat testicular extracts reveals similar levels of protamine 1 mRNAs in both species, but an approximately 30-fold reduction in the steady-state level of the mRNA coding for rat protamine 2 precursor (Bower et al. 1987). Although reduced in amount, the rat protamine 2 precursor mRNA is found on polysomes, suggesting the precursor rat protein is translated. The synthesis of rat protamine precursor has been confirmed by the immunological detection of small amounts of the precursor in nuclear extracts of elongating spermatids (Bunick et al. 1990a). The lower in vivo level of rP2 precursor mRNA may be partially a result of reduced transcription, since the rP2 gene is transcribed in vitro only about 30% as well as an equivalent mP2 construct (Bunick et al. 1990a).

The absence of any mature rat protamine 2 in spermatozoa is puzzling. It may be the result of a posttranslational processing defect in the rat. The coding sequence of rP2 is identical to the 63-amino-acid mature mouse protamine 2 except for two amino acids. Comparison of the amino acid sequences of the putative leader region of the mouse and rat precursor protamines 2 reveals three additional amino acid differences. The proline, glycine, and histidine at amino acid positions 12, 30, and 43 of the mouse are replaced by glutamine, glutamate, and glutamate in the rat. Processing is known to occur at amino acids 12 and 43 (R. Balhorn, pers. comm.). Assuming that the rat putative precursor requires processing at the same sites used by the mouse, such sequence alterations may explain why the precursor rat protamine 2 is translated but no mature rat protamine 2 is present in spermatozoa. The absence of the precursor form of rat protamine 2 in spermatozoa suggests it must be processed to be stably maintained in the mature spermatozoa.

Translational Regulation

The transition proteins and protamines are representative of a large number of testicular proteins that are synthesized after transcription has terminated during midspermiogenesis. As a result, understanding the mechanisms regulating their storage (from 3 to 7 days) and subsequent activation is likely to be of general importance for numerous postmeiotic genes.

Fractionation of testicular extracts by sucrose gradients reveals that 80–90% of the mTP1, mP1, and mP2 mRNAs are present in the ribonucleoprotein (nonpolysomal) fraction. Analysis of the distribution of these mRNAs in extracts of isolated cell types demonstrates that these mRNAs are not present in the meiotic pachytene spermatocytes but are first found in the postmeiotic round spermatids in the nonpolysomal fraction. Days later, in elongating spermatids, the mRNAs move onto polysomes and are translated.

The sizes of the mTP1, mP1, and mP2 mRNAs present in the polysomal and nonpolysomal fractions differ. In the ribonucleoprotein fraction, they are 600, 580, and 830 nucleotides, respectively. At the time of translation, the mRNAs shorten to approximately 480–600, 450, and 700 nucleotides, respectively. This shortening is a result of partial deadenylation to molecules containing about 30 poly(A)s. Such mRNA length changes are not unique to the mouse or to its chromatin proteins. Similar shortenings of protamine mRNA length have been reported for trout (Dixon et al. 1985), rat and hamster (Bower et al. 1987), rooster (Oliva et al. 1988), and for other testicular genes (Kleene 1989; Kleene et al. 1990).

Although the regulatory significance of the partial deadenylation of postmeiotic mRNAs is not understood, recent transgenic studies have demonstrated an essential role for the 3′-untranslated region of a protamine mRNA in temporal expression (Braun et al. 1989). Combining the protamine 1 promoter and a human growth hormone reporter gene, Braun et al. (1989) have shown that the 3′-untranslated region of the construct determines whether the mRNA is stored or immediately translated. When the 3′-untranslated region from human growth hormone is used in place of the 3′-untranslated region of protamine 1, the transgenic mRNAs are not stored, but are translated immediately in round spermatids. Replacement of the 3′-untranslated region of the human growth hormone with the 3′-untranslated region of protamine 1 delays the translation of the transgene mRNA until the time during spermatogenesis when the protamines are normally translated. These results demonstrate that the 3′-untranslated region of protamine 1 directs its temporal expression during spermiogenesis. The precise sequence in the protamine mRNA that controls this process remains to be determined. It is noteworthy that both protamine 1 and 2 mRNAs contain several groups of conserved sequences in the 3′-untranslated region. Recent studies also reveal protein binding to specific sequences in the 3′-untranslated mRNAs of protamine 2 (Y. Kwon and N.B. Hecht, unpubl. observations). The elucidation of regulatory factors that modulate the translation of the

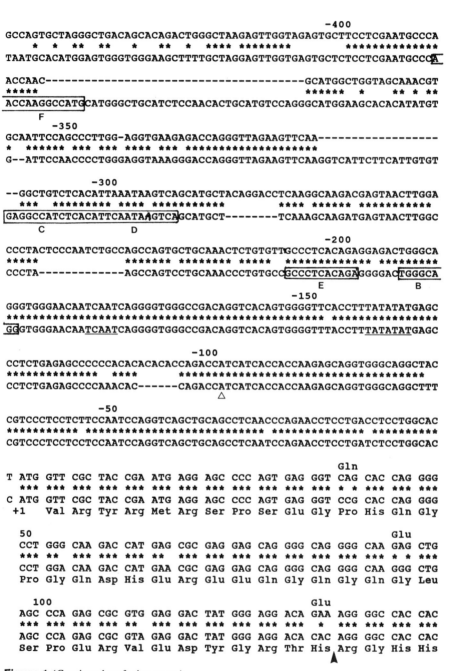

Figure 1 (*Continued on facing page.*)

```
                150
CAC --- AGA CAC AGG CGC TGC --- --- AAG AGG CTT CAC AGG ATC CAC
***     *** *** *** *** ***         *** *  *  **  **  *** *** ***
CAC CAC AGA CAC AGG CGC TGC TCT CGT AAG AAG CTA CAT AGG ATC CAC
His His Arg His Arg Arg Cys Ser Arg Lys Arg Leu His Arg Ile His
                        200
AAG AGG CGC CGG TCA TGC AGA AGG CGG AGG AGA CAC TCC TGC CGC CAC
*** *** **  *** *** *** *** *** *** *** *** *** *** *** *** ***
AAG AGG CGT CGG TCA TGC AGA AGG CGG AGG AGA CAC TCC TGC CGC CAC
Lys Arg Arg Arg Ser Cys Arg Arg Arg Arg Arg His Ser Cys Arg His
                        250
AGG AGG CGG CAT CGC AGA G taagcacccagtagccaagtccccgctacctctgct
*** *** *** *** *** *** * *********    **** *  ***  ** **** ****
AGC AGG CGG CAT CGC AGA G taagcaccccacagccgaccccctggccacctgtgct
Arg Arg Arg His Arg Arg
            300                                         350
gctgccacgaaggcagggtgcccctctccaaaccctgttgccttccgggcagcctcgcaaacct
*****                 ***  ** ******** ******** * *      *******
gctgctg---------------cccatctaaaccctgctgccttccagacgtaga--caaacct

                                                              Ser
gaactttctctacccag  GC TGC AGA AGA TCC CGA AGG AGG AGG AGC TGC
* *****  ** *      ***  *  *** *** *** *** *** *** *** **  ***
ggactttcctttgtacag GC TGC AGA AGA TCC CGA AGG AGG AGA TGC
                   Gly Cys Arg Arg Ser Arg Arg Arg Arg Arg Cys

            Trp     Tyr Tyr                         450
AGG TGC AGG AAA TGC AGG TGG CAC TAT TAT TAA GCCTCCCCAGGTCGATCCAT
*** *** *** *** **  *** **  *** **  *   *** **********  *  *****
AGG TGC AGG AAA TGT AGG AGG CAC CAT CAC TAA GCCTCCCCAGGCCTGTCCAT
Arg Cys Arg Lys Cys Arg Arg His His His
                                       500
TCTGCCTGGAGCTAAGGAAGTCACCTGCCTCAGGAA--GTCACCTGCCCAAGCAAAGTCATGTA
************ ************ **** *****  ****************** ******
TCTGCCTGGAGCCAAGGAAGTCACTTGCCCAAGGAATAGTCACCTGCCCAAGCAACATCATGTG
                        550
AGGCCACAACACCATTCCATGTTGATATCTGAGCCCTGAGCTGCCAAGGAGCCACGAAATCTAA
********  *********** ***  *********************************  * *
AGGCCACACCACCATTCCATGTCGATGTCTGAGCCCTGAGCTGCCAAGGAGCCACGAGATCTGA
            600
GTAATGAGCAAAGCCACCTGCTAAATAAAGCTT
***  ***************** **********
GTACTGAGCAAAGCCACCTGCCAAATAAAGCTT
```

Figure 1

Comparison of the nucleotide sequence of the rat and mouse protamine 2 genes. Rat (*top sequence*); mouse (*bottom sequence*). Potential upstream promoter sequences are indicated by boxes and letters. For mP2, the open arrow indicates the site of transcription initiation, +1 indicates the start of translation, and the bold arrowhead indicates the processing site at which the mature mP2 protein is cleaved from the precursor form. The nucleotide numbers of the rP2 gene are noted. Asterisks indicate identity, dashes indicate deletions. The dark underlining denotes the conserved sequence in the 3′-untranslated region. (Modified, with permission, from Bunick et al. 1990a.)

mTP1, mP1, and mP2 mRNAs will provide important insights into a primary level of gene regulation during spermiogenesis.

ACKNOWLEDGMENTS

The research presented in this paper from the author's laboratory was supported by National Institutes of Health and Environmental Protection Agency grants. The author thanks Carol Valente for her outstanding secretarial assistance.

REFERENCES

Balhorn, R. 1989. Mammalian protamines: Structure and molecular interactions. In *Molecular biology of chromosome function* (ed. K.W. Adolph), p. 366. Springer-Verlag, New York.

Balhorn, R., S. Weston, C. Thomas, and A. Wyrobek. 1984. DNA packaging in mouse spermatids. Synthesis of protamine variants and four transition proteins. *Exp. Cell. Res.* **150:** 298.

Behringer, R.S., J.J. Peschon, A. Messing, C.L. Gartside, S.D. Hauschka, R.D. Palmiter, and R.L. Brinster. 1988. Heart and bone tumors in transgenic mice. *Proc. Natl. Acad. Sci.* **85:** 2648.

Bellvé, A.R. 1979. The molecular biology of mammalian spermatogenesis. In *Oxford reviews of reproductive biology* (ed. C.A. Finn), p. 151. Oxford University Press, London.

Bellvé, A.R. and D.A. O'Brien. 1983. The mammalian spermatozoon: Structural and temporal assembly. In *Mechanism and control of animal fertilization* (ed. J.F. Hartman), p. 55. Academic Press, London.

Bower, P.A., P.C. Yelick, and N.B. Hecht. 1987. Both protamine 1 and 2 genes are expressed in the mouse, hamster, and rat. *Biol. Reprod.* **37:** 479.

Braun, R.E., J.J. Peschon, R.R. Behringer, R.L. Brinster, and R.D. Palmiter. 1989. Protamine 3'-untranslated sequences regulate temporal translational control and subcellular localization of growth hormone in spermatids of transgenic mice. *Genes Dev.* **3:** 793.

Bunick, D., R. Balhorn, L.H. Stanker, and N.B. Hecht. 1990a. Expression of the rat protamine 2 gene is suppressed at the level of transcription and translation. *Exp. Cell. Res.* **188:** 147.

Bunick, D., P. Johnson, T.R. Johnson, and N.B. Hecht. 1990b. Transcription of the testis-specific protamine 2 gene in a homologous *in vitro* transcription system. *Proc. Natl. Acad. Sci.* **87:** 891.

Calvin, H.I. 1976. Comparative analysis of the nuclear basic proteins in rat, human, guinea pig, mouse, and rabbit spermatozoa. *Biochim. Biophys. Acta* **434:** 377.

Dixon, G.H., J.M. Aiken, J.M. Jankowsky, D.I. McKenzie, R. Moir, and J.C. States. 1985. Organization and evolution of the protamine genes of salmonid fishes. In *Chromosomal proteins and gene expression* (ed. G. Reeck et al.), p. 287. Plenum Press, New York.

Eddy, E.M. 1988. The spermatozoon. In *The physiology of reproduction* (ed. E. Knobil and J. Neill), p. 27. Raven Press, New York.

Goldberg, E. 1977. Isozymes in testes and spermatozoa. In *Isozymes: Current topics in biological research* (ed. M.C. Rattazzi et al.), p. 79. A.R. Liss, New York.

Griswold, M.D., M. Collard, S. Hugly, and J. Huggenvik. 1988. The use of specific cDNAs to assay Sertoli cell functions. In *Molecular and cellular aspects of reproduction* (ed. D.S. Dhindsa and O.P. Bahl), p. 301. Plenum Press, New York.

Handel, M. 1987. Genetic control of spermatogenesis in mice. In *Results and problems in cell differentiation in spermatogenesis: Genetic aspects* (ed. W. Hernig), p. 1. Springer-Verlag, Berlin.

Hecht, N.B. 1986. Regulation of gene expression in mammalian spermatogenesis. In *Experimental approaches to mammalian embryonic development* (ed. J. Rossant and R.A. Pedersen), p. 151. Cambridge University Press, New York.

———. 1989a. Mammalian protamines and their expression. In *Histones and other basic nuclear proteins* (ed. L. Hnilica et al.), p. 347. CRC Press, Boca Raton, Florida.

———. 1989b. Molecular biology of structural chromosomal proteins of the mammalian testis. In *Molecular biology of chromosome function* (ed. K.W. Adolph), p. 396. Springer-Verlag, New York.

———. 1990. Gene expression during male germ cell development. In *Cell and molecular biology of the testis* (ed. L. Ewing and C. Desjardins). Oxford University Press, New York. (In press.)

Heidaran, M.A., C.A. Kozak, and W.S. Kistler. 1989. Nucleotide sequence of the *Stp-1* gene coding for rat spermatid nuclear transition protein 1 (TP1): Homology with protamine P1 and assignment of the mouse *Stp-1* gene to chromosome 1. *Gene* **75**: 39.

Johnson, P.A, J.J. Peschon, P.C. Yelick, R.D. Palmiter, and N.B. Hecht. 1988. Sequence homologies in the mouse protamine 1 and 2 genes. *Biochim. Biophys. Acta* **950**: 45.

Kistler, W.S. 1989. Structure of testis-specific histones, spermatid transition proteins, and their genes in mammals. In *Histones and other basic nuclear proteins* (ed. L. Hnilica et al.), p. 331. CRC Press, Boca Raton, Florida.

Kleene, K.C. 1989. Poly(A) shortening accompanies the activation of translation of five mRNAs during spermiogenesis in the mouse. *Development* **106**: 367.

Kleene, K.C., R.J. Distel, and N.B. Hecht. 1983. cDNA clones encoding cytoplasmic poly(A)$^+$ RNAs which first appear at detectable levels in haploid phases of spermatogenesis in the mouse. *Dev. Biol.* **98**: 455.

———. 1984. Translational regulation and coordinate deadenylation of a haploid mRNA during spermiogenesis in the mouse. *Dev. Biol.* **105**: 71.

———. 1985. The nucleotide sequence of a cDNA clone encoding mouse protamine 1. *Biochemistry* **24**: 719.

Kleene, K.C., A. Bozorgzadeh, J.F. Flynn, P.C. Yelick, and N.B. Hecht. 1988. Nucleotide sequence of a cDNA encoding mouse transition protein 1. *Biochim. Biophys. Acta* **950**: 215.

Kleene, K.C., J. Smith, A. Bozorgzadeh, M. Harris, L. Hahn, I. Karimpour, and J. Gerstel. 1990. Sequence and developmental expression of the mRNA encoding the seleno-protein of the sperm mitochondrial capsule in the mouse. *Dev. Biol.* **137**: 395.

Meistrich, M.L. 1989. Histone and basic nuclear protein transitions in mammalian spermatogenesis. In *Histones and other basic nuclear proteins* (ed. L. Hnilica et al.), p. 165. CRC Press, Boca Raton, Florida.

Meistrich, M.L., J. Longtin, W.A. Brock, S.J. Grimes, and M.L. Mace. 1981. Purification of rat spermatogenic cells and preliminary biochemical analysis of these cells. *Biol. Reprod.* **25**: 1065.

Mezquita, C. 1985. Chromatin proteins and chromatin structure in spermatogenesis. In *Chromosomal proteins and gene expression* (ed. G. Reeck et al.), p. 315. Plenum Press, New York.

Olds-Clarke, P. 1988. Genetic analysis of sperm function in fertilization. *Genet. Res.* **20:** 241.

Oliva, R., J. Mezquita, C. Mezquita, and G.H. Dixon 1988. Haploid expression of the rooster protamine mRNA in the postmeiotic stages of spermatogenesis. *Dev. Biol.* **125:** 332.

Peschon, J.J., R.R. Behringer, R.L. Brinster, and R.D. Palmiter. 1987. Spermatid-specific expression of protamine 1 in transgenic mice. *Proc. Natl. Acad. Sci.* **84:** 5316.

Poccia, D. 1986. Remodeling of nucleoproteins during gametogenesis, fertilization, and early development. *Int. Rev. Cytol.* **105:** 1.

Reeves, R.H., J.D. Gearhart, N.B. Hecht, P. Yelick, P. Johnson, and S.J. O'Brien. 1990. The gene encoding protamine 1 is located on human chromosome 16 near the proximal end of mouse chromosome 16 where it is tightly linked to the gene encoding protamine 2. *J. Hered.* (in press).

Stewart, T.A., N.B. Hecht, P.G. Hollinghead, P.A. Johnson, J.A.C. Leong, and S.L. Pitts. 1988. Haploid-specific transcription of protamine-myc and protamine-Tag fusion genes in transgenic mice. *Mol. Cell. Biol.* **8:** 1748.

Tanhauser, S. and N.B. Hecht. 1989. Nucleotide sequence of the rat protamine 2. *Nucleic Acids Res.* **17:** 4395.

Trasler, J.M., P.A. Johnson, L.E. Hake, A.A. Alcivar, C.F. Millette, and N.B. Hecht. 1990. DNA methylation and demethylation of post-meiotically expressed genes during meiotic prophase in the mouse testis. *Mol. Cell. Biol.* (in press).

Willison, K. and A. Ashworth. 1987. Mammalian spermatogenic gene expression. *Trends Genet.* **3:** 351.

Wolgemuth, D.J., E. Gizang-Ginsberg, C. Ponzetto, and Z.F. Zakeri. 1989. Genetic control of germ cell function: Developmentally regulated gene expression during spermatogenesis. In *Molecular biology of fertilization* (ed. G. Schatten and H. Schatten), p. 235. Academic Press, New York.

Yelick, P.C., Y.K. Kwon, J.F. Flynn, A. Borzorgzadeh, K.C. Kleene, and N.B. Hecht. 1989. Mouse transition protein 1 is translationally regulated during the postmeiotic stages of spermatogenesis. *Mol. Rep. Dev.* **1:** 193.

Yelick, P.C., R. Balhorn, P.A. Johnson, M. Corzett, J.A. Mazrimas, K.C. Kleene, and N.B. Hecht. 1987. Mouse protamine 2 is synthesized as a precursor whereas mouse protamine 1 is not. *Mol. Cell. Biol.* **7:** 2173.

COMMENTS

Sega: Is anything known about the proteins that are associated with mRNA?

Hecht: We don't know anything about the proteins. We know there are binding sites on the protamine mRNAs that are binding proteins. We are trying to figure out what they are now. There are several different sites along the protamine mRNA that bind protein rather nicely.

Lonnie Russell: Norm, you mentioned that RNPs can be stored in the cytoplasm a long time and they express much later. We know that the chromatoid body incorporates uridine, but what is necessary structurally for these things to be tied up for such a long period of time, to hold them, and be expressed later? I can see the advantage of that in the testis. We see all kinds of structures in the germ cells that we can't identify at this point.

Hecht: We know very little about where these mRNAs are, for instance. We don't know if they are randomly expressed in the testis and simply stored in the cytoplasm. We don't know if they are sequestered. We don't know the role of the chromatoid body. The chromatoid body occurs too early during spermatogenesis for only postmeiotic gene expression. But certainly, there are genes—for instance, PGK-2 is transcribed during meiosis and probably not expressed until spermiogenesis, so perhaps that could be one of the places that it could be stored. We know very little about the cytoskeleton of these cells and whether there could be attachment to the cytoskeleton. These are all things that certainly should be looked at.

Liane Russell: One of your early slides showing Southern blots had different lanes for type A and type B spermatogonia. Were they actually physically isolated?

Hecht: Yes.

Liane Russell: How pure are your samples?

Hecht: We use the testes from 8-day-old animals. On a good day, we can get probably 80% purity of type A and type B.

Liane Russell: So it's just different maturation stages?

Hecht: Yes.

Wiley: Is it possible to take the mRNA prior to the time that it is translated and to inject it into another cell type and see if it's translated there? Can it be activated, or are those factors end-specific?

Hecht: We don't know that. You mean injected into an oocyte or something like that?

Wiley: Yes.

Hecht: I think we would have a technical problem of injecting into a somatic cell. Those are awfully small cells. I guess it is possible to do, but we certainly don't have that expertise.

Wiley: Or a *Xenopus* oocyte.

Hecht: Yes. The problem with those studies is you isolate the RNP, then you inject it, and you don't know what happens to it. I mean, if the proteins stay bound and there is no translation, fine. However, if your injection, or just the entrance of the RNP into the oocyte milieu dissociates those factors, then it's going to translate. So, it is the kind of experiment that is difficult to interpret depending on what the results would be.

Woychik: Is there any change in the methylation of the RNA that correlates with the activation of translation, for example, in the cap structure?

Hecht: We haven't looked. The mRNAs are capped. As far as any other changes, we don't know. We also couldn't exclude the possibility that there is a minor processing event. We have been running the mRNAs on acrylamide gels, and we see no difference in the electrophoretic mobilities. Either it is the proteins that are sequestering their activity, or perhaps a localization to a particular part of the cell, or perhaps other factors. These changes are not specific to the protamines because you have an elaborate axoneme and sperm tail that have to be synthesized. My guess is that all those genes are going to be under very similar control because they are all made very late in spermatogenesis, and transcription has ceased around step 7 or step 8.

Handel: I think this whole idea of sequestration within the cell and how these things are maintained in active form impinges on whether or not there can be transcript sharing among cells.

Hecht: Yes.

Molecular Targets, DNA Breakage, and DNA Repair: Their Roles in Mutation Induction in Mammalian Germ Cells

GARY A. SEGA
Biology Division, Oak Ridge National Laboratory
Oak Ridge, Tennessee 37831-8077

OVERVIEW

Variability in genetic sensitivity among different germ-cell stages in the mammal to various mutagens could be the result of how much chemical reaches the different stages, what molecular targets may be affected in the different stages, and whether or not repair of lesions occurs. Several chemicals have been found to bind very strongly to protamine in late-spermatid and early-spermatozoa stages in the mouse. The chemicals also produce their greatest genetic damage in these same germ-cell stages. Although chemical binding to DNA has not been correlated with the level of induced genetic damage, DNA breakage in the sensitive stages has been shown to increase. This DNA breakage is believed to indirectly result from chemical binding to sulfhydryl groups in protamine, which prevents normal chromatin condensation within the sperm nucleus.

INTRODUCTION

In studies of the genetic effects of chemical agents in mice, the exposure given to the animals (by injection, inhalation, skin application, etc.) is accurately known. However, in general, little or nothing is known about the molecular dose of the chemical that actually reaches the different germ-cell stages (see Fig. 1).

The variability in genetic sensitivity among different germ-cell stages could be the result of how much chemical reaches the different stages, what molecular targets the chemical may interact with in different stages, and whether or not repair of lesions in specific germ-cell stages is possible.

In particular, the chemical mutagens ethyl methanesulfonate (EMS), methyl methanesulfonate (MMS), ethylene oxide (EtO), and acrylamide (AA) have all been shown, in genetic studies, to produce very many more dominant lethal mutations and heritable translocations in late spermatids and early spermatozoa than in other germ-cell stages in male mice (EMS [Ehling et al. 1968; Generoso and Russell 1969]; MMS [Ehling et al. 1968; Ehling 1980]; EtO [Generoso et al. 1980]; AA [Shelby et al. 1986, 1987]). We have, subsequently, used all four of these chemical mutagens in radioactive form to find out where they are binding within the germ

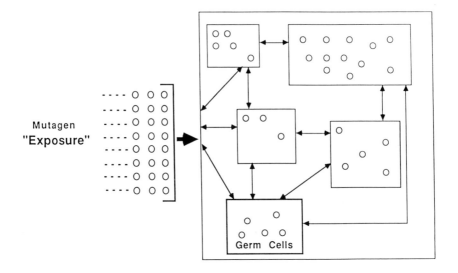

Figure 1
External exposure vs. molecular dose: The amount of chemical (circles) to which an animal is exposed can be accurately established, but its distribution within molecular targets in different tissues is more difficult to analyze. Smaller boxes shown within "mouse" (large box) represent different tissues. Levels of binding in each tissue can vary, depending on such factors as transport, metabolic conversion to more (or less) potent chemical species, and repair. The effect of chemicals on targets within the germ cells is our principal interest.

cells and to see what correlations may exist between the extent of chemical binding in different stages and the amount of induced genetic damage.

In addition, three of these chemicals have been studied for their ability to induce DNA breakage in different germ-cell stages of male mice using an alkaline elution technique (MMS [Sega et al. 1986]; EtO [Sega and Generoso 1988]; AA [Sega and Generoso 1990]). This technique makes use of filters to differentiate DNA strands on the basis of their single-strand size in a high pH solution (Kohn 1979). The filters act to physically impede the passage of long DNA strands. Thus, if single-strand breaks have been induced, the smaller DNA fragments will pass more quickly through the filter than will undamaged DNA when an alkaline solvent is slowly pumped through the filter.

All of the chemicals have also been studied for their ability to induce DNA repair in mouse germ cells (MMS and EMS [Sega and Sotomayor 1982]; EtO [Sega et al. 1988]; AA [G.A. Sega, in prep.]). The occurrence of DNA repair in the germ cells of treated animals is monitored following testicular injection of [^{3}H]thymidine

([^{3}H]dThd). If DNA repair has occurred in meiotic or postmeiotic stages, the [^{3}H]dThd will be incorporated into cells in which no scheduled DNA synthesis normally takes place. Thus, the repair is detected by an unscheduled DNA synthesis (UDS) response in the affected stages.

RESULTS

Chemical Binding to Whole Germ Cells

Mice from the same strains used in genetic experiments were exposed to radioactively labeled EMS, MMS, EtO, or AA. Sperm moving through the reproductive tracts were recovered from the animals at different times (up to 3 weeks) after exposure and assayed for how much chemical was bound (alkylation). With chemicals such as EMS, MMS, EtO, and AA, which are powerful mutagens in late spermatids and early spermatozoa, we found a dramatic increase in the amount of chemical bound to these stages (at least an order of magnitude more binding than in other stages). Thus, there was a correlation between increased genetic damage and increased levels of chemical binding to the sensitive stages.

Figure 2 shows the binding pattern of AA in sperm recovered from the vasa deferentia over a 3-week period after exposure. The same basic pattern of binding was found with EMS, MMS, and EtO. The binding reaches a peak in the recovered sperm between ~8 and 12 days after exposure and represents germ cells that were late spermatids to early spermatozoa at the time of treatment (Sega and Sotomayor 1982).

Chemical Binding to Germ-cell DNA

To further characterize the molecular nature of the lesions in the germ cells, DNA was extracted from the sperm recovered from the vasa deferentia at different times after treatment and assayed for the amount of chemical bound to it. It was found that the amount of chemical binding to DNA represented only a very small fraction of the total chemical binding to the germ cells. Furthermore, there was no increase in DNA alkylation in the most sensitive stages, late spermatids and early spermatozoa. Figure 2 also shows the pattern of binding of AA to the DNA in the developing spermiogenic stages. Similar results were found with EMS, MMS, and EtO.

Protamine as a Possible Target for Alkylation

These surprising results forced us to look for alternative molecular targets for mutagenesis within the germ cells. It was known that in mid- to late-spermatid stages of mammals, the usual chromosomal proteins (histones) are replaced with small, very

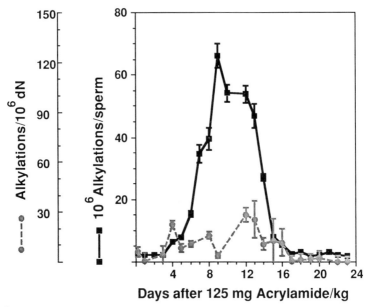

Figure 2
Alkylation of sperm heads and sperm DNA taken from the vasa deferentia during a 3-week period following i.p. injection of 125 mg [^{14}C]AA/kg. 10^6 alkylations per sperm head (boxes); alkylation/10^6 deoxynucleotides (circles). Error bars represent ± 1 S.D. (Reprinted, with permission, from Sega et al. 1989.)

basic proteins (protamines) that contain more than 50% of the amino acid arginine (Monesi 1965; Bellvé et al. 1975; Balhorn et al. 1977). In addition, mammalian protamines contain cysteine (Bedford and Calvin 1974; Calvin 1975), and it is the cross-linking of the cysteine amino acids in protamine (through disulfide-bond formation) that gives mammalian spermatozoa their keratin-like properties (similar to some of the properties of fingernails and hair). Because protamine contains nucleophilic sites (e.g., the –SH group in cysteine), we reasoned that it might be a target for attack by chemicals such as EMS, MMS, EtO, and AA.

Chemical Binding to Germ-cell Protamine

When we purified protamine from sperm recovered from the caudal epididymis (an area of the reproductive tract that immediately precedes the vas) at different times after chemical treatment, we were excited to find that the level of mutagen that was bound to protamine increased greatly in the most sensitive germ-cell stages (late spermatids–early spermatozoa). The amount of mutagen binding to protamine exactly paralleled the total amount of binding to the sperm. In fact, with chemicals such as EMS, MMS, EtO, and AA that have their greatest effect in late spermatids and early spermatozoa, we have found that almost all of the binding in the sensitive stages can be attributed to interaction with protamine.

Figure 3 shows the binding pattern of AA in sperm recovered from the caudal epididymides over a 3-week period after exposure. Since sperm from the caudal epididymides take about 2 days to reach the vas (Sega and Sotomayor 1982), the alkylation pattern of the sperm recovered from the caudal epididymides is shifted toward earlier times. Taking this time shift into account, the alkylation pattern that was seen with the vas sperm is well reproduced with the epididymal sperm.

Also shown in Figure 3 is the AA binding to sperm protamine. Final protamine purification was carried out on an HPLC column and detected at 214 nm. The amount of protamine recovered was then expressed in units of OD_{214}-min, i.e., the area under the protamine peaks recovered from the HPLC column. One OD_{214}-min unit represents the protamine from ~22 x 10^6 sperm. Within experimental uncertainty, all of the sperm alkylation can be accounted for by the protamine alkylation. The same basic observations were made with EMS, MMS, and EtO.

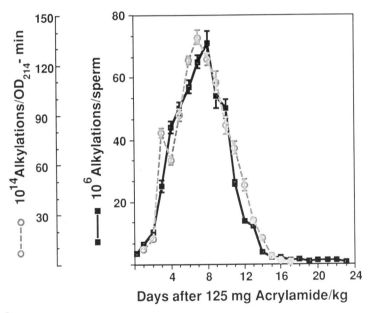

Figure 3
Alkylation of sperm heads and sperm protamine recovered from the caudal epididymides during a 3-week period following i.p. injection of 125 mg [^{14}C]AA/kg. 10^6 alkylations per sperm head (boxes); 10^{14} alkylations/OD_{214}-min (circles). Error bars represent ± 1 S.D. (Reprinted, with permission, from Sega et al. 1989.)

Chemical Binding to Cysteine within the Protamine

To determine where on the protamine the chemicals were binding, samples of alkylated protamine were acid-hydrolyzed. By analysis of the protamine hydrolyzates on an amino acid analyzer and by thin-layer chromatography, we have been able to show that these types of chemicals do, in fact, bind to the sulfur in cysteine. We have shown that with MMS at least 80% of the protamine adducts are S-methylcysteine (Sega and Owens 1983). AA has been found to yield the adduct S-carboxyethylcysteine, as well as a second, as yet unidentified adduct (Sega et al. 1989).

DNA Breakage

Germ-cell DNA from chemically treated and control animals was prelabeled with [^{3}H]dThd and [^{14}C]dThd, respectively. At various times after mutagen treatment, sperm are gently recovered from the vasa of one treated and one control animal, pooled, placed on a polycarbonate filter, and lysed. The DNA from the sperm is denatured using an alkaline buffer (pH, 12.2) and eluted through the filter over a 15-hour period. Radioactivity in collected fractions is determined by liquid scintillation counting (Sega et al. 1986; Sega and Generoso 1988). The difference between the amount of [^{3}H]dThd- and [^{14}C]dThd-labeled DNA eluted gives a relative measure of the amount of single-strand breaks in the treated DNA.

Typical of the pattern of DNA breakage induced in developing germ-cell stages by these chemicals is that shown for EtO in Figure 4. During the first week after exposure there is a gradual increase in the amount of DNA eluted from the sperm of the EtO-treated animals. The amount of DNA then peaks between days 7 and 13. Starting at about the end of the second week after exposure, the elution of the treated DNA begins to decrease. By 23 days posttreatment, there is barely a detectable difference between the amount of treated- and control-sperm DNA eluted.

DNA Repair

We have made several detailed studies of the meiotic and postmeiotic germ-cell stages that undergo DNA repair (as measured by UDS) when exposed to mutagens (EMS [Sega 1974]; cyclophosphamide and mitomen [Sotomayor et al. 1978]; MMS and X-rays [Sotomayor et al. 1979]; AA [G.A. Sega, in prep.]). In all of these studies, the basic finding has been that mutagens induce UDS in meiotic stages and in postmeiotic stages up to about midspermatids. No UDS has been detected in late spermatids and spermatozoa.

Figure 5 shows the pattern of the UDS response induced in different germ-cell stages of the mouse after an i.p.-injected dose of 250 mg EMS/kg (the [^{3}H]dThd

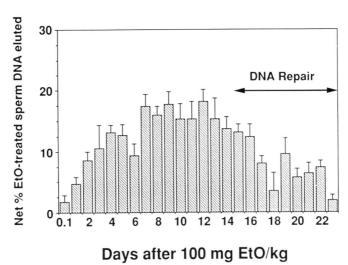

Figure 4
Pattern of sperm DNA elution over a 23-day period following exposure to 100 mg EtO/kg. The difference between the amounts of [^{3}H]dThd-labeled DNA and [^{14}C]dThd-labeled DNA eluted has been subtracted from all the data points. Beyond the end of the second week, spermatozoa recovered from the vasa were, at the time of treatment, in germ-cell stages capable of DNA repair (Sega and Generoso 1988). Error bars represent ± 1 S.E.M..

was given at the same time as the EMS). For about the first 2 weeks after treatment, there is no unscheduled presence of [^{3}H]dThd label in the sperm recovered from the vas. These sperm represent germ-cell stages that were treated as mature spermatozoa to late spermatids. In the third and fourth weeks after EMS treatment, the vas sperm show the unscheduled presence of [^{3}H]dThd. These sperm represent germ-cell stages that were treated as midspermatids to early meiotic stages.

The absence of UDS in mature spermatozoa to late spermatid stages is not due to failure of EMS (or other chemicals) to alkylate DNA in these stages. In our chemical dosimetry studies we have found that the DNA in these germ-cell stages is being alkylated (Sega and Owens 1978, 1983, 1987; Sega et al. 1989). The germ-cell stages that exhibit no UDS are those in which protamine either has replaced, or is in the process of replacing, the usual chromosomal histones. At the time of protamine synthesis, an extensive condensation of the spermatid nucleus begins. It is probable that the DNA lesions present in these cells have become inaccessible to the enzymatic system that gives rise to UDS in the earlier germ-cell stages. Also, much of the cytoplasm is lost from the spermatids as they develop from mid- to late-spermatid stages, and the enzymatic system that produces the UDS response may be lost at this time.

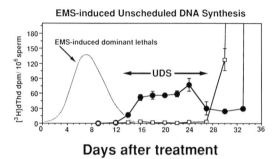

Figure 5
Unscheduled uptake of [^{3}H]dThd into meiotic and postmeiotic germ-cell stages of male mice after testicular injection of [^{3}H]dThd and i.p. injection of 250 mg EMS/kg. Uptake of [^{3}H]dThd was measured in sperm passing through the vas deferens. (Circles) [^{3}H]dThd activity in sperm heads from EMS-treated animals; (boxes) [^{3}H]dThd activity in sperm heads from control animals. Also indicated is the dominant-lethal frequency pattern obtained with EMS (the maximum dominant lethality is around 50–60% with 250 mg EMS/kg (Generoso and Russell 1969). (Adapted from Sega and Sotomayor 1982.)

DISCUSSION

All of these findings have led us to postulate a model for the binding of certain chemicals to mouse chromatin in germ-cell stages that are sensitive to the induction of chromosome breakage and dominant lethality (Sega et al. 1989). In normal nuclear condensation of the developing spermatids, the sulfhydryl groups of cysteine cross-link to form disulfide bridges in the chromatin. However, if chemical binding to a nucleophilic sulfhydryl group occurs before a disulfide bond has formed, cross-linking of the sulfhydryl groups may not take place. This could lead to stresses in the chromatin structure that eventually produce either a single- or double-strand DNA break. The end result would be a dominant lethal mutation or chromosome translocation.

The patterns of DNA-strand breakage observed in mouse sperm over a 3-week period after exposure to MMS (Sega et al. 1986), EtO (Sega and Generoso 1988), and AA (Sega and Generoso 1990) all support the above model. In each case, the DNA breakage is greatest in germ-cell stages having the highest levels of protamine alkylation. (Increased DNA breakage in the late-spermatid and early-spermatozoa stages is not correlated with greater DNA alkylation.)

Sperm recovered in the third week after chemical treatment show a relatively rapid decrease, for successive days, in the amount of DNA that eluted through the filters (refer to Fig. 4 for the elution pattern observed after EtO exposure). Our molecular dosimetry experiments with all four chemicals have also shown a great reduction in protamine alkylation by the middle of the third week after exposure.

We attribute the decrease in sperm DNA elution during this period to two factors. First, the sperm protamine sampled in the third week was synthesized after the chemical mutagens had disappeared from the testes. Thus, the protamine in these sperm was not heavily alkylated, and chromatin breakage was subsequently reduced. Second, the sperm recovered in this time period were derived from stages capable of DNA repair at the time of treatment (Sega and Sotomayor 1982), so that any DNA lesions that might otherwise have given rise to DNA-strand breaks were repaired before the alkaline elution assay.

Clearly, not all mutagenic chemicals act by the above mechanism, and other molecular targets may be important in other germ-cell stages. However, our observations of how some chemicals bind strongly to sperm protamine in mammals give a new dimension to our understanding of mutational processes in mammalian germ cells.

ACKNOWLEDGMENTS

This research was jointly sponsored by the Office of Health and Environmental Research, U.S. Department of Energy, under contract DE-AC05-84OR21400 with Martin Marietta Energy Systems, Inc.; by the National Institute of Environmental Health Sciences under IAG 222Y01-ES-10067; by the National Institute of Child Health and Human Development under grant 5-R01HD17345; and by the U.S. Environmental Protection Agency under IAG DW-930141-01-1.

REFERENCES

Balhorn, R., B.L. Gledhill, and A.J. Wyrobek. 1977. Mouse sperm chromatin proteins: Quantitative and partial characterization. *Biochemistry* **16**: 4074.

Bedford, J. and H. Calvin. 1974. The occurrence and possible functional significance of -S-S- crosslinks in sperm heads, with particular reference to eutherian mammals. *J. Exp. Zool.* **188**: 137.

Bellvé, A., E. Anderson, and L. Hanley-Bowdoin. 1975. Synthesis and amino acid composition of basic proteins in mammalian sperm nuclei. *Dev. Biol.* **47**: 349.

Calvin, H. 1975. Keratinoid proteins in the heads and tails of mammalian spermatozoa. *Biol. J. Linn. Soc.* **7**: 257.

Ehling, U.H. 1980. Induction of gene mutations in germ cells of the mouse. *Arch. Toxicol.* **46**: 123.

Ehling, U.H., R.B. Cumming, and H.V. Malling. 1968. Induction of dominant lethal mutations by alkylating agents in male mice. *Mutat. Res.* **5**: 417.

Generoso, W.M. and W.L. Russell. 1969. Strain and sex variations in the sensitivity of mice to dominant-lethal induction with ethyl methanesulfonate. *Mutat. Res.* **8**: 589.

Generoso, W.M., K.T. Cain, M. Krishna, C.W. Sheu, and R.M. Gryder. 1980. Heritable translocation and dominant-lethal mutation induction with ethylene oxide in mice. *Mutat. Res.* **73**: 133.

Kohn, K.W. 1979. DNA as a target in cancer chemotherapy: Measurement of macromolecular DNA damage produced in mammalian cells by anticancer agents and carcinogens. In *Methods in cancer research* (ed. V.T. DeVita, Jr. and H. Busch), vol. XVI, p. 291. Academic Press, New York.

Monesi, V. 1965. Synthetic activities during spermatogenesis in the mouse: RNA and protein. *Exp. Cell Res.* **39**: 197.

Sega, G.A. 1974. Unscheduled DNA synthesis in the germ cells of male mice exposed in vivo to the chemical mutagen ethyl methanesulfonate. *Proc. Natl. Acad. Sci.* **71**: 4955.

Sega, G.A. and E.E. Generoso. 1988. Measurement of DNA breakage in spermiogenic germ-cell stages of mice exposed to ethylene oxide, using an alkaline elution procedure. *Mutat. Res.* **197**: 93.

———. 1990. Measurement of DNA breakage in specific germ-cell stages of male mice exposed to acrylamide, using an alkaline-elution procedure. *Mutat. Res.* (in press).

Sega, G.A. and J.G. Owens. 1978. Ethylation of DNA and protamine by ethyl methanesulfonate in the germ cells of male mice and the relevancy of these molecular targets to the induction of dominant lethals. *Mutat. Res.* 52: 87.

———. 1983. Methylation of DNA and protamine by methyl methanesulfonate in the germ cells of male mice. *Mutat. Res.* **111**: 227.

———. 1987. Binding of ethylene oxide in spermiogenic germ cell stages of the mouse after low-level inhalation exposure. *Environ. Mol. Mutagen.* **10**: 119.

Sega, G.A. and R.E. Sotomayor. 1982. Unscheduled DNA synthesis in mammalian germ cells—Its potential use in mutagenicity testing. In *Chemical mutagens: Principles and methods for their detection* (ed. F.J. de Serres and A. Hollaender), vol. 7, p. 421. Plenum Press, New York.

Sega, G.A., E.E. Generoso, and P.A. Brimer. 1988. Inhalation exposure-rate of ethylene oxide affects the level of DNA breakage and unscheduled DNA synthesis in spermiogenic stages of the mouse. *Mutat. Res.* **209**: 177.

Sega, G.A., R.P. Valdivia Alcota, C.P. Tancongco, and P.A. Brimer. 1989. Acrylamide binding to the DNA and protamine of spermiogenic stages in the mouse and its relationship to genetic damage. *Mutat. Res.* **216**: 221.

Sega, G.A., A.E. Sluder, L.S. McCoy, J.G. Owens, and E.E. Generoso. 1986. The use of alkaline elution procedures to measure DNA damage in spermiogenic stages of mice exposed to methyl methanesulfonate. *Mutat. Res.* **159**: 55.

Shelby, M.D., K.T. Cain, C.V. Cornett, and W.M. Generoso. 1987. Acrylamide: Induction of heritable translocations in male mice. *Environ. Mutagen.* 9: 363.

Shelby, M.D., K.T. Cain, L.A. Hughes, P.W. Braden, and W.M. Generoso. 1986. Dominant lethal effects of acrylamide in male mice. *Mutat. Res.* **173**: 35.

Sotomayor, R.E., G.A. Sega, and R.B. Cumming. 1978. Unscheduled DNA synthesis in spermatogenic cells of mice treated in vivo with the indirect alkylating agents cyclophosphamide and mitomen. *Mutat. Res.* **50**: 229.

———. 1979. An autoradiographic study of unscheduled DNA synthesis in the germ cells of male mice treated with X-rays and methyl methanesulfonate. *Mutat. Res.* **62**: 293.

COMMENTS

Hecht: Gary, it would be very interesting to see, since we know very little about how the protamine 1 and protamine 2 interact, whether both of the protamines are being affected or whether one is accessible and the other is not accessible to these external compounds.

Sega: With methyl methanesulfonate, we looked at it a little. We didn't try to separate the MP1 and MP2, but we could see the pattern of binding. We could see enough resolution in our chromatographs of the MP1 and MP2 to know there was activity associated with both of the components. We knew that both MP1 and MP2 are being alkylated. In the future, I would like to find out what cysteines in particular are being alkylated—if they are all being alkylated, or if there are specific ones that may be more important in alkylation.

Hecht: It would be interesting to experiment with marsupials. Marsupial protamine does not contain cysteine. Either you are not going to see this effect, or it is going to be working through a very different mechanism.

Smithies: Gary, you mentioned in passing that all these three mutagens caused translocations.

Sega: They all can produce translocation in the same stages. The same ones that cause dominant lethal effects also cause other chromosome breakage that sometimes restitutes and forms translocations.

Smithies: Yes, but they're all DNA breakages and not base substitutions?

Sega: I can't say that. There is always the possibility that they could cause base substitutions. I think most of the genetic effects are caused by gross chromosome aberrations.

Smithies: That fits your mechanism more closely, doesn't it?

Sega: Yes.

Favor: If you look at the dose response for dominant lethals MMS or EMS, MMS is linear, EMS is shoulder. On the basis of your model, would you interpret the shoulder as due to a detoxification mechanism rather than repair, since the DNA lesions should be the same for both?

Sega: At what kind of a dose do you get the shoulder?

Favor: Up to about 150. At 150 you start to see an effect. I repeated the EMS a number of times and Ehling has done the same with MMS. I know you have a shoulder effect with EMS. Do you interpret that as detoxification?

Sega: I don't think it's detoxification. You may be getting to a point where you are saturating some of the sites in the protein, for example, at a certain level of EMS, and beyond that it takes a higher concentration of EMS to further alkylate. It may be harder to alkylate sites in the protamine.

Favor: But there's no effect up to 150. The effect starts at 150.

Sega: I think it's a matter of detection. I just can't believe that EMS doesn't produce any dominant lethals below 150.

Favor: I repeated it. I think we had 60 females.

Sega: I don't know how to explain that.

Bridges: Gary, can I draw to your attention the fact that the stork population in Germany over the last 50 years has exactly paralleled the birth rate in Germany. The conclusions you might draw from that might not be right. I am very much concerned that you seem to be using a similar logic here. Have you really got any evidence, apart from a similar temporal pattern of binding and effect, that your initial lesion is not in DNA?

Sega: We know that DNA is alkylated with these chemicals. If you remember some of the earlier graphs I showed, the level of alkylation of the DNA is usually highest right after the exposure, and then you can usually see a gradual decrease in the DNA alkylation with time. It is a very slow thing. It may be just hydrolysis of some of these alkylated sites in the DNA. If you are going to argue that it's this dealkylation of the DNA that is causing some breakage of the DNA, then you would expect to get a continually increasing dominant lethal effect. What happens, in fact, is you get dominant lethals increasing up to a point, and then they fall off.

Bridges: I wouldn't expect them to increase necessarily, because all sorts of repair and restitution processes can occur.

Sega: There's no DNA repair going on in those stages that we can measure by the unscheduled DNA synthesis test. In that regard, at least, we can say that there is no repair.

Bridges: That merely tells you that there's nothing very gross; it doesn't tell you anything about breaks. You just didn't detect break repair.

Sega: We know that DNA breaks are not being repaired. As I showed in that last slide with ethylene oxide, the DNA breakage as a function of days after exposure, and those breaks are increasing, follows the same pattern exactly as the pattern of binding to the protamine. So it is not being repaired until the third week after the exposure, where we know that DNA repair competency is in those cells. The pattern of DNA breakage then falls off. We think it is a combination of the fact that there is repair going on in those stages, plus the fact that the protamine that was in those stages was never exposed to the chemical.

Fritz: Another thing that certainly is critical to understanding why you can't evoke protamines is the effect of EDS on Leydig cells in the rat; as you know, EDS completely destroys Leydig cells. You get sterility because there are no androgens, and for some reason it doesn't work in the mouse. Is there any kind of correlation there?

Sega: I didn't know that. I don't know of any correlation there. That's interesting.

Dellarco: Gary, have you looked at mutagens that cause dominant lethals or heritable translocations in earlier germ-cell stages for any alkylation pattern?

Sega: We just got some labeled chlorambucil. What Lee (Russell) has found is that with chlorambucil the maximum genetic effect is found in the third week after exposure, which corresponds to the early spermatids, which would be before the protamine synthesis occurs. We are just starting a study to see what the binding pattern with chlorambucil may look like compared with the genetic effects being produced. That will be the first chemical we will look at.

Detecting Specific-locus Mutations in Human Sperm

ANDREW J. WYROBEK, MARGE CURRIE, JACKIE L. STILWELL, ROD
BALHORN, AND LARRY H. STANKER
Biomedical Sciences Division
Lawrence Livermore National Laboratory
Livermore, California 94550

OVERVIEW

We present a strategy for detecting gene mutations in human sperm, using protamine as the prototypic mutational target. Protamines are well-characterized sperm-specific proteins with attractive features for mutational analysis. We have developed highly specific fluorescently tagged antibodies against human protamines for use in flow cytometry to detect sperm with phenotypic variations representing gene-expression-loss or amino-acid-substitution mutations in the protamine genes. To date, we have developed the conditions for labeling protamine P1 of human sperm nuclei, and we have detected approximately 200–400 nonfluorescent particles with scatter signals similar to those of sperm nuclei per million brightly fluorescing nuclei. This is a higher proportion of nonfluorescing particles than predicted by estimates of spontaneous mutation rates. We are now determining what fraction of these nonfluorescing particles contain sperm-equivalent amounts of DNA and bind antibodies against P2, the sister protamine. Variant nuclei will be sorted and analyzed for both protein and DNA defects using accelerator-mass-spectrometry and polymerase-chain-reaction methods, respectively.

INTRODUCTION

The current methods for measuring germinal mutations in human beings have not detected increased mutation frequencies in mutagen-exposed individuals. The two best-studied groups of exposed individuals (Japanese atomic-bomb survivors and patients receiving chemotherapy, as reviewed by OTA 1986) received substantial doses of agents known to be germinal mutagens in the mouse. The apparent discrepancy between the mouse and human responses raises two questions with implications for genetic risk assessment: (1) Are the current methods for measuring germinal mutations in people too insensitive or inefficient to detect the types of genetic lesions induced by these agents? (2) Is the human germ line less sensitive to induced mutations than that of the mouse?

Current detection methods may be inefficient in identifying human germinal mutations. These methods are typically based on pregnancy and progeny evaluations and include analysis of inherited chromosomal abnormalities, measurement of inherited alterations in protein mobilities, detection of well-characterized sentinel phenotypes, and a variety of epidemiologic indicators of abnormal reproductive outcomes (OTA 1986). Since the spontaneous mutation frequencies for these indicators are low (e.g., 6 offspring with chromosomal abnormalities per 10,000 newborn and about 2 offspring with autosomal dominant diseases per 100,000 newborn), large numbers of offspring are required to detect a statistically significant increase above baseline rates. Therefore, it is possible that the inability to detect mutations with the current methods is simply a signal-to-noise problem. This problem is difficult to circumvent because of the immense costs and practical limitations associated with obtaining large numbers of progeny for analysis.

Although the recent BEIR V report (BEIR 1990) reviewed the radiation sensitivities of male germ cells of specific species, it did not explain the apparent discrepancy between the murine and human responses to ionizing radiation. One interpretation of this apparent discrepancy is that the human and murine assays detect different genetic lesions, which also differ in their rate of induction. Another possibility is that there is a true species difference in the rate of induction of the same radiation-induced lesion. Clearly, improved methods are needed for detecting similar molecular lesions in both mice and human beings so that we can better evaluate the mouse as a model for genetic risk assessment.

GENETIC FACTORS AFFECTING THE INDUCTION OF GERMINAL MUTATIONS BY CHEMICALS

Results with chemical mutagens broaden the uncertainties involved in using the mouse as an indicator of potential human genetic risk. The chemical induction and persistence of germinal mutations are known to depend on a variety of factors, including the nature of chemical exposure, metabolism, and pharmacokinetic factors.

The following are two examples of species differences in spermatogenic cytotoxicity and germinal mutagenicity after chemical exposures. Dibromochloropropane (DBCP) is an example of a chemical to which the human seminiferous epithelium is more sensitive than that of the mouse (Whorton and Foliart 1983; Oakberg and Cummings 1984). Human exposure to DBCP resulted in extensive cell-killing and sterility. In contrast, no detectable cell-killing, dominant lethality, or induced specific-locus mutations were recorded even in heavily dosed mice (Teramoto and Shirasu 1989). The spermatogenic cytotoxicity of exposed rats and rabbits, however, is closer to that of man (Rao et al. 1983). Interestingly, DBCP also induced postimplantation losses in rats in untreated females mated with treated males (Teramoto and Shirasu 1989). These observations suggest that both induced cell-killing and genetic damage require metabolic processing in vivo and that neither cytotoxic nor mutagenic intermediates are present in the testes of treated mice. Adriamycin is an example of the opposite cytotoxicity relationship; it induces extensive germ-cell killing and infertility in mice (Meistrich et al. 1985), whereas human germ cells appear to be resistant (da Cunha et al. 1983). The germinal mutagenicity of adriamycin in mice remains controversial. Meistrich et al. (1985) reported induced dominant lethality using an in vitro assay, but Generoso et al. (1989) found no increase using the in vivo assay. These examples underscore the difficulties encountered in using animal data to extrapolate cytotoxicity and mutational risk to man, especially for xenobiotic agents that are processed in vivo. They also suggest that until we have a better understanding of the relationship between animal and human germinal mutagenicity, more emphasis should be placed on the development of improved methods for detecting germinal mutations directly in people.

In addition, spermatogenic cells can show major differences in mutational sensitivity at different stages of differentiation. This sensitivity pattern can differ substantially from one chemical to the next and from one species to the next. The majority of chemicals known to induce mutations in mouse germ cells are effective only in postmeiotic cells, although several chemicals induce mutations in both meiotic and postmeiotic cells (Russell and Shelby 1985; L.B. Russell et al., this volume). Furthermore, few of the chemicals that induce mutations in differentiating cells also induce mutations in stem cells. Ethylnitrosourea is an example of a chemical mutagen that predominantly induces stem-cell mutations in mice. If human spermatogenic cells also display the chemical and stage-specific differences in mutational sensitivities, we predict that the rate of induced human germinal mutations is dependent on the specific nature of the chemical exposure and on the length of time between exposure and conception. Murine data further suggest that stem-cell mutagenicity (or lack thereof) does not successfully predict the mutagen sensitivity of differentiating cells. This finding may apply to pregnancy outcomes following cancer chemotherapy involving stem-cell exposures (Mulvihill and Byrne 1985). Thus, the lack of detection of induced germinal mutations in chemotherapy patients after stem-cell exposures should not be considered evidence for the lack of mutagenicity in differentiating male germ cells. Cyclophosphamide, for example, is a chemotherapeutic drug that is mutagenic in differentiating germ cells but not stem cells of mice (Mohn and Ellenberger 1976; L.B. Russell et al., this volume). Although only stem-cell mutations are thought to persist with time, mutations in differentiating cells may be important for people with repeated or chronic exposures to chemicals that are minimally cytotoxic to germ cells.

The current progeny-based methods for detecting human germinal mutations offer little hope of deciphering germ-cell-stage differences in germinal mutation rates. New detection methods are needed that have the resolution to identify induced mutations in differentiating germ cells as well as in stem cells.

NEW METHODS FOR DETECTING HUMAN GERMINAL MUTATIONS

There are several promising new approaches for detecting germinal mutations (OTA 1986; CLS 1989). De novo molecular changes may be detected in the DNA of an offspring from parents exposed to a potential mutagen by comparing the offspring's DNA with the DNA of both its biological parents (OTA 1986; Mohrenweiser et al., this volume). Using this strategy, sensitive molecular methods are being developed to maximize the amount of DNA that is analyzed per offspring, thereby reducing the numbers of offspring needed to determine germinal mutation rates.

Another approach is to detect mutations directly in sperm. Although this limits the analysis to mutations in males, sperm-based methods have several advantages. By expanding the analysis to the millions of sperm normally found per ejaculate, mutant frequencies can be determined, in principle, using single semen samples. Based on various estimates of the spontaneous specific-locus germinal mutation rate (Table 1), an average semen sample (containing about 3×10^8 sperm) contains approximately 3000 sperm carrying mutations that result in the lack of expression of a specific gene. Sperm with chromosomal aneuploidy occur at higher frequencies (1–2%) and would be correspondingly more prevalent in semen.

Sperm analyses also provide the temporal resolution not possible in offspring studies (i.e., to determine the effect of time between exposure and appearance of germinal mutations). In addition, sperm assays do not require tissue from offspring

Table 1
Prevalence of Chromosomal Abnormalities and Gene Mutations among Human Sperm

Based on direct measurements	
Structural chromosomal aberrations	3 – 13%
Abnormal chromosome numbers	
hyperhaploidy only	1 – 2%
Based on indirect estimates	
Specific-locus gene mutations	
expression loss	
(somatic cells in vivo)	$1–10 * 10^{-6}$
autosomal dominant	
(sentinel phenotypes)	$20 * 10^{-6}$
Specific DNA mutations	
base pair exchanges	
(other systems)	$\sim 10^{-8}$/bp or $\sim$30 events/sperm

or their female parent so that, in principle, germinal mutation rates can be determined whether or not a man has fathered children.

METHODS FOR DETECTING SPERM MUTATIONS

Chromosomal Mutations in Sperm

Human sperm can be incorporated into enzymatically prepared hamster eggs and therein processed to yield human metaphase chromosomes that can be evaluated by standard cytogenetic methods. The human-sperm/hamster-egg method (see Brandriff and Gordon, this volume) has been used to determine (1) the person-to-person variation in the frequencies of sperm chromosomal aberrations, (2) the lack of a major paternal-age effect in the frequency of sperm chromosomal abnormalities, (3) that aneuploidies for all chromosomes are represented in sperm, (4) the frequency of specific types of chromosomal imbalances in sperm from men with constitutive rearrangements, and (5) that radiotherapy involving the testes increased the proportion of sperm with chromosomal aberrations. This method is also being used to identify the early postfertilization events in the processing of the human male genome. However, it is a difficult and time-consuming procedure, requiring fresh semen samples, cell cultures, and banding analysis. Presently, it is still impractical for large-scale studies of mutagen-exposed men. However, it has been used to obtain baseline levels of cytogenetic defects, which now serve as benchmarks for the development of easier procedures for measuring chromosomal abnormalities in sperm.

Alterations in Sperm DNA

In mice, alkali-labile sites, adducts, and other alterations have been measured in sperm DNA after chemical treatments (CLS 1989). Singh et al. (1989) described an unusually high spontaneous occurrence of alkali-labile sites in human sperm using alkaline elution and whole-cell electrophoresis, but this observation has not yet been confirmed. A distinction should be made between measurements applicable to individual sperm cells and those based on bulk analyses. Methods based on bulk analysis may provide compelling evidence for induced DNA alterations and, therefore, may be useful indicators of exposure. However, such data are not directly convertible to germinal mutation frequencies. The polymerase chain reaction (PCR) was recently adapted for amplifying specific DNA segments from individual human sperm (see Arnheim, this volume). However, PCR remains impractical for application to the thousands of sperm required to identify the few cells containing mutations in the specific region amplified.

Detection of Sperm Carrying Gene Mutations

Any method for detecting sperm containing gene mutations hinges critically on the ability to identify mutant cells among the vast preponderance of normal cells. The approach we are developing is based on preselecting sperm with variant protein phenotypes using flow cytometry to identify sperm with specific phenotypes that can arise by gene-expression-loss or amino-acid-substitution mutations. Individual variant sperm are then sorted to determine the nature of the underlying molecular lesion.

DETECTING SPERM MUTATIONS IN THE PROTAMINE GENES

Our approach to measuring sperm mutations is limited to genes expressed only in germ cells because transfer of gene product from somatic cells to sperm might obscure the detection of mutant sperm. Among the genes specific to the male germline, protamine genes P1 and P2 are well characterized in several species, including man. The following is a brief review of the features that make protamines attractive candidate genes for the detection of sperm mutations and a report of our progress toward detecting sperm carrying protamine mutations.

Molecular Characteristics of the Protamine Genes

Protamines are small proteins rich in arginine (~60%) and cysteine (~15%) that replace the histone and nonhistone chromosomal proteins late in spermiogenesis (Ando et al. 1973). Mature sperm of many mammals contain only one protamine (P1). P1 is generally 50 amino acids long and contains a strictly conserved 6-amino-acid amino terminus (Hecht 1989). Balhorn (1982) proposed a model describing how protamine and DNA interact in sperm of species that contain only P1. According to this model, protamines bind in the minor groove of the DNA helix so that the arginines are in a position to neutralize the charges along the phosphate backbone of the DNA. Numerous mammals, including mouse, hamster, and man, contain a second protamine (P2), in addition to protamine P1 (Balhorn 1989). Protamine P2 is longer and more heterogeneous both in length and sequence than P1, ranging from 54 amino acids in humans to 63 in mice.

Protamine genes have been well characterized in mice. Single copies of the protamine P1 and P2 genes are located on mouse chromosome 16 (Hecht et al. 1986). There is low homology between the two gene sequences, indicating substantial evolutionary separation (Kleene et al. 1985; Johnson et al. 1988; Yelick et al. 1987; Hecht 1989). However, the flanking sequences of the two genes contain several identical short DNA sequences (Johnson et al. 1988; Hecht 1989), suggesting commonalities in regulation. There are also numerous similarities in the transcription and translation controls of both mouse protamine genes. Both are haploid expressed, and their messages remain inactive for about 7 days before the poly(A) tail is shortened (Bower et al. 1987) and translation begins (Kleene et al. 1984). However, the two mouse protamines are regulated differently after translation. Mouse P2 is synthesized as a precursor containing 106 amino acids and is processed by the stepwise removal of several amino-terminal fragments to yield a final protein containing 63 amino acids (Yelick et al. 1987). Mouse P1, however, is synthesized and deposited onto DNA without further alteration to its sequence. The way in which P1 and P2 interact with each other and with sperm DNA is still largely unknown (Rodman et al. 1984).

Protamines are also the predominant nuclear protein of human sperm, first appearing in condensing spermatids at steps 5–6 (Dadoune and Alfonsi 1986). In man, there are two variants of protamine P2: P2a and P2b. P2a is 57 amino acids long and includes the entire 54-amino-acid sequence of P2b plus an additional 3 amino acids at the amino terminus (McKay et al. 1985, 1986; Ammer et al. 1986; Balhorn et al. 1987). Protamine P2a may be a precursor form of human P2b, because in addition to their amino acid similarity, the three amino-terminal amino acids of P2a are identical to the 3 amino acids on the amino-terminal side of the last processing site of mouse P2 (Yelick et al. 1987). The cDNA sequences for human protamine P1 (Lee et al. 1987) and P2 (Domenjoud et al. 1988) are consistent with the peptide sequences. A fourth minor protamine variant, P4, found by Gusse et al. (1986), appears to be a closely related or modified form of protamine P2.

In sum, the attractive features of protamine genes are: (1) They are germ-cell-specific, not known to be expressed in any somatic cell type; (2) they are haploid expressed; (3) they are expressed by single-copy genes; and (4) there are at least two protamine variants expressed together in sperm nuclei, thus permitting the development of internal cellular controls.

Strategy for Detecting Sperm Carrying Protamine Mutations

Figure 1 illustrates our approach for detecting sperm carrying either gene-expression-loss mutations (top) or amino-acid-substitution mutations (bottom) in the protamine gene. Both approaches require highly specific antibodies. The detection of gene-expression-loss mutations uses two fluorescently tagged antibodies, each specific for one of the two major human protamine variants. Gene-expression-loss mutations would appear phenotypically as sperm that have lost the ability to bind one of the two antibodies. The detection of cells containing amino-acid-substitution mutations requires an antibody that does not recognize either protamine P1 or P2 unless a specific amino acid alteration has occurred. Cells with new amino-acid-substitution mutations would appear phenotypically as those that have acquired the ability to bind an antibody among normal cells that do not bind the antibody.

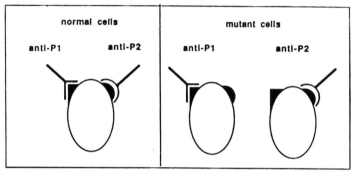

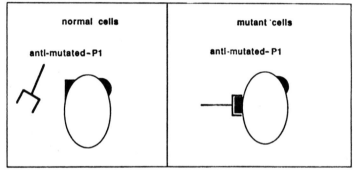

Figure 1
Strategy for detecting gene-expression-loss (*top*) and amino-acid-substitution mutations (*bottom*) in human sperm. These approaches require highly specific monoclonal antibodies against protamine 1 (P1), protamine 2 (P2), or against slightly modified forms of these proteins. Antibodies are reacted with isolated human sperm nuclei.

Production of Antiprotamine Monoclonal Antibodies

Human protamines are highly immunogenic (Stanker et al. 1987; Roux et al. 1988), and we have developed several highly specific monoclonal antibodies (MAbs) against the human protamines. These antibodies fall into five broad reactivity classes as assessed by enzyme-linked immunosorbent assay (ELISA), on immunoblots and, to a limited extent, with cytometry of isolated sperm nuclei. These reactivity classes are: (1) antibodies specific for human protamine P1 with no reactivity with any other human or nonhuman protamines (e.g., MAb hup 1m), (2) antibodies that react with protamine P1 but not protamine P2 of all species tested (e.g., MAb hup 1n), (3) antibodies that react with both human protamines P1 and P2 but not with protamines of other tested species (MAb hup B), (4) antibodies that react with human protamines P1 and P2, and also react with protamines of other species (MAb hup A), and (5) antibodies that react with protamine P2 but not P1 of all tested species (e.g., MAb hup 2b).

Labeling of Human Sperm Nuclei

We developed a procedure for isolating sperm nuclei free of tail, acrosomal, and cytoplasmic components (Balhorn et al. 1977) and used this technique to prepare sperm nuclei that react with monoclonal antibodies. A mixture of bull and human nuclei is shown in Figure 2; the bull nuclei are larger and more homogeneous in shape than are the human nuclei (top). Mixtures labeled with fluorescently tagged MAb hup 1m show that human nuclei are labeled and bull nuclei are not (bottom), which is consistent with the ELISA results for this antibody.

Flow Cytometric Analysis of Antibody-labeled Sperm Nuclei

The fluorescence intensity of sperm nuclei labeled with fluorescently tagged antibody was quantified on a cell-by-cell basis using dual-beam flow cytometry. Figures 3, 4, and 5 show the results of typical analyses of human sperm nuclei labeled with hup 1m, using bull nuclei as a control for antibody specificity. Bull and human sperm nuclei have differing scatter signals that can be used to identify the respective populations in flow graphs (Fig. 3). hup 1m showed a homogeneous labeling of human nuclei (Fig. 4), with no detectable labeling of bull nuclei in mixing experiments (Fig. 5).

Figure 5 shows that the modal brightness of human sperm labeled with hup 1m is related to the antibody concentration. hup 1m antibody did not cross-react with bull nuclei even at the highest concentration, and there was an approximate 1500-fold difference in their respective brightnesses.

As shown in Figure 4 (lower left panel), there are very few particles in the human sperm nuclei suspension that scatter laser light as sperm nuclei do but do not bind antibody. In prototypic experiments, we detected approximately 200–400 unlabeled particles (i.e., particles with less than 10% of the intensity of the main fluorescence peak) per million brightly labeled nuclei. This proportion of unlabeled particles is at least 20-fold higher than would be expected for spontaneous germinal mutation frequencies (Table 1). We hypothesize that many of these nonfluorescing particles are not intact sperm nuclei, and we are refining our detection strategy by addressing two questions: (1) What fraction of the particles with sperm-nuclear scatter also contain the amount of DNA expected in a normal sperm nucleus? (2) What fraction of these nuclei express the second protamine typically found in human sperm nuclei?

Sperm nuclei contain a haploid complement of genomic DNA, and sorted phenotypic variants can be analyzed (e.g., by PCR) to identify the DNA defects that

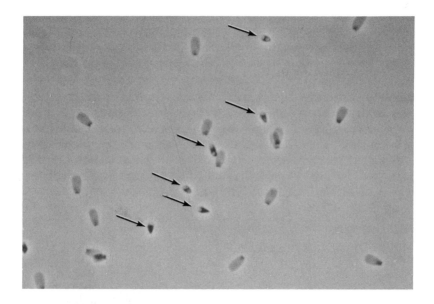
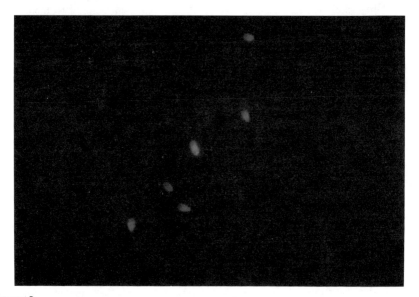

Figure 2
Photomicrographs of mixture of human and bull nuclei labeled with fluoresceinated antibody hup 1m with phase-contrast microscopy (*top*) and fluorescence microscopy (*bottom*). The average length of the major axis of the human sperm nuclei (marked with arrows) is ~ 5 μm.

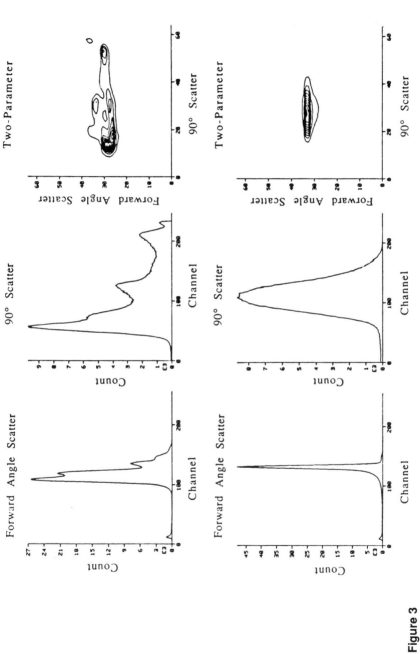

Figure 3
Scatter distributions of unstained human (*bottom*) and bull (*top*) sperm nuclei obtained by dual-beam flow cytometry. Column 1 represents forward-angle scatter, column 2 represents 90° scatter, and column 3 shows two-parameter graphs for both types of scatter.

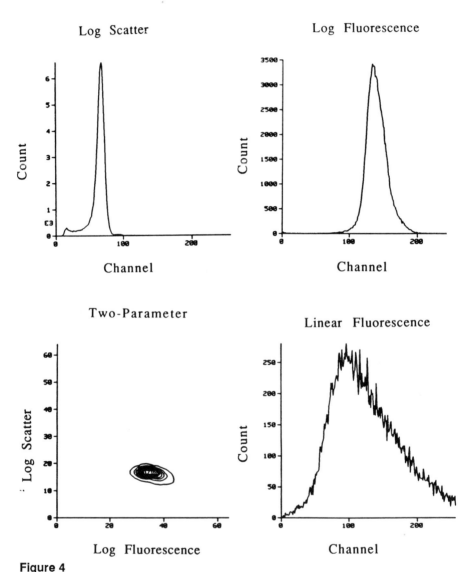

Figure 4
Forward-angle scatter and fluorescence graphs of human sperm stained with fluoresceinated hup 1m antibody. Graphs in the top row represent forward-angle scatter (*left*) and fluorescence (*right*). Graphs on the bottom row represent two-parameter forward-scatter vs. fluorescence (*left*) and fluorescence distribution using linear amplification (*right*).

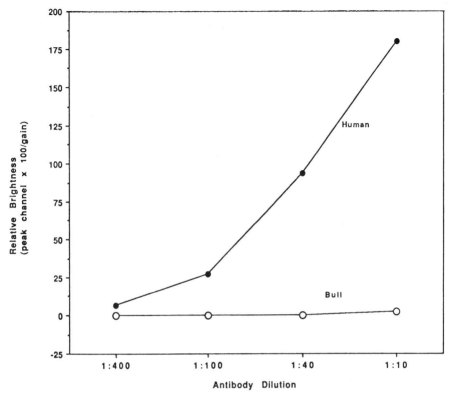

Figure 5
Effect of increasing hup 1m antibody concentration on brightness of human and bull sperm nuclei measured in flow cytometry. Each point represents the peak intensity on a corresponding linear analysis in co-mixtures of human and bull sperm nuclei using linear signal amplification.

underlie the protamine phenotype. In addition, we propose to use accelerator-mass spectrometry (Elmore and Phillips 1987) to determine whether sperm nuclei selected for lack of protamine labeling have indeed lost one of the protamines while retaining the other.

DISCUSSION

We have demonstrated that human sperm nuclei can be labeled with fluorescently tagged antibodies and that flow cytometry can be used to quantify the amount of protamine labeling on a cell-by-cell basis. Using this approach, we have found a small proportion of particles that do not fluoresce but that scatter laser light as sperm nuclei do. Our current efforts are to determine what fraction of these particles are sperm carrying mutations in the protamine genes.

There are several unresolved biological questions related to the development of sperm mutational assays. First, we do not known the role of diffusion via the intercellular bridges, which are known to connect human spermatogenic cells (Dym and Fawcett 1971). Protamine diffusion can alter sperm phenotype. Complete diffusion would nullify the gene-expression-loss analysis and would slightly amplify the frequency of amino-acid-substitution mutations resulting from stem-cell mutations. A study in transgenic mice using a protamine promoter and a recorder gene suggested that the protamine promoter does not inhibit diffusion (Braun et al. 1989). However, there are no direct data regarding diffusion of native protamine or the location of protamine message during storage prior to translation. Additional studies with transgenic mice are under way to resolve these questions.

Another relevant question is whether or not a sperm deficient in one of the protamines will survive to reach the ejaculate. Available evidence suggests that it would. The sperm of many species contain only protamine P1, which indicates that protamine P2 is not absolutely required for making viable sperm (Balhorn 1989). However, no species has been found with sperm containing only protamine P2, so we do not know whether such sperm can be made. In addition, some sterile men produce nonfunctioning and nonviable seminal sperm that apparently contain no protamines at all (Silvestroni et al. 1976). Therefore, we have examples of both null-protaminic and uni-protaminic ejaculated sperm.

Despite the fact that we know very little about the mutational sensitivity of the protamine genes, they are known on an evolutionary scale to be among the fastest evolving proteins (Balhorn 1989).

Other genes have been described that are expressed exclusively in spermatogenic cells (Hecht 1986). Several of the protein products are sperm-specific, haploid expressed, and have had antibodies developed specific to them. Our strategy for detecting sperm carrying protamines may also be applicable to these other sperm-specific genes.

One problem in this area of research concerns the need for animal models. Although we have stressed the current uncertainties with the mouse as an animal model for genetic risk assessment, we believe that this problem can be resolved. Surprisingly few chemicals have been analyzed for germinal mutagenicity in mice, presumably because the available tests are very expensive. For example, only about 25 chemicals have been evaluated by the mouse specific-locus test, a number that is dwarfed by the approximately 60,000 commercially registered chemicals in the U.S.A. Except for ionizing radiation and certain chemotherapies, we have essentially no human data for comparison. Therefore, improved methods are urgently needed to compare the same types of genetic lesions in both mice and man so that we can compare the mutational responsiveness of these two species and develop better risk assessment methods. It is hard to imagine that human-based methods will ever be efficient enough to allow us to evaluate the germinal mutagenicity of the 60,000 registered chemicals. Therefore, we must continue to develop the mouse and other species as test systems for human genetic risk assessment. In addition, the mouse will continue to play a major role in experimental studies of mechanisms of mutation induction and for evaluating the relationship between specific lesions and health of the offspring.

ACKNOWLEDGMENT

This work was performed under the auspices of the U.S. Department of Energy by the Lawrence Livermore National Laboratory under contract number W-7405-ENG-48.

REFERENCES

Ammer, H., A. Henschen, and C.H. Lee. 1986. Isolation and amino acid sequence analysis of human sperm protamines P1 and P2. *Biol. Chem. Hoppe-Seyler* **367**: 515.

Ando, T., M. Yamasaki, and K. Suzuki. 1973. Protamines. Isolation, characterization, structure and function. *Mol. Biol. Biochem. Biophys.* **12**: 1.

Balhorn, R. 1982. A model for the structure of chromatin in mammalian sperm. *J. Cell Biol.* **93**: 298.

———. 1989. Mammalian protamines: Structure and molecular interactions. In *Molecular biology of chromosome function* (ed. K.W. Adolph), p. 366. Springer Verlag, New York.

Balhorn, R., B.L. Gledhill, and A.J. Wyrobek. 1977. Mouse sperm chromatin proteins: Quantitative isolation and partial characterization. *Biochemistry* **16**: 4074.

Balhorn, R., M. Corzett, J. Mazrimas, L.H. Stanker, and A.J. Wyrobek. 1987. High-performance liquid chromatographic separation and partial characterization of human protamines 1, 2, and 3. *Biotechnol. Appl. Biochem.* **9**: 82.

Biological Effects of Ionizing Radiation Committee (BEIR). 1990. *Health effects of exposure to low levels of ionizing radiation.* National Academy Press, National Research Council, Washington, D.C.

Bower, P., P.C. Yelick, and N.B. Hecht. 1987. Both protamine 1 and 2 genes are expressed in mouse, hamster, and rat. *Biol. Reprod.* **37**: 479.

Braun, R.E., R.R. Behringer, J.J. Peschon, R.L. Brinster, and R.D. Palmiter. 1989. Genetically haploid spermatids are phenotypically diploid. *Nature* **337**: 373.

Committee on Life Sciences (CLS). 1989. *Biologic markers in reproductive toxicity.* National Academy Press, National Research Council, Washington, D.C.

da Cunha, M.F., M.L. Meistrich, H.L. Ried, L.A. Gordon, G. Watchmaker, and A.J. Wyrobek. 1983. Active sperm production after cancer chemotherapy with doxorubicin. *J. Urol.* **130**: 927.

Dadoune, J.P. and M.F. Alfonsi. 1986. Ultrastructural and cytochemical changes of the head components of human spermatids and spermatozoa. *Gamete Res.* **14**: 33.

Domenjoud, L., C. Fronia, F. Uhde, and W. Engel. 1988. Sequence for human protamine 2 cDNA. *Nucleic Acids Res.* **16:** 7733.

Dym, M. and D.W. Fawcett. 1971. Further observations on the number of spermatogonia, spermatocytes, and spermatids connected by intercellular bridges. *Biol. Reprod.* **4:** 195.

Elmore, D. and F.M. Phillips. 1987. Accelerator mass spectrometry for measurement of long-lived radioisotopes. *Science* **236:** 543.

Generoso, W.M., K.T. Cain, L.A. Hughes, and L.B. Foxworth. 1989. A restudy of the efficacy of adriamycin in inducing dominant lethals in mouse spermatogonia stem cells. *Mutat. Res.* **226:** 61.

Gusse, M., P. Sautiere, D. Belaiche, A. Martinage, C. Roux, J.P. Dadoune, and P. Chevaillier. 1986. Purification and characterization of nuclear basic proteins of human sperm. *Biochim. Biophys. Acta* **884:** 124.

Hecht, N.B. 1986. Regulation of gene expression during mammalian spermatogenesis. In *Experimental approaches to mammalian embryonic development* (ed. J. Rossant and R. Pederson), p. 151. Cambridge University Press, New York.

———. 1989. Molecular biology of structural chromosomal proteins of the mammalian testis. In *Molecular biology of chromosome function* (ed. K.W. Adolph), p. 396. Springer Verlag, New York.

Hecht, N.B., K.C. Kleene, P.C. Yelick, P.A. Johnson, D.D. Pravtcheva, and F.H. Ruddle. 1986. Mapping of haploid expressed genes: Genes for both mouse protamines are located on chromosome 16. *Somatic Cell Mol. Genet.* **12:** 203.

Johnson, P.A., J.J. Peschon, P.C. Yelick, R.D. Palmiter, and N.B. Hecht. 1988. Sequence homologies in the mouse protamine 1 and 2 genes. *Biochim. Biophys. Acta* **950:** 45.

Kleene, K.C., R.J. Distel, and N.B. Hecht. 1984. Translational regulation and deadenylation of a protamine mRNA during spermiogenesis in the mouse. *Dev. Biol.* **105:** 71.

———. 1985. Nucleotide sequence of a cDNA clone encoding mouse protamine 1. *Biochemistry* **24:** 719.

Lee, C.H., S. Hoyer-Fender, and W. Engel. 1987. The nucleotide sequence of a human protamine 1 cDNA. *Nucleic Acids Res.* **15:** 7639.

McKay, D.J., B.S. Renaux, and G.H. Dixon. 1985. The amino acid sequence of human sperm protamine P1. *Biosci. Rep.* **5:** 383.

———. 1986. Human sperm protamines: Amino-acid sequences of two forms of protamine P2. *Eur. J. Biochem.* **156:** 5.

Meistrich, M.L., L.S. Goldstein, and A.J. Wyrobek. 1985. Long-term infertility and dominant lethal mutations in male mice treated with adriamycin. *Mutat. Res.* **152:** 53.

Mohn, G.R. and J. Ellenberger. 1976. Genetic effects of cyclophosphamide, ifosfamide, and trofosfamide. *Mutat. Res.* **32:** 331.

Mulvihill, J.J. and J. Byrne. 1985. Offspring of longterm survivors of childhood cancer. *Clin. Oncol.* **4:** 333.

Oakberg, E.F. and C.C. Cummings. 1984. Lack of effect of dibromochloropropane on the mouse testis. *Environ. Mutagen.* **6:** 621.

Office of Technology Assessment (OTA). 1986. *Technologies for detecting heritable mutations in human beings*, OHA-H-298. U.S. Government Printing Office, Washington, D.C.

Rao, K.S., J.D. Burek, J.A., John, B.A. Schwetz, F.J. Murray, T.J. Bell, J.E. Battjes, W.J.

Potts, and C.M. Parker. 1983. Toxicologic and reproductive effects of inhaled 1,2-dibromo-3-chloropropane in rats. *Fundam. Appl. Toxicol.* **3:** 104.

Rodman, T.C., F.H. Pruslin, and V.G. Allfrey. 1984. Protamine-DNA association in mammalian spermatozoa. *Exp. Cell Res.* **150:** 269.

Roux, C., M. Gusse, P. Chevaillier, and J.P. Dadoune. 1988. An antiserum against protamines for immunohistochemical studies of histone to protamine transition during human spermiogenesis. *J. Reprod. Fertil.* **82:** 35.

Russell, L.B. and M.D. Shelby. 1985. Tests for heritable genetic damage and the evidence for gonadal exposure in mammals. *Mutat. Res.* **154:** 69.

Silvestroni, L., G. Frajese, and M. Fabrizio. 1976. Histone instead of protamines in terminal germ cells of infertile, oligospermic men. *Fertil. Steril.* **27:** 1428.

Singh, N.P., D.B. Danner, R.T. Tice, M.T. McCoy, G.D. Collins, and E.L. Schneider. 1989. Abundant alkali-sensitive sites in DNA of human and mouse serum. *Exp. Cell Res.* **184:** 461.

Stanker, L.H., A.J. Wyrobek, and R. Balhorn. 1987. Monoclonal antibodies to human protamines. *Hybridoma* **6:** 293.

Teramoto, S. and Y. Shirasu. 1989. Genetic toxicology of 1,2-dibromo-3-chloropropane (DBCP). *Mutat. Res.* **221:** 1.

Whorton, M.D. and D.E. Foliart. 1983. Mutagenicity, carcinogenicity and reproductive effects of dibromochloropropane (DBCP). *Mutat. Res.* **123:** 13.

Yelick, P.C., R. Balhorn, P.A. Johnson, M. Corzett, J. Mazrimas, K.C. Kleene, and N.B. Hecht. 1987. Mouse protamine 2 is synthesized as a precursor whereas mouse protamine 1 is not. *Mol. Cell. Biol.* **7:** 2173.

COMMENTS

Solter: Could you give us some feeling for the usefulness of the gene mutation test? For example, what would happen if you examined somebody and found 10 sperm with protamine mutations out of 1 million? Would that be prognostic or what would be the purpose?

Wyrobek: The main purpose for these sperm mutation tests is similar to that of the sister chromatid exchange, chromosomal aberration, specific amino acid substitution gene mutation tests, and the micronuclei tests in somatic cells. The usefulness of those tests is in the context of epidemiologic studies where you define an exposed population and a control population and do a cross-sectional study, or you identify individuals before, during, and after an exposure. Another usefulness is to help us identify agents that could be causing germinal mutations so that we could protect other people from those kinds of exposures.

Handel: Would you have to have males before exposure to get a baseline for each individual male?

Wyrobek: Not necessarily. There are two major study designs. One is as you described. The most commonly used study design, however, is the cross-sectional one where a certain number of controls and a certain number of exposed are identified. Using the cross-sectional approach, researchers attempt to assign men to exposure categories—like nonsmokers, high smokers, low smokers. After a few baseline studies to determine variation, we can calculate the number of samples and people needed in each category to determine whether there is an effect of the exposure.

Generoso: Do you have a feeling now for nonspecific binding? Is that going to be a problem?

Wyrobek: We are using bull and other species where we expect a protamine 2 not to be present. Bull sperm nuclei shouldn't bind protamine 2 antibodies because they don't have protamine 2. Bull sperm should bind protamine 1 antibodies that cross-react with the bull, but not those that don't. We are using these controls to work the specificity problem, which is really critical.

Woychik: I would like to comment at this point regarding the new transgenic technology in the mouse and get your views. We are making hundreds of different lines of transgenic mice with the pronuclear microinjection procedure. In many of these cases, we are getting very high copy number integrations, where, for example, the average size of the transgene would be 500 kb. With these animals essentially what we are doing is making a chromosome-specific repetitive sequence that should be relatively straightforward to detect in the individual sperm. Furthermore, if we go on and we map metaphase spreads, we can derive high copy number integrations unique to chromosome 1, chromosome 2, 3, 4, and on to X and Y. Have you thought about setting this up with mice so you can take the same kind of experiments that are being done where there is an in vivo mutagenic frequency or a defect frequency and then correlate that with an in situ hybridization of individual sperm and correlate the actual incidence with which you are seeing defects using in situ hybridization with in vivo defects?

Wyrobek: In fact, I was just at Research Triangle Park where there was lots of interest in animal counterparts to the human sperm mutation tests including the sperm tests for aneuploidy. The scenario I see is that we may have initial human sperm studies which are suggestive of a germinal effect (or even highly statistically significantly positive) and then we are faced with the problem of whether to go back and do another human population study to verify the result. Human populations with well-controlled exposures are hard to identify. An equally good approach might be to have an animal system or even two animal systems and test the putative compound in the dose-response mode in the animals. I suggest that we make studies in animal models of germinal mutagenicity part of our response to these human exposure situations.

Woychik: What do you think of taking, say, protamine 2 promoter and hooking it up to the *lacZ* gene and then making a single-copy transgenic, or possibly even a homozygote, where individual sperm would have one copy of the *lacZ*, so you could then assay for mutation events in the *lacZ* simply by X-Gal staining of individual sperm instead of having to go through antibody staining of the individual sperm?

Wyrobek: We have done some preliminary mouse work and we found that the hup 1m doesn't react with the mouse either. In fact, we did the flow work to confirm that. There is about a 1500- to 2000-fold difference in brightness between mouse protamine 1 and human protamine 1, and we can't detect the mouse sperm. So, we have actually proposed and we have in place an effort, together with Norm Hecht, to try to make transgenic mice using the mouse promoter and the human cDNA, and then to go ahead and use the antibodies that we have.

Woychik: Do you think antibody staining is a better method than X-Gal staining?

Wyrobek: I think so, yes.

Woychik: How many of the mutations are likely to actually affect antibody binding? Are all of them, since it's a relatively small molecule?

Wyrobek: It depends on the design of the mutation test. If you're looking for gene expression loss, then if that protein is gone there is no binding. Anything that knocks out that gene expression will result in no antibody binding. The other construct that we really haven't evaluated very much because the rates are likely to be lower is the one for specific amino acid substitution mutations. For that, you design the antibody to detect a very specific epitope and only that epitope. A normal sperm will not bind whereas a mutated sperm will bind. Unlike gene expression loss mutations, the approach for amino acid substitution mutations is very dependent on epitope. If you want to see another epitope, you have to design yet another antibody.

Aberrant Chromosome Structure/Behavior

Clastogenic Effects of Acrylamide in Different Germ-cell Stages of Male Mice

ILSE-DORE ADLER
GSF-Institut für Säugetiergenetik, D-8042 Neuherberg
Federal Republic of Germany

OVERVIEW

Acrylamide (AA) is an industrial chemical for which exposure limits have been set because of its neurotoxic effect. However, in view of its genotoxic potential, it seemed important to characterize the germ-cell clastogenicity of AA. The attempt was made to determine particularly sensitive germ-cell stages in a qualitative way and also to give a quantitative comparison.

Cytogenetic studies of meiotic chromosomes were performed after treatment of male mouse germ cells during meiotic prophase and as spermatogonia. Furthermore, a mouse heritable translocation assay was conducted with sampling progeny from AA-treated spermatids and spermatogonia. The results show that spermatids are very sensitive to the clastogenic effect of AA, spermatocytes are affected, and spermatogonia are insensitive. The quantitative comparison was based on the doubling dose, that is the dose that induces as many chromosomal aberrations per generation as occur spontaneously. In the absence of dose-response data, point estimates were used to calculate the doubling dose. For translocation induction in spermatids, it was about one order of magnitude higher than for chromosomal aberrations observed in meiosis after treatment of prophase stages. The quantitative comparison is provisional, since different clastogenicity endpoints were compared and relevant dose-response data are still pending. However, it points to the need for further experimental data and is of relevance to exposed human individuals.

INTRODUCTION

The differential spermatogenic response to chemical mutagens has been described first with data of the dominant lethal assay (Ehling et al. 1972). The authors found that the yield of dominant lethal effects was different for the various mating intervals that represented particular germ-cell stages sampled after treatment with *n*-propyl methanesulfonate (PMS) and isopropyl methanesulfonate (IPMS). PMS, like ethyl methanesulfonate (EMS) and methyl methanesulfonate (MMS), induced dominant lethal mutations only in spermatozoa and spermatids, whereas IPMS affected also spermatocytes (Ehling et al. 1968, 1972). Subsequently, it was shown with a variety of other mutagens that the response of the individual germ-cell stages was characteristic for a given chemical and could differ from compound to compound (Ehling 1977). Since it has been demonstrated for MMS that the basis for dominant lethal mutations is chromosomal aberrations (Brewen et al. 1975), it was concluded that the differential spermatogenic response is characteristic for the clastogenic effect of chemicals. Thus, it should be possible to determine the particular sensitivity of germ-cell stages with cytogenetic methods. Chromosome analyses can be performed in mitotic divisions of differentiating spermatogonia and in meiotic divisions of spermatocytes (Adler 1982, 1984; Adler and Brewen 1982). Postmeiotic stages of germ-cell maturation that do not divide any further can be analyzed cytogenetically in first-cleavage divisions after fertilization (Matter and Jaeger 1975). However, this procedure is difficult and time consuming, so that dominant lethal tests may be preferred for the characterization of the clastogenic effect in spermatids and spermatozoa. With a combination of cytogenetic analyses and dominant lethal studies, one can cover the whole scale of male germ-cell development to assess the clastogenic response to a chemical mutagen. Recently, the analysis of synaptonemal complex irregularities in pachytene spermatocytes has been added to determine effects on premeiotic and meiotic prophase stages of germ cells (Allen et al. 1987).

A large proportion of the aberrations observed in the germ line of treated animals will result in cell lethality or in lethality of the fertilization product as evidenced in the dominant lethal test. Of major concern, however, is the chromosomal effect transmitted to the viable progeny. Only balanced structural rearrangements of chromosomes are compatible with survival. These are basically reciprocal translocations, inversions, and Robertsonian translocations. They can be assessed in the heritable translocation assay (Generoso et al. 1973), the inversion test (Roderick 1971), and a variant of the heritable translocation assay that only uses meiotic analyses for translocation identification in male progeny of treated animals (Adler 1978).

It became of particular interest to apply as many cytogenetic techniques as possible to the characterization of the spermatogenic response with AA because the chemical had been suspected to be a rare example of a germ-cell-specific clastogen. Early investigations of AA found a clastogenic effect not in mouse bone marrow cells, but in differentiating spermatogonia (Shiraishi 1978). A strong dominant lethal effect of AA was reported in spermatids of mice (Shelby et al. 1986), which was accompanied by the induction of heritable translocations in the same germ-cell stage (Shelby et al. 1987). Similarly, dominant lethal effects were reported in spermatids of rats (Smith et al. 1986). In this trial, which involved animal exposure to AA in drinking water for 72 days, no aberrations were found in differentiating spermatogonia. Similarly, in an acute i.p. study in mice, spermatogonial mitoses did not reveal an increased frequency of chromosomal aberrations (Adler et al. 1988). However, a clastogenic effect in bone marrow cells could be demonstrated by analysis of chromosomal aberrations and in the micronucleus test (Adler 1987; Adler et al. 1988). The clastogenic effect in bone marrow cells was also observed by Cihak and Vontorkova (1988). Consequently, AA cannot be termed a unique germ-cell mutagen, but there may be differences in the effectiveness to induce chromosomal breakage between somatic cells and germ cells as well as between germ-cell stages.

The presently reported studies contribute data to a quantitative comparison of the clastogenicity of AA in germ cells. Analyses of chromosomal aberrations have been performed in first meiotic divisions after treatment of stem-cell spermatogonia and meiotic prophase stages. Furthermore, a heritable translocation assay was conducted to confirm the findings of Shelby et al. (1987) and to extend the collection of progeny from treated spermatids to spermatogonial stages.

RESULTS

Meiotic Chromosome Studies

The analysis of chromosomal aberrations in spermatocytes at diakinesis-metaphase I was performed at various intervals after i.p. treatment of male (102/E1 x C3H/E1)F_1 mice, aged 10–12 weeks, weighing 25–28 g. The animals were injected with single doses of an aqueous solution of AA (Serva, Heidelberg, FRG). For each male, 100 diakineses were scored for univalents of the sex chromosomes and autosomes, chromatid and isochromatid gaps, and fragments as well as rearrangements.

In the experiments to determine a sensitive stage during meiotic prophase, testes were sampled 1, 5, 9, 11, 12, and 15 days after treatment with 100 mg/kg of AA. These intervals represent cells that were treated at diplotene, pachytene, zygotene, leptotene, and preleptotene, respectively. The last interval represented treated differentiating spermatogonia. The results are presented in Table 1. The univalent frequencies were within normal control ranges. Only chromatid or isochromatid gaps and isochromatid fragments were observed; rearrangements were not found. No particularly sensitive stage during prophase was recognized; the clastogenic effects occur at all stages except during leptotene and in differentiating spermatogonia. However, although the differences in the frequencies of fragments between the controls and the treated groups were significant ($p < 0.05$, chi-square test), the effect was not very high compared to the more stage-specific effects observed with mitomycin C (Adler 1976), which induced aberrations primarily during preleptotene.

In a second set of experiments, the male mice were treated with single doses of 100 and 125 mg/kg of AA. Testes were sampled after 90 days, such that the cells analyzed at diakinesis-metaphase I had been stem-cell spermatogonia at the time of treatment. Additionally, ten of the animals from the heritable translocation assay, which were treated with five daily doses of 50 mg/kg of AA were sacrificed 116 and 117 days after the end of treatment. Details of the heritable translocation assay are given below. The cells at diakinesis-metaphase I were scored for univalents of the sex chromosomes and autosomes, chromatid or isochromatid fragments, and

Table 1
Chromosomal Aberrations in Diakinesis-metaphase I Cells of Male (102/E1 x C3H/E1)F_1 Mice Treated during Prophase of Meiosis with 100 mg/kg of Acrylamide

Interval (days)	Stage treated	Univalents XY (%)	Univalents autosomal (%)	Gaps (%)	Fragments (% ± S.E.)
1	diplotene	4.2	3.8	0.4	1.6 ± 0.4[a]
5	pachytene	6.4	3.0	0.2	1.2 ± 0.5[b]
9	zygotene	6.0	4.0	0.2	2.2 ± 0.7[a]
11	leptotene	8.3	2.8	0.3	0.5 ± 0.3
12	preleptotene	7.2	2.3	0	1.2 ± 0.3[b]
15	differentiating spermatogonia	4.0	2.8	0.3	0.3 ± 0.2
Control		4.1	1.9	0	0.3 ± 0.1

Sample size: 5-6 males per group, 100 cells per male in the treated group; 11 males, 100 cells per male in the control combined from concurrent solvent control animals at each sampling interval.
[a]$p < 0.01$ (chi-square test).
[b]$p < 0.05$.

translocation multivalents, such as chains or rings of four or more chromosomes. Again, the univalent frequencies of the treated groups were within the normal control range (Table 2). Similarly, fragments did not occur more frequently after AA treatment than in controls, and only one multivalent, a chain of four chromosomes, was found in the group treated five times with 50 mg/kg of AA. Thus, in a total sample of 2700 cells treated with AA as stem-cell spermatogonia, no increased translocation yield was revealed.

Heritable Translocation Assay

In the present experiment, C3H/E1 males, aged 10–12 weeks, weighing 25–28 g, were treated. The males were i.p. injected daily on 5 consecutive days with 50 mg/kg of AA. This treatment regimen was also used by Shelby et al. (1987). Each male was mated to two virgin 102/E1 females 7–11 days after the end of treatment. A second mating was performed 36–42 days after the end of treatment. The experiment was repeated to sample enough progeny for the translocation analysis.

The results of the parental matings are given in Table 3. A reduction in the number of litters born as compared to females that had copulated is evident. Also, during the first mating interval, the litter size is reduced due to the dominant lethal effect of the AA treatment. A slight loss of young between birth and weaning occurred in all mating groups.

Table 2
Chromosomal Aberrations in Diakinesis-metaphase I Cells of Male (102/E1 x C3H/E1)F_1 Mice after Treatment of Spermatogonial Stem Cells

Dose (mg/kg)	Interval (days)	No. of cells[a]	Univalents XY (%)	Univalents autosomal (%)	Fragments (% ± S.E.)	Multivalents (% ± S.E.)
100	90	1100	6.9	1.8	0.3 ± 0.2	0
125	90	600	8.2	2.5	0.2 ± 0.2	0
5 x 50	116–117	1000	7.0	2.6	0.3 ± 0.2	0.1 ± 0.1
Control		900	10.1	2.1	0.3 ± 0.2	0

[a] 100 cells scored per animal.

Male and female progeny derived from the first mating interval were mated at the age of 10 weeks to (102/E1 x C3H/E1)F_1 animals. Up to three litters were observed to select semisterile or sterile animals according to the criteria described by Adler (1980). Male progeny from the second mating interval were mated to female progeny of the same experimental group, avoiding the pairing of siblings. Suspect male translocation carriers on the basis of litter size reduction were unilaterally orchidectomized, and meiotic chromosome preparations were obtained to confirm the presence of a reciprocal translocation (Evans et al. 1964). The suspect translocation females were mated to (102/E1 x C3H/E1)F_1 males to produce male progeny that could then be subjected to the meiotic proof of the reciprocal translocation. All confirmed translocation heterozygotes, except the steriles, were mated to the same hybrid stock animals to maintain the line through male carriers. At the end of these matings, the animals were sacrificed to perform a karyotype analysis on G-banded mitoses (Gallimore and Richardson 1973) of bone marrow preparations.

A total of 23 translocation heterozygous animals were obtained in 105 progeny of the first mating interval. Table 4 presents the results of the karyotype analysis for

Table 3
Heritable Translocation Assay with Acrylamide

Mating interval (days)	Treated males	Mated females	Plugged females	Litters	Mean litter size ± S.E. birth (%)	Mean litter size ± S.E. weaning
7–11	48	96	89	36	2.1 ± 0.2	1.6 ± 0.3
7–11	51	101	67	30	1.9 ± 0.2	1.6 ± 0.2
36–42	48	100	95	76	6.2 ± 0.2	5.8 ± 0.2
36–42	51	144	129	105	5.8 ± 0.2	5.4 ± 0.2

C3H/E1 males, 102/E1 females, age 10–14 weeks. Five daily doses of 50 mg/kg, i.p.

Table 4
Semisterile Translocation Heterozygous Males

Animal code	Translocations	Presumed breaks	Marker chromosomes
AA-1	T(3;11)	3F1, 11B5	
AA-2	T(7;12)	7E1, 12B	sm 12^7
AA-4	T(9;12)	9B, 12D1	
AA-18	T(15;18)	15C, 18B2	
AA-20	T(2;15)	2C1, 15E	
AA-21	T(10;18)	10D2, 18C	sm 18^{10}
AA-24	T(8;12)	8B2, 12C2	
AA-27	T(2;9)	2B, 9F3	lm 9^2, sm 2^9
AA-300	T(1;10) + T(3;19)	1H1, 10B2	lm 10^1
		3H1, 19C2	
AA-306	T(3;11)	3H3, 11A4	
AA-313	T(1;7) + T(3;14)	1E2, 3E2	
		7E1, 14D1	
AA-316	T(11;16)	11B3, 16B1	
AA-320	T(5;7)	5G2, 7B5	
AA-323	T(4;16)	4D3, 16C2	
AA-325	T(3;6)	3H1, 6C2	
AA-607	T(4;13)	4A2, 13D1	lm 13^4, sm 4^{13}

(lm) Long marker; (sm) small marker.

16 semisterile male translocation heterozygotes. Several of the chromosomal rearrangements produced marker chromosomes, i.e., T(7;12), T(10;18), T(8;12), and T(4;13). The latter male carrier was conceived in the second mating interval on day 39 after the end of the paternal treatment. Two semisterile males actually carried two reciprocal translocations, i.e., AA-300 and AA-313. Another two translocation heterozygotes involved the same chromosomes, namely 3 and 11, but the breaks were located at different positions so that the translocations were not identical and thus not preexisting.

Table 5 gives a further characterization of the various semisterile male translocation carriers. Litter size reductions were most dramatic for the double heterozygotes. Also, translocation T(3;11) of male AA-306 had extremely small litters, probably due to the transposition of almost the entire chromosome 11 to the distal end of chromosome 3. This is similar to the c/t-types of translocations described by Cacheiro et al. (1974), which were associated with male sterility. The other translocation heterozygous males showed a litter size reduction of 30–60%. Low litter size reduction was associated with low multivalent frequency and a small

Table 5
Characterization of Translocation Heterozygous Males

Translocations	Mean litter size ± S.E. (no. litters)		Multivalent frequency	Translocation males/ total male progeny (%)	
T(1;7) + T(3;14)	1.3 ± 0.2	(9)	100% (80% 2TM)	2/2	(100)
T(1;10) + T(3;19)	1.9 ± 04	(10)	100% (60% 2TM)	6/6	(100)
T(2;9)	3.6 ± 0.6	(8)	100% C	4/13	(31)
T(2;15)	4.6 ± 0.8	(7)	84% C	7/14	(50)
T(3;6)	4.6 ± 0.6	(7)	100% R	8/15	(53)
T(3;11)	4.6 ± 0.5	(7)	100% C	7/16	(44)
T(3;11)	1.9 ± 0.4	(10)	88% R	7/8	(88)
T(4;13)	4.5 ± 1.2	(4)	100% C	3/6	(50)
T(4;16)	5.0 ± 0.6	(6)	75% R	6/15	(40)
T(5;7)	5.3 ± 0.6	(7)	95% R	3/13	(23)
T(7;12)	3.6 ± 0.6	(8)	88% C	8/15	(53)
T(8;12)	5.3 ± 0.8	(7)	60% R	10/18	(56)
T(9;12)	4.0 ± 0.7	(7)	28% C	4/6	(67)
T(10;18)	3.2 ± 1.0	(6)	64% C	7/9	(78)
T(11;16)	4.0 ± 0.5	(7)	85% R	5/10	(50)
T(15;18)	6.4 ± 0.4	(7)	32% C	5/18	(28)

(C) Predominantly chains of four; (R) predominantly rings of four; C:R = 8/6.

number of translocation heterozygote progeny in T(15;18). The predominance of ring and chain multivalents indicated in Table 5 was equally distributed among the various translocations.

The number of female translocation carriers was comparatively small. Table 6 gives the details of the karyotype analysis. Again, two double heterozygous individuals were identified, AA-51 and AA-54. One female carried an inversion of chromosome 1 in addition to a translocation between chromosomes 2 and 4. Several marker chromosomes resulted from the rearrangements in the females; particularly noticeable was the minute chromosome 18^{14} in female AA-51.

The average litter sizes of the translocation females are presented in Table 7. The multivalent frequencies given in Table 7 refer to the translocation heterozygote sons of the female carriers. It was of particular interest to note that the female with the double translocation, T(2;10) and T(14;18), had a son with a tertiary trisomy additionally to T(2;10), i.e., 41 $(+18^{14})$. This male and the carriers of T(14;18) as well as the double heterozygote son were sterile. Therefore, unfortunately, the T(14;18) translocation is lost.

The characterization of the sterile males observed in the present translocation assay is given in Table 8. One double heterozygous male carried a reciprocal trans-

Table 6
Semisterile Translocation Heterozygous Females

Animal code	Translocations	Presumed breaks	Marker chromosomes
AA-48	T(6;10)	6F3, 10B2	
AA-51	T(2;10) + T(14;18)	2H1, 10B4, 14E3, 18B2	sm 18^{14}
AA-54	T(5;11) + T(6;7)	5B, 11B1, 6G2, 7B1	lm 6^7, sm 7^6
AA-328	T(2;4) + Inv 1	2A3, 4B3; 1B and F	
AA-335	T(7;15)	7A1, 15F3	lm 15^7 sm 7^{15}
AA-344	T(1;15)	1F, 15E	

(lm) Long marker; (sm) small marker.

location involving three chromosomes, i.e., T(1;9;19)+T(7;8). One male carried a c/t-type translocation, T(2;17). Particularly interesting is the translocation between chromosome 12 and the Y chromosome. This male was conceived in the second mating interval on day 39 after the end of the paternal treatment. The fourth sterile male, also conceived in the second mating interval, showed a 41 (XXY) karyotype.

The summary of results from the translocation assay with AA is given in Table 9. During the first mating interval, in which the progeny were derived from AA-treated spermatids, the frequency of translocation carriers was 22% (23/105). It is noticeable that significantly fewer female than male translocation carriers were recovered (6/48 females vs. 17/57 males, $p < 0.05$).

Counting each translocation of double heterozygotes individually, the frequencies are 27% (28/105) for the total number of progeny, 35% (20/57) for

Table 7
Characterization of Translocation Heterozygous Females

Translocations	Mean litter size ± S.E. (no. litters)	Multivalent frequency	Translocation males/ total male progeny (%)
T(1;15)	4.4 ± 0.8 (5)	60% R	6/10 (60)
T(2;4) + Inv 1	2.4 ± 0.5 (5)	96% R	7/11 (64)
T(2;10) + T(14;18)	3.3 ± 0.6 (8)	60% (20% 2TM)	8/10 (80)
T(5;11) + T(6;7)	1.7 ± 0.3 (7)	100% (76% 2TM)	3/5 (60)
T(6;10)	3.4 ± 0.4 (7)	88% C	3/6 (50)
T(7;15)	4.9 ± 0.6 (7)	100% C	4/15 (27)

(C) Chains of four; (R) rings of four; C:R = 2/2.

Table 8
Characterization of Sterile Males

Translocations	Presumed breaks	Multivalent frequency	Testis weight (mg)	Age at mating (no. plugged females/ no. litters)
T(1;9;10) + T(7;8)	1C2, 9D, 10C3, 7E1, 8B3	96% (56% 2TM)	33.1	2.5–6 months (8/0)
T(2;17)	2H3, 17A2 (c/t type)	90%	35.5	2.5–6 months (8/0)
T(12;Y)	12D2, YD/E	no cells at diakinesis	24.0	2.5–7 months (8/0)
41 (XXY)	—	no prophase stages	13.5	2.5–7 months (8/0)

males, and 17% (8/48) for females. During the second mating interval, which represents treated spermatogonial stem cells, 1004 progeny were sired and 2 male translocation heterozygotes were found. The total translocation frequency of 0.2% is not significantly different from the historical control accumulated over 10 years in our laboratory. If only male progeny are considered, the difference between controls and treated animals is significant (control 0/3000 vs. treated group 2/556, $p = 0.03$, Fisher's exact test). However, the sterile male T(12;Y) is probably not treatment-derived, since it was sired by a premeiotically treated germ cell. It would have been lost during meiosis because the carrier male showed arrest of spermatogenesis during pachytene.

DISCUSSION

The present experiments reveal that among the various developmental germ-cell stages of male mice, spermatids are the most susceptible to the clastogenic effect of AA. Distribution studies of labeled AA in male mice and binding studies to DNA and protamines in mouse germ cells corroborate this result (Marlowe et al. 1986; Sega et al. 1989). The spermatocytes during meiotic prophase are affected but do not show a particularly sensitive phase. This is in accord with the results on AA-induced synaptonemal complex irregularity studies in pachytene cells of mice (Backer et al. 1989). Differentiating and stem-cell spermatogonia are insensitive.

The results of the heritable translocation assay are seemingly at variance from those reported by Shelby et al. (1987). The total frequency of translocation carriers derived from treated spermatids is significantly lower in our experiment (22 vs. 39%, $p < 0.01$, chi-square test). However, this is due to the fact that females with

Table 9
Summary of the Heritable Translocation Assay with Acrylamide (5 x 50 mg/kg i.p.)

Mating interval (days)	F_1 males	F_1 females	Semisterile males	Semisterile females	Sterile males	Sterile females	Frequency of translocation carriers (%)
7–10	57	48	15(15)	7(6)	2(2)	—	21.9[a]
36–42	556	449	1(1)	2(0)	2(1)	—	0.2[a]
Control	3000	3000	0	2(2)	10[b]	—	0.03[c]

Numbers in parentheses represent confirmed translocation carriers.
[a]Male frequency 29.8% (pg) and 0.36% (g).
[b]2 XYY males and 1 XXY male.
[c]Female frequency 0.07%.

many fewer translocations recovered are included in the present study. If the comparison is restricted to male progeny, the difference is not significant.

It is of interest and reflects the clastogenic potency of AA in spermatids that 5 of the 23 translocation heterozygotes (22%) were double heterozygotes, and one animal carried an inversion additionally to a reciprocal translocation. Furthermore, one of the translocations in a double heterozygote involved three chromosomes. The high yield of translocations may be related to the 5-daily-treatment regimen. The pertinent question, whether the individual doses act additively or synergistically, can only be solved by a further experiment with single treatments. At present, under the assumption of additivity and using the point estimate of the total dose of 250 mg/kg, a doubling dose of 0.3 mg/kg of AA for translocation induction in spermatids was calculated (Adler 1990). The doubling dose for clastogenic effects in somatic cells, i.e., the bone marrow micronucleus test (Adler et al. 1988), can be calculated as 23 mg/kg of AA based on the linear-dose response. For the clastogenic effect in spermatocytes, again only a point estimate at 100 mg/kg of AA can be used to calculate a doubling dose that results in 14 mg/kg AA.

Since these estimates of doubling doses are burdened with uncertainties of lacking dose-response data for spermatids and spermatocytes, they can only be compared in orders of magnitude. However, it appears that the doubling dose for spermatids is one order of magnitude lower than the doubling doses for somatic cells and spermatocytes. It is of relevance to workers exposed to AA, particularly to the staff members of molecular biology laboratories who use the chemical often with little precaution when preparing polyamide gels, to establish the dose-response relationship for translocation induction in spermatids and to obtain data on other genetic endpoints such as specific-locus mutations, which allow an estimate of the genetic risk.

ACKNOWLEDGMENTS

Parts of the experiments on meiotic chromosome analysis after acrylamide treatment of meiotic prophase stages were performed by Dr. A. El-Tarras, Cairo University, Egypt, when he was a guest scientist at the GSF in Neuherberg. Full details of his experiments will be published separately. Ruth Schmöller was responsible for the breeding studies in the heritable translocation assay, and Ingrid Ingwersen performed the meiotic verification of translocation heterozygosity in suspect carriers. All their efficient help is gratefully acknowledged.

REFERENCES

Adler, I.-D. 1976. Aberration induction by mytomycin C in early primary spermatocytes of mice. *Mutat. Res.* **35**: 247.

———. 1978. The cytogenetic heritable translocation test. *Biol. Zentralbl.* **97**: 441.

———. 1980. New approaches to mutagenicity studies in animals for carcinogenic and mutagenic agents. I. Modification of the heritable translocation test. *Teratog. Carcinog. Mutagen.* **1**: 75.

———. 1982. Male germ cell cytogenetics. In *Cytogenetic assays of environmental mutagens* (ed. T.C. Hsu), p. 249. Allanheld, Osmund & Co., New York.

———. 1984. Cytogenetic tests in mammals. In *Mutagenicity testing: A practical approach* (ed. S. Venitt and J.M. Parry), p. 275. IRL Press, Oxford.

———. 1987. Clastogenic effect of acrylamide in mouse bone marrow. *Environ. Mol. Mutagen.* (suppl. 9) **8**: 3.

———. 1990. Cytogenetic studies in male germ cells, their relevance for the prediction of heritable effects and their role in screening protocols. In *Chromosomal aberrations: Basic and applied aspects* (ed. G. Obe and A.T. Natarajan). Springer-Verlag, Heidelberg. (In press.)

Adler, I.-D. and J.G. Brewen. 1982. Effects of chemicals on chromosome aberration production in male and female germ cells. In *Chemical mutagens. Principles and methods for their detection* (ed. A. Hollaender and F.J. de Serres), vol. 7, p. 1. Plenum Press, New York.

Adler, I.-D., I. Ingwersen, U. Kliesch, and A. El-Tarras. 1988. Clastogenic effects of acrylamide in mouse bone marrow cells. *Mutat. Res.* **206**: 379.

Allen, J.W., G.K. De Weese, J.B. Gibson, P. Poorman, and M.J. Moses. 1987. Synaptonemal complex damage as a measure of chemical mutagen effects on mammalian germ cells. *Mutat. Res.* **190**: 19.

Backer, L.C., K.L. Dearfield, G.L. Erexson, J.A. Campbell, B. Westbrook-Collins, and J.W. Allen. 1989. The effects of acrylamide on mouse germ-line and somatic cell chromosomes. *Environ. Mol. Mutagen.* **13**: 218.

Brewen, J.G., M.S. Payne, K.P. Jones, and R.J. Preston. 1975. Studies on chemically induced dominant lethality. I. The cytogenetic basis of MMS-induced dominant lethality in postmeiotic male germ cells. *Mutat. Res.* **33**: 239.

Cacheiro, N.L.A., L.B. Russell, and M.S. Swartout. 1974. Translocations, the predominant cause of total sterility in sons of mice treated with mutagens. *Genetics* **76**: 73.

Cihak, R. and M. Vontorkova. 1988. Cytogenetic effects of acrylamide in the bone marrow of mice. *Mutat. Res.* **209**: 91.

Ehling, U.H. 1977. Dominant lethal mutations in male mice. *Arch. Toxicol.* **38**: 1.

Ehling, U.H., R.B. Cumming, and H.V. Malling. 1968. Induction of dominant lethal mutations by alkylating agents in male mice. *Mutat. Res.* **5**: 417.

Ehling, U.H., D.G. Doherty, and H. Malling. 1972. Differential spermatogenic response of mice to the induction of dominant lethal mutations by n-propyl methanesulfonate and isopropyl methanesulfonate. *Mutat. Res.* **15**: 175.

Evans, E.P., G. Breckon, and C.E. Ford. 1964. An air-drying-method for meiotic preparations from mammalian testes. *Cytogenet.* **3**: 289.

Gallimore, P.M. and C.R. Richardson. 1973. An improved banding technique exemplified in the karyotype analysis of two strains of rats. *Chromosoma* **41**: 259.

Generoso, W.M., W.L. Russell, and D.G. Gosslee. 1973. A sequential procedure for the detection of translocation heterozygotes in male mice. *Mutat. Res.* **21**: 220.

Marlowe, C., M.J. Clark, R.W. Mast, M.A. Friedman, and W.J. Waddell. 1986. The distribution of [^{14}C]acrylamide in male and pregnant Swiss-Webster mice studied by whole-body autoradiography. *Toxicol. Appl. Pharmacol.* **86**: 457.

Matter, B.E. and I. Jaeger. 1975. Premature chromosome condensation, structural chromosome aberrations, and micronuclei in early mouse embryos after treatment of paternal postmeiotic germ cells with triethylenemelamine possible mechanisms for chemically induced dominant-lethal mutations. *Mutat. Res.* **33**: 251.

Roderick, T.H. 1971. Producing and detecting paracentric inversion in mice. *Mutat. Res.* **11**: 59.

Sega, G.A., R.P.V. Alcota, C.P. Tancongco, and P.A. Brimer. 1989. Acrylamide binding to DNA and protamine of spermatogenic stages in the mouse and its relationship to genetic damage. *Mutat. Res.* **216**: 221.

Shelby, M.D., K.T. Cain, C.V. Cornett, and W.M. Generoso. 1987. Acrylamide: Induction of heritable translocations in male mice. *Environ. Mutagen.* **9**: 362.

Shelby, M.D., K.T. Cain, L.A. Hughes, P.W. Braden, and W.M. Generoso. 1986. Dominant lethal effects of acrylamide in male mice. *Mutat. Res.* **173**: 35.

Shiraishi, Y. 1978. Chromosome aberrations induced by monomeric acrylamide in bone marrow and germ cells of mice. *Mutat. Res.* **57**: 313.

Smith, K.M., H. Zenick, R.J. Preston, E.L. George, and R.E. Long. 1986. Dominant lethal effects of subchronic acrylamide administration in the Long-Evans rat. *Mutat. Res.* **173**: 273.

COMMENTS

Dearfield: The USEPA has been taking regulatory action concerning several uses of acrylamide and, accordingly, they are very interested in how to propose a possible quantitative risk assessment model on acrylamide. This question might go to Gary Sega as much as to you. After looking at your data, it sounds as though acrylamide generally induces a postmeiotic effect. We are mostly concerned about what type of target might be the most important aspect of its heritable risk. We don't know whether it's going to be protein or DNA. Do you have any comments about that?

Adler: I tend to believe in Gary Sega's data that protamines are the target in this particular germ-cell stage. Other information suggests that tubulin is another target and that acrylamide has the potential to induce aneuploidy by distorting the spindle effect. We haven't analyzed that yet. There is proof to be obtained of that. To really quantify the genetic risk from acrylamide exposure of spermatids, one would have to have individual dose data. We have been extrapolating, assuming that the effect is additive from five different individual doses, but there is no proof of that yet. Single exposures would be necessary to see whether we can show additivity.

Favor: Have you calculated doubling dose in females?

Adler: I don't have any data on females.

Favor: Oh, it's treated males?

Adler: It's treated males.

Favor: And it's transmission to females?

Adler: The interesting thing is that among the offspring of the treated male germ cells, there were fewer female carriers than male carriers. It was significantly different, like 1 to 3.

Favor: Why do you think that is?

Adler: I have no idea, but it's a general observation. For the first time now I have enough data to do the statistics on it. I have observed that with mitomycin C, with MMS, with ENU, with procarbazine. Whenever I did an experiment, I recovered fewer translocation females than translocation males. But still, in these six females there were three that carried more than one translocation.

Wyrobek: Was the frequency of double heterozygotes in complex translocations higher than you would expect from probability alone?

Adler: Oh, yes, it was 20%. It was 5 in 23, which is tremendous.

Wyrobek: Any thoughts on why that might be?

Adler: I have no idea.

Bishop: In the process by which you scored males and females, was the sequence of events that you did different? You didn't do the females for fertility first?

Adler: I did the females for fertility first, but then I proved the presence of the translocations in their sons, and then I went back to the female and karyotyped the female. Whenever I had suspect females where I didn't have enough offspring, I karyotyped the female right away. I karyotyped a lot more animals than I showed you in the data. I don't think I've lost females, if that's what you're driving at.

Bishop: No. My question is: Did you karyotype more females mitotically than you did males?

Adler: Yes. I karyotyped maybe two more males which did not eventually show up to be translocation heterozygotes. I karyotyped maybe twice as many females.

Bishop: But you did the males only if they were double-translocation males, basically?

Adler: No. I karyotyped every translocation carrier.

Lonnie Russell: I'm interested in why you thought that acrylamide might act on microtubules. It has been shown to be an intermediate filament inhibitor. Intermediate filaments are in both germ cells and Sertoli cells.

Adler: There is a paper by Friedman and co-workers is which they showed tubulin binding.

Lonnie Russell: To acrylamide?

Adler: Yes. Acrylamide to tubulin. That's 1986.

Hecht: Was that in the testis?

Adler: No, that was in somatic cells.

Lonnie Russell: It's well known that intermediate filaments can be depolymerized by acrylamide.

Liane Russell: Have you considered the possibility that translocation-bearing females are less apt to be semisterile, maybe because their segregation is different?

Adler: Actually, the proven translocation-carrying females that we had usually had a lower fertility rate than the males. In previous experiments, just to make sure I didn't lose any females, because the litter size reduction was not that dramatic, I have karyotyped a lot more females than I actually had to, and I could not find one that I had missed by the litter size reduction according to our criteria. So, I don't know; maybe it has something to do with embryonic survival.

Sega: It's interesting that you don't find translocations or dominant lethals in, for example, pachytene spermatocytes or spermatids. We looked shortly after exposure to acrylamide, using radical elution procedures, with which we can measure substantial DNA breakage in pachytene spermatocytes and spermatid stages. But, apparently, by the time these cells have matured, the DNA repair processes have taken care of all this damage and the chromosomes have gotten back to normal, as far as we can tell.

Adler: Well, we did find chromosome breakage in spermatocytes. It was not related to any specific stage of prophase, but there was a general positive trend, and the aberrations that we saw were double chromatid fragments, isochromatid fragments, but no rearrangement.

Dearfield: I was just wondering, then, if there could be a role for both targets in, say, acrylamide-induced heritable effects. There might be some DNA component that you cannot really rule out at this time; or, since there was a large amount of protamine there and we know that acrylamide binds quite readily to proteins, that might provide a pool which mops up a lot of the acrylamide that might get into the testis and limit the amount of DNA being attacked. Yet, would that little bit of acrylamide that might be creeping through still have an effect for, say, a translocation or anything like that?

Adler: I'm not sure. I have talked to Jim about the effect of acrylamide on the synaptonemal complexes and changes. Are you going to talk about that?

Allen: Actually, I'm not. It turned out to be a surprise because we didn't see very much. Although we found acrylamide to induce high levels of micronuclei in spleen cells, no structural aberrations were observed in premeiotic or meiotic chromosomes—and there were very minimal increases in asynapsis and breakage of synaptonemal complexes.

Adler: Right. Then, again, the complex contains a lot of protein that is susceptible to acrylamide interaction.

Dearfield: Maybe DNA repair and selection really take care of all that, and then maybe there are protamine effects? That would be very interesting.

Adler: No, I think it's just open breakage that you see.

Smithies: Would there be any evidence that protamine isn't more than an indicator? I mean, if the protamine is affected, presumably other proteins with SH groups are equally affected, but not so easy to measure. Is it just an indicator or is it the primary target? Is there any argument to say that it's not just an indicator rather than the primary target?

Sega: I think it's the primary target because the sulfhydryl groups in the cysteine in the protamine are very important. When they are alkylated, they can affect the whole structure of the chromatin.

Smithies: But then, presumably, a DNA repair enzyme, let's say for the sake of argument, has some SH groups that are equally important, and they are alkylated; but one doesn't know because there isn't enough of it to assay, and so the two will be completely in parallel, but you don't know which would be the prime target.

Sega: We do know that acrylamide does break the DNA in earlier stages where protamine is not present, but this is repaired. For some reason, the cells are not capable of repairing this breakage that occurs in the spermatid stages; even in the oocyte this is not repaired. We found that when we did a UDS study with acrylamide, we did not get a maximum induction of UDS response in the germ cells until six hours after the treatment with the acrylamide. This suggests to me that the acrylamide may have been metabolized, possibly to some other component, some other chemical, which then alkylates the DNA. Also, in binding studies we did with acrylamide, we looked at the testis DNA binding to ^{14}C-labeled acrylamide, and that paralleled the UDS response. We found a maximum binding level of acrylamide six hours after the injection. On the other hand, the liver DNA showed maximal binding right after the injection of the acrylamide, then it started to go down after that. Thus, there are very different kinetics in the way the acrylamide binds to liver DNA and testis DNA.

Hecht: We have to be very careful with respect to the disulfide situation. Certainly, DNA polymerase α is exquisitely sensitive to N-ethylenemelamine, and if that is involved in repair, then you could easily also be inactivating that. I think the evidence from which we can say that protamine is a primary target is quite weak. It certainly is a target, but if you have a particular enzyme which has an essential function, a very low level of the acrylamide could be knocking that out and creating the problems.

Sega: What are you hypothesizing now for this particular enzyme you mentioned? What is that going to do?

Hecht: You said that you felt that the protamine was a primary target. Well, perhaps, a primary target because the damage occurs at that point and there is no repair mechanism which is operative at that point.

Sega: Right.

Hecht: I'm saying perhaps the reason the repair mechanism is not working is that, let's say, the repair enzyme is present at a very low level which does not have very good accessibility at that point because of the protamine's presence. I think one has to be careful.

Sega: But, then, I'm still puzzled by what happens after fertilization. If there are repair enzymes in the oocyte, why is the damage not capable of being repaired there in the oocyte? If it occurs in the late spermatids, it is apparently so severe that it just can't be handled very well even by the repair system in the oocyte where a lot of repair enzymes are present.

Hecht: Is that clear?

Adler: That's where the rejoining of the open breaks occur when translocations are formed. There has to be some sort of repair in the oocytes, otherwise we wouldn't have translocations.

The Synaptonemal Complex as an Indicator of Induced Chromosome Damage

MONTROSE J. MOSES,[1] PATRICIA POORMAN-ALLEN,[2]
JAMES H. TEPPERBERG,[3] JAMES B. GIBSON,[4] LORRAINE C. BACKER,[5]
AND JAMES W. ALLEN[6]

[1]Department of Cell Biology
Duke University Medical Center
Durham, North Carolina 27710
[2]Wellcome Research Laboratories
Research Triangle Park, North Carolina 27709
[3]Genetics and Developmental Biology Program
Division of Plant and Soil Sciences
College of Agriculture and Forestry
West Virginia University
Morgantown, West Virginia 26506
and Department of Cell Biology
Duke University Medical Center
Durham, North Carolina 27709
[4]Children's Hospital
Philadelphia, Pennsylvania 19104
[5]Environmental Health Research and Testing, Inc.
Research Triangle Park, North Carolina 27709
[6]Genetic Toxicology Division
Health Effects Research Laboratory
U.S. Environmental Protection Agency
Research Triangle Park, North Carolina 27709

OVERVIEW

The synaptonemal complex (SC) is a transient, axial chromosomal structure that is unique to meiotic prophase, representing the specialized events of homologous synapsis and desynapsis with such fidelity as to equate SC formation with synapsis. The proven capability of the SC for detecting and defining synaptic aberrations, such as chromosome rearrangements, proved an asset in detecting and assessing structural and synaptic aberrations induced by alkylating agents (mitomycin C, cyclophosphamide) acting at premeiotic S, and antimitotic agents (AMAs) (colchicine, vinblastine, et al.) acting at leptotene. The similarities and differences between alkylator and AMA effects correlate with their molecular modes of action and indicate dependence of SC integrity and homologous synapsis on interactions between DNA and specific proteins. Induced SC damage correlates generally with abnormalities in MI chromosomes. The greater amount of SC damage in contrast to MI damage makes the SC a more "sensitive" indicator of mutagen effect. The difference may be explained by concomitant cytotoxicity that prevents damaged cells from reaching MI, either through loss from the seminiferous tubule or arrest at prophase. The elimination of damaged cells thus amounts to a protective effect. The correlated use of residual MI chromosome damage as an endpoint for predicting potential heritable effects, and SC abnormalities as a sensitive endpoint for detecting and assessing mutagenicity, constitutes a strong approach for evaluating potentially heritable effects in the germ line.

INTRODUCTION

This paper concerns the induction of potentially heritable germ-line damage during meiosis, the period of gametogenesis in which the germ-line genome is processed for distribution to the gametes. Mutations occur spontaneously during this period, which is also sensitive to mutation induction by external agents. The processing involves events that are common to somatic mitosis, such as DNA replication and repair, and cell division; but in meiosis these events are integrated with others that are unique, such as synapsis, crossing-over, chiasma formation, and chromosome reduction.

Evaluating the risk of induced mutation during this period requires understanding of the mechanisms of the meiotic processes. Although well documented for many years, these mechanisms remain poorly understood. Nevertheless, empirical estimation of the risk of heritable damage by mutagenic agents and of the meiotic stage of greatest sensitivity is theoretically possible, knowing the stage at which the agent acts and assessing the damage induced at a subsequent endpoint where the effects can be observed and measured.

Meiosis is marked by at least five sensitive events, represented diagrammatically in Figure 1A. Interference with any one of these events could produce chromosomal and genetic alterations that would be heritable if carried by a functional gamete. The sensitive events are (1) replication of gonial stem cells that are the source of gametocytes entering meiosis; (2) synapsis, the process of intimate homologous chromosome pairing that is a prerequisite for both meiotic chromosome disjunction/reduction and crossing-over; (3) crossing-over, the consummation of reciprocal recombinational exchanges between chromatids of homologous chromosomes that lead to interhomolog gene reassortment; (4) disjunction of homologs at meiotic division I that provides for reduction of chromosome number and reassortment of paternal and maternal chromosomes; and (5) separation of sister chromatids at meiotic division II that yields a complete haploid chromosome complement for each daughter cell destined to become a gamete.

Early studies on the effects of mutagenesis in the germ line involved ionizing radiation. Chromosome sensitivity to structural damage by X rays was shown to

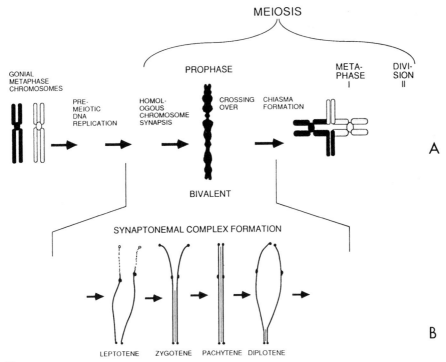

Figure 1
Schematic representation of premeiotic and meiotic prophase chromosomal events in typical mammalian spermatocytes. (A) The selected meiotic events shown include those known to be affected by mutagens: premeiotic DNA replication, synapsis of homologs, and crossing-over/chiasma formation. Primary damage is known to be induced during premeiotic DNA replication and may be variously induced throughout meiotic prophase. The usual endpoints for assessing damage are metaphase I, where chromosome breakage, univalency, etc., may be scored, and division II, where further structural damage and hyperploidy/aneuploidy may be scored. (B) SC formation and disposition is shown schematically for a typical mammalian sub-metacentric chromosome during meiotic prophase. See text for brief description of process. Axial/lateral elements have terminal thickenings or plaques, by which they are attached to the nuclear envelope where synapsis (SC formation) is usually initiated. Kinetochores are seen as thickenings on the axial/lateral elements.

depend on the stage of meiosis irradiated. An example is the extensive evidence from the work of Sparrow et al. (1952) on Trillium microsporogenesis, a process that is protracted over many weeks; stages are well synchronized in the anthers and follow a predictable time schedule. When cells were irradiated at different meiotic prophase stages and scored for breaks at anaphase I (AI), pachytene appeared to be the most sensitive stage. In contrast, diplotene was five times more sensitive than pachytene when metaphases were scored at the microspore division following meiosis. These new lesions must have been present at the preceding pollen mother cell division (AI) as unexpressed potential breaks that required passage through the intervening microspore replication phase for their expression.

Such results constitute a caveat: The kind and extent of chromosome damage induced by the mutagen and the stage of maximum sensitivity to induction may depend on the endpoint at which the damage is scored. It must be concluded that between primary damage and endpoint assessment, metabolic/molecular conditions may intervene that modify the expression of the primary damage qualitatively and quantitatively, placing qualifications on the endpoint used as an indicator of sensitivity.

Comparable information has been more difficult to obtain in mammalian gametogenesis because of a general lack of synchronization of stages compared with plant material, and with assessment of induced structural damage limited mainly to chromosome analysis at meiotic divisions I and II endpoints. More recent investigations of the effects of mutagens on the synaptonemal complex in mouse and hamster have provided a refined schedule of meiotic events that, together with new endpoints, permit identification of the stage and specific target of the primary lesion and more sensitive assessment of the consequent chromosome damage. Most of the evidence presented in this paper comes from studies on mouse spermatogenesis, which will thus become the focus of the discussion to follow.

RESULTS AND DISCUSSION

The Synaptonemal Complex: Nature and Relation to Synapsis

The SC is a major chromosomal component that is unique to meiosis (Moses 1981). It is a transient structure, axial along the entire length of each bivalent. It is formed during synapsis (zygotene) and disassembles at desynapsis (diplotene). It appears to be a prerequisite, although not a sufficient one, for crossing-over and is a participant in, if not part of the mechanism of, homologous chromosome recognition, synaptic initiation, and synaptic progression. It may also be the means by which the chromatids are kept from separating during meiotic prophase. SC formation is operationally synonymous with synapsis.

Each SC is thus a chromosomal structure made up principally of the axes of the two synapsed homologous chromosomes (Fig. 1B). Prior to synapsis (at leptotene), a single dense, proteinaceous axial element assembles along the length of each homolog. During the process of synapsis at zygotene, the axes pair and become the lateral elements of the SC. They appear to be held in parallel alignment by fine filaments that cross the interaxial space and participate in forming a third, less discrete central element that is aligned in parallel with the lateral elements. At the end of the synaptic period (pachytene), the process is reversed: The lateral elements separate again, becoming the axial elements of the individual homologs, and then disperse as the chromosomes move into prometaphase (diakinesis) and thence into meiotic divisions I and II.

The SCs thus provide axial frameworks for the meiotic bivalents and as such reveal the behavior of the chromosomes during synapsis in clearly visible, reproducible, diagrammatic images. They do so with greater accuracy and higher resolution than usually possible in whole chromosome preparations. They are most conveniently studied in quickly and easily prepared whole mounts of surface-spread, paraformaldehyde-fixed, air-dried gametocyte nuclei in which entire SC complements are preserved (Moses 1977).

Following silver staining (Dresser and Moses 1979), the SCs may be analyzed with the light and/or electron microscope (EM). The latter is necessary for observations requiring resolution of the individual axial/lateral elements. Scanning with the light microscope is sufficient to visualize the SC complements and is useful in determining stages, detecting rearrangements, and for the superficial assessment of induced SC and other damage. However, synaptic perturbations can only be assessed accurately with the EM where lateral elements are resolvable. Spread preparations are also amenable to study of the SC using a growing number of molecular probes, such as tritiated thymidine (^{3}H-Tdr) or bromodeoxyuridine incorporation to detect and localize DNA synthesis, and antibodies to specific molecules immunocytochemically localized in situ (Moses et al. 1984; Dresser 1987).

Bulk isolation of SCs by Heyting et al. (1989) has opened the door to identifying molecular components, and application of the spreading technique to the SCs of yeast by Dresser and Giroux (1988) and Dresser (this volume) makes possible molecular studies of the genetic control of synapsis, recombination, and other meiotic events, and holds promise of unlocking their underlying mechanisms.

A Meiotic Timetable in the Male Mouse

The SC is an integral chromosomal participant in meiotic events. As such, it has clarified much about the rules and mechanics of meiosis, particularly of synapsis. This has been made possible largely by the ability to define meiotic prophase stages according to successive morphological changes in SCs of the autosomes and the XY pair, nucleoli, heterochromatin, and sex body, together with other characteristic structures (Fig. 2).

Leptotene, zygotene, five pachytene, and two diplotene substages have been characterized by these criteria in mouse spermatocytes (Moses 1981). The relative duration of each stage has been estimated from its proportionality to the frequency of the stage in spread preparations scored by light microscopy. Using published data on the length of meiotic prophase in the mouse (Oakberg 1957), the relative stage

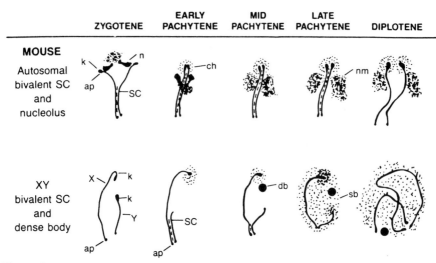

Figure 2
Abridged, diagrammatic representation, taken from electron micrographs of surface spread preparations, of the principal morphological changes in the SC and associated structures during meiotic prophase of the mouse. Such changes are used to characterize specific stages. (ap) Attachment plaque; (ch) centromeric heterochromatin; (db) dense body; (k) kinetochore; (n) nucleolus; (nm) nucleolar material; (sb) sex body; (X) X chromosome axis; (Y) Y chromosome axis. (Reprinted, with permission, from Dresser et al. 1987.)

duration data were converted to hours and days (Fig. 3). These values were verified experimentally by administering doses of ^{3}H-Tdr to mice and scoring the stages that were labeled autoradiographically at successive days. The first spermatocytes to be labeled must have incorporated ^{3}H-Tdr at the ultimate premeiotic S phase, and thus the time course of the progression of stages from that point could be determined (Moses et al. 1984). The results agreed with the estimates derived from frequency data and showed that incorporation at premeiotic S precedes leptotene by about 24 hours. Table 1 summarizes the durations of the major stages.

Such a time line partially compensates operationally for the heterogeneity of spermatogenic stages in whole mount spreads. In cross sections of intact seminiferous tubules, the regular succession of syncytial stages has long provided the basis for a timetable of spermatogenesis in mammals. Real-time estimates of stage duration have been based on the frequency of radial patterns of mouse spermatogenic stages and have been verified by ^{3}H-Tdr labeling (Clermont and Trot 1969). Such histological progression patterns are of course lost in spreads of testicular cell suspensions. However, the spreading technique's advantages of speed, ease, and convenience have been enhanced by the usefulness of the SC time line as applied to the detailed analysis of meiosis.

Table 1
Duration of Meiotic Prophase Stages in Male Mice from Frequencies of SC-characterized Stages

Stage	Duration (days)
Leptotene/zygotene	1.9
Pachytene	6.8
Diplotene	2.4
Divisions I and II	0.9
Total	12.0

Because stage is defined by a number of different morphological markers, there is concern as to whether perturbations in the schedule of one of them may not also affect the others. However, our experience suggests that they are not necessarily interdependent. For example, the rapid synapsis of the Y chromosome with the homologous segment of the X, and the subsequent gradual desynapsis that occurs during the first half of pachytene is normally out of phase with autosomal synapsis at zygotene and desynapsis at diplotene (Fig. 2). Yet XY and autosomal schedules are each reproducible and consistently out of phase. Interfering with one does not automatically alter the other. Thus, an X-autosome translocation may affect the synaptic pattern of the XY, but it does not necessarily affect the synaptic schedule of the autosomes, nucleoli, or other structures. However, in practice, when there is a suspected perturbation in the progression or morphology of any one of the markers—e.g., of the XY pair—it can be set aside as a staging criterion and the correct sequence and schedule can be established on the basis of the other criteria. In any case, our experimental procedure is to verify schedules routinely with stage frequency distributions and to check deviations with ^{3}H-Tdr chase assays.

The Precision of Homologue Synapsis Is Defined by the SC

Reproducibility and regimentation appear to be characteristics of mammalian meiotic prophase. Early studies on the SC of Chinese hamsters showed constancy of relative SC lengths and arm ratios throughout pachytene equivalent to those of chromosomes at somatic metaphase (Moses et al. 1977b). Thus, the pachytene SCs can be identified by their relative lengths and arm ratios to the same degree as their metaphase chromosome counterparts, using the same characteristics.

Evidence that the SC truly reflects the precision with which homologous synapsis is initiated (usually, but not always, at the terminal attachments of the SC to the inner face of the nuclear envelope) came from studies of mouse strains heterozygous for rearrangements that included autosomal and X-autosome reciprocal translocations (Moses et al. 1977a), a duplication (Poorman et al. 1981b), a number of inversions (Poorman et al. 1981a), and a deletion and duplication found in an insertion heterozygote (Mahadevaiah et al. 1984). In all rearrangements, complete, homologous SC formation gave predicted configurations, often with textbook diagram clarity. Duplications and deletions showed buckled segments of lateral elements that lacked a homologous partner; translocations gave quadrivalents, and inversions produced characteristic loops. These configurations were almost invariably present if observed at early pachytene. Furthermore, measurements at that stage showed breakpoints that as a rule approximated those calculated from banded somatic metaphase chromosomes (Moses 1981). These characterizations of known rearrangements with the SC confirmed that homologous synapsis, consisting of homologous recognition, synaptic initiation, and progressive homologous synapsis, occurs during zygotene and is completed by early pachytene.

Synaptic Adjustment

Normally, once the SC forms, homologous synapsis appears to be locked in until desynapsis begins at diplotene, followed by dispersion of the SC. However, the behavior of the rearrangement figures during pachytene is exceptional and reveals an unexpected phenomenon. Fully homologous synapsis, as demonstrated in configurations such as inversion loops and deletion buckles, is altered during the first half of pachytene. The homology-directed configuration (e.g., inversion loop) is regularly eliminated, and the pairing of the lateral elements is adjusted to produce simple, "straight" SCs containing nonhomologously synapsed segments (Moses et al. 1982). Differences in lateral element lengths (as in duplication and deletion buckles) are equalized (usually by shortening of the longer lateral element). This phenomenon, called "synaptic adjustment," is completed by mid-pachytene in the mouse (Moses et al. 1984) and evidently involves desynapsis of homologously synapsed segments, immediately followed by their nonhomologous resynapsis. It is not clear why the zygotene drive for homologous synapsis is replaced by a pachytene drive for topological simplification and indifference to homology in synapsis.

There is presently no proof of what consequences, if any, synaptic adjustment may have on the meiotic process. One possibility is that nonhomologous synapsis could prevent crossing-over in the rearranged segment. In the case of an inversion, the deletions, duplications, dicentrics, and acentrics that would otherwise result from crossovers within the homologously synapsed, inverted segment would be avoided or reduced, thus diminishing deleterious effects of the rearrangements on subsequent generations. However, the inversions studied were all detected by the presence of dicentric bridges at anaphase I (Roderick and Hawes 1974), indicating that a substantial amount of crossing-over had occurred within the inversions despite adjustment. This implies that the crossovers must have occurred during the first half of pachytene before adjustment, when the inversion segment was homologously synapsed.

This fact alone is notable because it indicates that crossing-over occurs, or at least is consummated, during the first half of pachytene. It follows that heterosynapsis of the inversion would have to occur earlier in order to reduce significantly the deleterious consequences of crossing-over. Early heterosynapsis may in fact occur under some conditions according to Ashley's evidence (1988) that when an inversion or translocation breakpoint falls within, or on the edge of, a dark G-band, nonhomologous synapsis follows from the breakpoint during zygotene.

Induction of SC Damage: Alkylating Agents

The foregoing sections are relevant to this paper because, beyond their biological implications, they provide evidence of the high degree to which the SC, in contrast with the chromosome, constitutes a clear, sensitive, and quantifiable indicator of the state of synapsis and thus can reveal perturbations in the synaptic process. Although our studies were centered on the biological aspects of synapsis, they also opened the way to using SC analysis for examining induced chromosome damage in germ-line cells at meiotic prophase. We hoped, for instance, to study the formation of mutagen-induced rearrangements by SC analysis at meiotic prophase and to determine the sensitive periods for their induction in vivo.

The chemotherapeutic alkylating agent, mitomycin C (MC), was chosen first to induce chromosomal mutations because its cross-linking effects on DNA were well documented and its induction of clastogenic and cytogenetic damage, including rearrangements, in germ-line cells was known. The results, subsequently confirmed with another chemotherapeutic alkylating agent, cyclophosphamide (CP), revealed yet another, unexpected dimension to the usefulness of the SC as a sensitive indicator of induced damage (Moses et al. 1985; Allen et al. 1987).

Substantial SC damage appeared in early pachytene nuclei at 3 days following intraperitoneal (i.p.) administration of MC at doses ranging over two orders of concentration (0.05–5.0 mg/kg body weight). Peak damage was observed at 5 days.

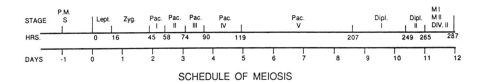

Figure 3
Linear time schedule of morphologically characterized meiotic substages, progressing from the beginning of leptotene through the end of division II. Relative stage durations were estimated from stage frequency and converted to hours and days using Oakberg's (1957) figure of 287 hours for the entire period. The schedule was verified by tracking ^{3}H-Tdr autoradiographically through the series of substages at successive time intervals.

According to the meiotic schedule (Fig. 3), primary damage must have been induced at premeiotic S. This assumption was borne out by administering ^{3}H-Tdr i.p. simultaneously with the drug (both MC and CP were tested). Cells containing SC damage were observed to be labeled autoradiographically, whereas most undamaged cells were not. Because the cross-linking action of alkylating agents is known to interfere with DNA synthesis, the principal target presumably is DNA in replication and/or in repair synthesis. It follows that the initial lesion is then translated into SC damage.

Alkylator damage to the SC is manifested by two classes of aberrations: structural, consisting of fragmentation of the SC and its parts, and synaptic, consisting of interference with homologous synapsis leading to asynapsis and nonhomologous synapsis of the SC and its elements.

Breaks in the individual axes appear as discontinuities in the unpaired axial elements, in individual lateral elements of the SC, or in the SC itself (both lateral elements simultaneously). Broken fragments may or may not be displaced and may associate endwise with other fragments, or with telomeric ends of intact elements. Endwise associations may represent true translocations. Inversions, deletions, and insertions would not necessarily be detected in the absence of homologous synapsis.

Pairing aberrations result from homosynaptic failure leading to partial (most common) or total asynapsis (single axial or lateral elements) and heterosynapsis (partial or total SC formation with mismatched lateral elements and multiaxial configurations). Single elements at early pachytene appear not to originate from desynapsis; evidently, once homologous synapsis has occurred, it remains, and the induced damage does not easily release it.

Both structural and synaptic damage results from MC and CP treatment. MC is more effective than CP in producing SC damage, the detectable alterations being found at doses of MC as low as 0.05 mg/kg i.p., in contrast to a low dose of 10.0 mg/kg for CP. In both cases, amounts of damage are dose-dependent within the range tested, as scored either according to numbers of abnormalities per cell or as fractions of cells containing damage.

Higher doses of both drugs showed cytotoxic effects, temporarily blocking cells entering meiosis and indicating the primary target to be spermatogonial S phase. For example, 7 days after 5.0 mg/kg of MC, leptotene/zygotene and P1 cells were absent. Repopulation of these stages was apparent at 9 days, and the frequency distribution of stages became normal in about 2 weeks.

Long-term spermatogonial damage was evident as much as 50 days (the longest time tested) after drug administration. Low levels of SC and lateral element breakage were observed at all pachytene stages. The lesions giving rise to this SC damage were evidently induced in the DNA of gonial stem cells. The progeny of these cells would be expected to perpetuate the damaged DNA over the reproductive lifetime of the individual. The continued appearance of SC damage is thus a measure of long-term, potentially heritable effects.

Experiments by Backer et al. (1988) (see also discussion by Allen et al., this volume) demonstrated correlations between CP-induced SC damage and MI and MII chromosome aberrations. Pulse-chase ^{3}H-Tdr labeling confirmed that at least some cells of the damaged SC cohort progressed on to meiotic divisions. The observed fourfold excess of pachytene SC breaks over MI chromosome aberrations points to the greater sensitivity of the SC to damage, but at the same time, does not account for the difference. There are at least three possible explanations: (1) Much of the SC damage is repaired during pachytene; (2) only some of the SC damage translates to MI/MII chromosome damage; and (3) most of the SC-damaged cohort is unable to progress to meiotic divisions and is lost.

Early experiments with MC and CP indicated a dose-dependent, cytotoxic effect at pachytene: Labeled, drug-damaged cells decreased in frequency as they were tracked through pachytene. Although these observations do not rule out the possible concurrent effect of (1) and/or (2) above, they implicate (3) as a major factor in reducing the amount of induced damage that is carried forward to the meiotic divisions, and thence to germ-cell differentiation.

Induction of SC Damage: Antimitotic Agents

Antimitotic agents (AMAs) also have significant effects on the SC. The induced perturbations have features in common with those induced by alkylator treatment, but there are essential and informative differences in the action of this different class of drug. Colchicine has long been known to induce univalency and to reduce crossing-over (chiasma formation) following treatment of plant meiocytes at meiotic prophase (Levan 1939; Shepard et al. 1974; Bennett et al. 1979).

Treatment at Leptotene Induces Asynapsis

The effects of colchicine, colcemid, nocodazol, and vinblastine sulfate were examined in mice (Gibson and Moses 1986; Allen et al. 1988; Gibson 1988). Although there were minor differences in details of induced SC perturbations, the results were essentially alike for all four AMAs: The major effect was interference with homologous SC formation (i.e., synapsis), giving rise to univalents (single axial elements) and partial asynapsis (incompletely synapsed SCs) in early to mid-pachytene nuclei. Failure of XY synapsis was most frequent. A lesser effect was the occasional induction of SC, lateral element, and axial element breaks, particularly at higher drug doses.

Most significant in these results is the preponderance of synaptic aberrations induced by AMAs in contrast to the mixture of synaptic and fragmentation damage caused by the alkylators MC and CP. The divergences are probably attributable to differences in the molecular targets of the two classes of drugs and the stages at which they induce primary lesions. Whereas MC and CP acted at premeiotic S to produce damage that showed up 4–5 days later as SC aberrations in early pachytene cells, AMAs had no demonstrable effect at premeiotic S. Instead, SC aberrations appeared in early pachytene (P1, P2, P3) cells after 48 hours, indicating that the primary lesion was induced at leptotene/zygotene. These observations confirm, in a mammalian system, what has been well documented in plants (e.g., Shepherd et al. 1974; Bennett et al. 1979): that the representative AMA, colchicine, induces asynapsis and univalents by acting at the time of SC formation (leptotene/zygotene) and earlier. In the mouse, colchicine-induced, pachytene, axial univalents, like those induced by MC and CP, evidently result from interference with homologous synapsis and not from desynapsis of already formed SCs. Furthermore, nonhomologous (and homology-indifferent) synapsis appears to persist unaffected throughout pachytene, producing unequal and multiaxial complexes, also similar to those induced by MC and CP, and hairpin foldbacks that are exclusively common to AMAs, but not to alkylators.

Comparison of Alkylator and AMA Effects

It is of no small biological interest that two distinctly different classes of mutagenic agents, one that primarily produces DNA lesions and the other that acts on a specific protein target molecule, tubulin, induce qualitatively similar, though quantitatively different, SC effects. The putative close association (binding ?) of DNA with protein subunits of the SC lateral elements, taken together with the assembly of fine filaments connecting the lateral elements to each other and to the central region material at the time of SC formation (Moens 1968; Solari 1972), suggests (1) that the linear assembly of the axial/lateral elements depends on subunit-to-subunit interactions and subunit-to-DNA binding, and (2) that the alignment and synapsing of lateral elements results from interactions of central region filaments with lateral element subunits and possibly with other proteins. From this hypothesis it is conceivable that homologous recognition is achieved by specific (unique ?) DNA sequence binding with one of the "synaptic" proteins. According to this reasoning, interruption of the SC's structural continuity (fragmentation) following alkylator interference with DNA replication and/or repair may be the consequence of localized failures of proper DNA/lateral element binding that could cause discontinuities in the assembly of lateral element subunits. In a similar manner, induced S-phase DNA lesions could interfere with the DNA-central-region-protein binding necessary for homologous recognition, initiation, and homologous SC formation leading to asynapsis.

From this kind of postulate, it would follow that induced, generalized DNA damage could lead to failure of two (or more) classes of DNA-protein interactions. The effect on one class could lead to SC fragmentation, whereas the effect on the other class of DNA-protein interactions could lead to blockage of homologous recognition and synaptic initiation. Thus, a drug may produce a specific type of lesion in DNA generally (e.g., cross-linking by an alkylating agent with consequent interference in DNA synthesis/repair) that could interfere subsequently with both SC structural integrity and homologous synapsis, as, in fact, is observed with MC and CP.

Colchicine's principal reaction, on the other hand, is restricted to a single protein molecule: tubulin. Our observation that colchicine interferes with synapsis whereas its derivative, lumicolchicine, which does not bind to tubulin, has no effect, points strongly to the involvement of tubulin in the synaptic mechanism, even in the face of the lack of any direct evidence for the presence of tubulin associated with the synaptic apparatus (Dresser 1987). The small proportion of SC breakage also caused by colchicine could be explained as a consequence of some unsuspected secondary or tertiary effect of the drug, such as interference with DNA synthesis by blocking incorporation of nucleotide (Creasey and Markiw 1964).

Cytotoxicity and Relationship of SC and MI Damage

Cytotoxicity that prevents damaged prophase cells from entering meiotic division was also observed following treatment with AMAs. First experiments using vinblastine sulfate (VS) (Gibson 1988) showed an 87% loss of damaged cells between 3 and 7 days following exposure to 5 µg VS intratesticular (i.t.) at leptotene (Table 2). Colchicine is known to be highly cytotoxic. Experiments were therefore carried out (J.H. Tepperberg, unpubl.) to determine the lowest dose of colchicine that would give significant SC damage with minimum cytotoxic effects in order to observe damage at MI and MII.

The minimal effective dose (0.1 µg colchicine i.t.) produced a wide range of synaptic aberrations in roughly half of the cells that had been marked by ^{3}H-Tdr 2 days before colchicine treatment. For example (Table 3), at 3 days following colchicine injection, 20% of the labeled cells contained XY univalents, and 10% contained univalent autosomal axial elements. Of the latter cells, 1% contained all univalents (20 pairs of single axes), and 9% contained from 1 to 19 pairs of uni-

Table 2
Frequency of Cells with AMA-induced SC Abnormalities

AMA	Dose (µg i.t.)	Control (%)	No. days after exposure to AMA (%)	
			3	7
Vinblastine sulfate	5.0	2.0	30.0	4.0
Colchicine	0.1	0	18.0	20.0

Table 3
Frequency of Cells with Colchicine-induced Axial (SC) and Chromosomal (MI) Univalents

Stage	Treatment	XY (%)	Autosomal (%)
Pachytene SC	control	<1	<1
	3 days post-injection	20.0	10.0
MI chromosome	control	15.0	2.0
	10 days post-injection	32.0	9.0

valent axes. All damaged cells were labeled; unlabeled cells had either missed the drug or were too lightly affected to show detectable damage. A comparison at 3 and 7 days showed no change in the frequency of colchicine-damaged cells (Table 2), suggesting that there was no significant loss of damaged cells at this dose.

These experiments were designed to test whether axial univalents correlated with MI univalents. ^{3}H-Tdr-labeled cells with SC damage were tracked to MI 12 days after labeling and 10 days after colchicine. It was predicted that autosomal and XY univalent chromosomes would occur at that time if SC univalents were precursors. Preliminary results show (Table 3) that 10 days following colchicine, 7% of the 10-day metaphases (experimental less control frequencies) contained autosomal univalents, and 17% contained XY univalents. Although there is an encouraging approximation between the SC and MI frequencies, a major discrepancy remains to be explained. A substantial number of the pachytene cells with univalents contained 3–20 pairs of univalent autosomal axes, whereas at MI, mostly one and no more than two pairs of univalents were found in any metaphase chromosome complement.

One possible explanation invokes prophase arrest of the more damaged cells. If it is assumed that the number of pachytene univalent axes induced is a measure of the intensity of the colchicine effect on the cell containing them, and that an intensity-dependent toxic consequence is to block cells from leaving prophase, then only lightly affected cells, such as those with XY or one or two autosomal pairs of univalents would reach MI. It would further have to be presumed that there are three levels of cytotoxic effect: severe pachytene cytotoxicity caused by higher doses, which would result in loss of damaged cells from the population; less severe toxic effect, which would block the cells from further development; and the more mildly affected cells, which would progress more or less on schedule.

Alternative explanations cannot be completely ruled out. One is that most of the synaptic damage is repaired. Repair of asynapsis means restitution of homologous synapsis, requiring homosynaptic initiation and SC formation at late pachytene, which is not known to occur. In fact, evidence indicates that synapsis of partially synapsed regions at late pachytene is indifferent to homology (Ashley et al. 1981; Moses et al. 1982) and that synapsis of univalents at this time would be expected to be random and nonhomologous, according to evidence from *Bombyx* (Rasmussen 1977). Heterosynapsis would presumably preclude crossing-over and chiasma formation and, theoretically, would produce univalents or abnormal disjunction at MI in any case. The second possible explanation is that axial univalency is not a precursor of MI univalency. If true, the correspondence between pachytene and MI frequencies of univalents would have to be coincidental; moreover, it would be difficult to explain the presence of any colchicine-induced XY or autosomal univalents at MI at all.

A possible explanation for the high-dose, damaged-cell dropout induced by AMAs in our experiments is suggested by Russell's observation (Russell et al. 1981) that colchicine has a pronounced effect on Sertoli cells, disrupting the cytoskeleton and interrupting intracellular transport. A major consequence is the sloughing-off of substantial layers of spermatogenic cells—mostly spermatids, but not precluding other damaged stages—into the tubule lumen with consequent loss from the spermatogenic population.

CONCLUDING COMMENTS

The sensitivity of the SC to aberrations induced by chemical agents adds a new meiotic prophase endpoint to the well-established chromosomal endpoints at meiotic divisions I and II. The apparent failure of damaged cells at late pachytene to reach MI following drug treatment, either through dropout from the population as a result of severe cytotoxicity, or from pachytene arrest as a result of milder toxic damage, suggests that the meiotic division endpoints may underestimate the real extent of primary damage induced and may give information on the type and amount of damage from a selected population of cells sufficiently unaffected by the drug to progress past meiotic prophase. Because SC damage is incurred prior to these cytotoxic consequences, SC sensitivity would seem to be the endpoint of choice for assessing induced damage (including long-term damage) in screening for chemical compounds as potential chromosomal mutagens. On the other hand, from the standpoint of risk assessment, the meiotic division endpoint may be a better choice, as it represents chromosomal damage in surviving cells that are further along the spermatogenic track and are more apt to produce germ cells capable of carrying the damage to offspring. Used concomitantly, however, the two endpoints provide a new and powerful means for assessing chromosome mutagenicity in germ-line cells in vivo.

ACKNOWLEDGMENTS

The research reported herein was supported by the National Science Foundation (PCM-8308651) and the United States Environmental Protection Agency (cooperative agreement CR-812736). Although the research described in this article has been supported by the United States Environmental Protection Agency, it has not been subjected to Agency review and therefore does not necessarily reflect the views of the Agency, and no official endorsement should be inferred. Mention of trade names or commercial products does not constitute endorsement or recommendation for use. The authors gratefully acknowledge the skilled efforts of John Graves for typing and preparing this manuscript for publication.

REFERENCES

Allen, J.W., G.K. DeWeese, J.B. Gibson, P.A. Poorman, and M.J. Moses. 1987. Synaptonemal complex damage as a measure of chemical mutagen effects on mammalian germ cells. *Mutat. Res.* **190:** 19.

Allen, J.W., J.B. Gibson, P.A. Poorman, L.C. Backer, and M.J. Moses. 1988. Synaptonemal complex damage induced by clastogenic and anti-mitotic chemicals: Implications for nondisjunctions and aneuploidy. *Mutat. Res.* **201:** 313.

Ashley, T. 1988. G-band position effects on meiotic synapsis and crossing over. *Genetics* **118:** 307.

Ashley, T., M.J. Moses, and A.J. Solari. 1981. Fine structure and behavior of a pericentric inversion in the sand rat (*Psammomys obesus*). *J. Cell Sci.* **50:** 105.

Backer, L.C., J.B. Gibson, M.J. Moses, and J.W. Allen. 1988. Synaptonemal complex damage in relation to meiotic chromosome aberrations after exposure of male mice to cyclophosphamide. *Mutat. Res.* **203:** 317.

Bennett, M.D., L.A. Toledo, and H. Stern. 1979. The effect of colchicine on meiosis in *Lilium speciosum* cv. "Rosemede." *Chromosoma* **72:** 175.

Clermont, Y. and M. Trott. 1969. Duration of the cycle of the seminiferous epithelium in the mouse and hamster determined by means of ^{3}H-thymidine and radioautography. *Fertil. Steril.* **20:** 805.

Creasey, W.A. and M.E. Markiw. 1964. Biochemical effects of the vinca alkaloids. I. Effects of vinblastine on nucleic acid synthesis in mouse tumor cells. *Biochem. Pharmacol.* **13:** 135.

Dresser, M.E. 1987. The synaptonemal complex and meiosis: An immunological approach. In *Meiosis* (ed. P.B. Moens), p. 245. Academic Press, Orlando.

Dresser, M.E. and C. Giroux. 1988. Meiotic chromosome behavior in spread preparations of yeast. *J. Cell Biol.* **106:** 567.

Dresser, M.E. and M.J. Moses. 1979. Synaptonemal complex karyotyping in spermatocytes of the Chinese hamster (*Cricetulus griseus*). IV. Light and electron microscopy of synapsis and nucleolar development by silver staining. *Chromosoma* **76:** 1.

Dresser, M.E., D. Pisetsky, R. Warren, G. McCarty, and M.J. Moses. 1987. A new method for the cytological analysis of autoantibody specificities using whole-mount, surface-spread meiotic nuclei. *J. Immunol. Methods* **104:** 111.

Gibson, J.B. 1988. "Effects of anti-mitotic agents on meiosis in spermatocytes and oocytes of the mouse." Ph.D thesis, Duke University, Durham, North Carolina.

Gibson, J.B. and M.J. Moses. 1986. Effects of vinblastine sulfate on the synaptonemal complex in *Mus musculus*. *Genetics* **113**: 566. (Abstr.).

Heyting, C., J.J. Dietrich, P.B. Moens, R.J. Dettmers, H.H. Offenberg, E.J.W. Redeker, and A.C.G. Vink. 1989. Synaptonemal complex proteins. *Genome* **31**: 81.

Levan, A. 1939. The effect of colchicine on meiosis in Allium. *Hereditas* **25**: 9.

Mahadevaiah, S., V. Mittwoch, and M.J. Moses. 1984. Pachytene chromosomes in male and female mice heterozygous for the Is(7;1)40H insertion. *Chromosoma* **90**: 163.

Moens, P.B. 1968. The structure and function of the synaptonemal complex in Lilium longiflorum sporocytes. *Chromosoma* **23**: 418.

Moses, M.J. 1977. Synaptonemal complex karyotyping in spermatocytes of the Chinese hamster (*Cricetulus griseus*). I. Morphology of the autosomal complement in spread preparations. *Chromosoma* **60**: 99.

———. 1981. Meiosis, synaptonemal complex and cytogenetic analysis. In *Bioregulators of human reproduction* (ed. J. Jagiello and H. Vogel), p. 187. Academic Press, New York.

Moses, M.J., M.E. Dresser, and P.A. Poorman. 1984. Composition and role of the synaptonemal complex. *Symp. Soc. Exp. Biol.* **38**: 245.

Moses, M.J., L.B. Russell, and N.L. Cacheiro. 1977a. Mouse chromosome translocations: Visualization and analysis by electron microscopy of the synaptonemal complex. *Science* **196**: 892.

Moses, M.J., P.A. Poorman, T.H. Roderick, and M.T. Davisson. 1982. Synaptonemal complex analysis of mouse chromosomal rearrangements. IV. Synapsis and synaptic adjustment in two paracentric inversion. *Chromosoma* **84**: 457.

Moses, M.J., G. Slatton, T. Gambling, and F. Starmer. 1977b. Synaptonemal complex karyotyping in spermatocytes of the Chinese hamster (*Cricetulus griseus*). III. Quantitative analysis and evaluation. *Chromosoma* **60**: 127.

Moses, M.J., P.A. Poorman, M.E. Dresser, G.K. DeWeese, and J.B. Gibson. 1985. The synaptonemal complex in meiosis: Significance of induced perturbations. In *Aneuploidy— Etiology and mechanisms* (ed. V.L. Dellarco et al.), p. 337. Plenum Press, New York.

Oakberg, E.F. 1957. Duration of spermatogenesis in the mouse. *Nature* **180**: 1137.

Poorman, P.A., M.J. Moses, and T.H. Roderick. 1981a. Synaptonemal complex analysis of mouse chromosomal rearrangements. III. Cytogenetic observations on two paracentric inversions. *Chromosoma* **83**: 419.

Poorman, P.A., M.J. Moses, L.B. Russell, and N.L.A. Cacheiro. 1981b. Synaptonemal complex analysis of mouse chromosomal rearrangements. I. Cytogenetic observations on a tandem duplication. *Chromosoma* **81**: 507.

Rasmussen, S.W. 1977. Chromosome pairing in triploid females of *Bombyx mori* analyzed by three-dimensional reconstruction of synaptonemal complexes. *Carlsberg Res. Commun.* **42**: 163.

Roderick, T.H. and N.L. Hawes. 1974. Nineteen paracentric chromosomal inversions in mice. *Genetics* **76**: 109.

Russell, L.D., J.P. Malone, and D.S. MacCurdy. 1981. Effect of the microtubule disrupting agents, colchicine and vinblastine, on seminiferous tubule structure in the rat. *Tissue Cell* **13**: 349.

Shepard, J., E.R. Boothroyd, and H. Stern. 1974. The effect of colchicine on synapsis and chiasma formation in microsporocytes of Lilium. *Chromosoma* **44:** 423.

Solari, A.J. 1972. Ultrastructure and composition of the synaptonemal complex in spread and negatively stained spermatocytes of the golden hamster and the albino rat. *Chromosoma* **39:** 237.

Sparrow, A.H., M.J. Moses, and R. Steele. 1952. III. A cytological and cytochemical approach to an understanding of radiation damage in dividing cells. *Br. J. Radiol.* **25:** 182.

COMMENTS

Lonnie Russell: Can you tell by sections alone if you're in diplotene?

Moses: By sections alone? Yes, I believe so.

Lonnie Russell: How?

Moses: I think the work of Holm and Rasmussen has demonstrated the morphological differences between zygotene and diplotene in sections.

Lonnie Russell: No, not zygotene, but pachytene and diplotene.

Moses: Diplotene is recognized by the separation of the lateral elements.

Lonnie Russell: But they're only partly separated.

Moses: Yes, that's at the beginning of diplotene. Then there are secondary characteristics visible after that. For example, at some time in diplotene the entire synaptonemal complex is disassembled and disintegrates. The time at which it disintegrates varies with different species. In some it persists longer; in some it starts to disintegrate earlier. After the breakdown (in mouse), the last things to go are the centromeric attachment points, and they can be visualized long after the rest is gone.

Smithies: Just as a point of information on the resolution of the inversion loops, at the time those have resolved and you've got the linear array, is the synaptonemal complex in the long homologous region indistinguishable from that in the whole molecule?

Moses: Yes. There is no way that we have been able to distinguish them.

Handel: Do you think the colchicine effects are direct effects on lateral element formation or do you think they are effects on the cytoskeleton which may be mediating homologous chromosomes finding each other and pairing?

Moses: It's a good question. When we know more about the molecular processes that are going on during the time of synapsis, what the elements are or the components are of the synaptonemal complex, if they play a role, and what that role may be, then we can begin to look at drugs that have specific molecular targets, such as colchicine for tubulin. If colchicine is acting on tubulin, I can't imagine where that action would take place, because there is no tubulin in the nucleus that we can identify and there is very little tubulin in the cytoplasm. We are trying to chase that down.

Lonnie Russell: If you pursue the tubulin theory, you could use taxol, which disrupts tubulins another way by hyperpolymerizing them, and you may get the same thing.

Moses: It's on the list.

Adler: With the univalency frequency in MI, did you follow this up into MII as to aneuploidy?

Moses: Yes. There was not any significant hyperploidy at MII.

Adler: We have worked with colchicine lately and we have treated at 22 hours before we counted MII, and we do get hyperploidies in MIIs, but we have not gone back all the way through prophase.

Moses: Our counts so far don't show it, but these are just the first experiments. We need larger samples.

Adler: I guess my question is more general. Do you believe or do you have any evidence that a higher rate of univalency at metaphase I is a causal thing for aneuploidy in metaphase II?

Moses: I don't have any evidence to support it, although I would like to believe it.

Adler: There's a lot of theory about it. I had an animal that had about 60% X/Y univalency; this was one individual animal from the colony that I just happened to analyze for some other reasons. I counted MIIs, and there were 50/50 X/Y MIIs. So there was no influence whatsoever, even though X/Y univalency was way up, 60%: 16% is the regular.

Moses: Are those cells lost too?

Adler: They were not—well, not as far as I can tell.

Moses: Well, if the MI with univalents never went on, would that have resulted in reduced fertility?

Adler: I should think a depletion of 60% of the cells would be noticeable.

Favor: Probably not. That's more or less a model of the SXR mouse where you get these high proportions of cells with X and Y univalents.

Adler: If 60% of the cells don't go on, I might have a hard time counting enough MIIs.

Favor: You get reduced testis sizes, and with SXR you get the 1:1:1:1 segregation of the different genotypes.

Adler: That animal had normal testis size.

Lonnie Russell: When you do i.t. injection, is this dose comparable to what we used a long time ago with colchicine and vinblastine? The apical two-thirds of the Sertoli cell sloughs within just a few hours of injection into the testis near the site of injection into the seminiferous tubules, so maybe you have a very sick general region where you have injected.

Moses: That's entirely possible; it's another point that concerns us. We know, because we are making spreads, we are throwing away the Sertoli cell relationships, we are throwing away the configuration of the whole environment, and until we stop and cut sections and compare what happens in the sections—and not just one or two sections, but sufficient sampling of the testis—I don't think we can really understand what is going on. I don't know what role the Sertoli cell is playing. Now, if the barrier works, it's easy to understand that agents that act at premeiotic S are probably going to have no trouble with access to the target cells, because that's just about the time that the basal spermatogonia are moving through to the adluminal compartment. But, if colchicine and bleomycin, for example, are able to damage cells in pachytene, those cells are in the adluminal compartment and are presumably protected by a Sertoli cell barrier. So, is it getting through some way? Is it going through the Sertoli cell?

Lonnie Russell: Well, it definitely gets into the Sertoli cell.

Moses: Is it damage to the Sertoli cell which is then transmitted in some way to the adjacent spermatocyte, and are we looking at a secondary effect? I'd like to answer these questions. I need help.

Lonnie Russell: You could give something like actinomycin D in a dose that will kill an awful lot of pachytene spermatocytes and just see if there are some general effects. Or, there are a number of other chemotherapeutic agents that are really toxic, with which to see if what you are looking at is a general phenomenon rather than a specific one.

Moses: These are questions that need to be examined because we have a powerful and sensitive index for agents that will produce damage, but until we can establish how that damage is produced and whether it is truly directly on the chromosome and affects the chromosome, in the same way, for example, in the ovary as it does in the testis, I think we are walking a very narrow path.

Adler: Jim, this one sort of cross-like configuration in the synaptonemal complex between two pairs of chromosomes, let's say, that looked like a translocation that was induced by the bleomycin—and you said that might be the equivalent of a translocation configuration, while at metaphase I you don't see any translocations—what do you think happened? That is supposedly one type of aberration that you see at a level that should go on and should be visible at metaphase I.

Allen: You're referring to that sublateral element bridge?

Adler: No. I'm referring to this sort of translocation configuration, the cross or quadrivalent.

Allen: Yes, that was bleomycin. Those cells, when damaged at prophase, probably die.

Adler: The damage doesn't seem to be that dramatic.

Allen: They are dramatic configurations. You don't see them at a high frequency, but you do see a significant amount of damage that is very dramatic like that and you can't pick it up in a later stage, so the cells must die.

Handel: What were the effects of colchicine on the nucleolus?

Moses: Damaged cells seem to stop where they were hit if synapsis is substantially incomplete. In less severely damaged cells, the SCs that were not affected appear to go on, together with their nucleoli, and take on some of the characteristics of late pachytene, while the SCs that were damaged in the same nucleus may retain the characteristics of zygotene. Thus, there may be a range of nucleolar stages in one nucleus.

Handel: So, the nucleolus shows a variable morphology?

Moses: Yes.

Synaptonemal Complex Analysis of Mutagen Effects on Meiotic Chromosome Structure and Behavior

JAMES W. ALLEN,[1] PATRICIA POORMAN-ALLEN,[2] LORRAINE C. BACKER,[3]
BARBARA WESTBROOK-COLLINS,[1] AND MONTROSE J. MOSES [4]
[1]Genetic Toxicology Division
Health Effects Research Laboratory
U.S. Environmental Protection Agency
Research Triangle Park, North Carolina 27711
[2]Wellcome Research Laboratories
Research Triangle Park, North Carolina 27709
[3]Environmental Health Research and Testing, Inc.
Research Triangle Park, North Carolina 27709
[4]Duke University Medical Center
Durham, North Carolina 27710

OVERVIEW

Homologous chromosome synapsis and crossing-over at meiosis are basic to mammalian gamete development: They achieve genetic recombination, regulate chromosome segregation, and are believed to function in repair and maturation. Synaptonemal complexes (SCs) are axial correlates of meiotic chromosome bivalents and develop in conjunction with homologous chromosome synapsis. It is shown here that various mutagens/anti-mitotic agents—cyclophosphamide (alkylating agent), colchicine (anti-tubulin alkaloid), amsacrine or *m*-AMSA (topoisomerase inhibitor), bleomycin (radiomimetic agent), and gamma radiation—induce diverse structural and synaptic errors in SCs of treated mice and hamsters. Conventional types of clastogenic effects as well as damage unique to meiotic prophase appear to be manifested in the SC. Distinctive patterns of damage are associated with specific mutagenic agents/mechanisms. Some SC aberrations are suggestive of a site specificity possibly related to crossing-over. Following treatments at premeiotic (ultimate) S phase, higher levels of damage are typically recovered from prophase SCs than from meiotic metaphase chromosomes, thereby indicating that intervening cell loss or repair processes may have occurred. The sensitivity of SC analysis for studies of meiotic prophase chromosomes allows the detection of structural/behavioral abnormalities that are otherwise unapparent. The

---

Although the research described in this article has been supported by the United States Environmental Protection Agency, it has not been subjected to Agency review and therefore does not necessarily reflect the views of the Agency and no official endorsement should be inferred. Mention of trade names or commercial products does not constitute endorsement or recommendation for use.

Banbury Report 34: Biology of Mammalian Germ Cell Mutagenesis
Copyright 1990. Cold Spring Harbor Laboratory Press. 0-87969-234-0/90.$1.00 + .00

significance of SC abnormalities lies in their implications for developmental and genetic impairment, which may result in gamete loss or heritable mutations.

INTRODUCTION

In mammals, homologous chromosomes at meiotic prophase normally pair, synapse, and cross over. It is widely held that this process of aligning and recombining the homologs also serves to help regulate their segregation. Various experimental lines of evidence indicate that anomalous synapsis/recombination can lead to reproductive or heritable consequences. Certain mutant stocks of mice with chromosome rearrangements experience failed or aberrant synapsis, which appears linked to gametogenic impairment and infertility (Burgoyne and Baker 1984; Handel 1987). Studies in various organisms, including mice (Henderson and Edwards 1968), *Drosophila* (Baker and Hall 1976), and individuals with trisomy 21 (Down syndrome; Warren et al. 1987), suggest that a reduced frequency of meiotic recombination gives rise to chromosomal nondisjunction. Recombination may also function in gamete repair and maturation (Holliday 1984); thus, abnormalities of chromosome structure/behavior during meiotic prophase could have a wide range of deleterious effects on germ-line development.

There is increasing evidence to suggest that many heritable syndromes and pathologies, e.g., certain aneuploidies, often arise as new mutations in parental meiotic cells (Juberg and Mowrey 1983; U.S. Congress 1986). This has generated interest in the effects of environmental genotoxicants on such cells and the mechanisms by which different agents might cause nondisjunction or other chromosomal anomalies at meiosis. Homolog pairing and crossing-over precondition subsequent chromosome segregation and invoke specific DNA replication and transcription as part of this orienting process (Hotta et al. 1985). These prophase events may present unique targets for mutagens and may mediate eventual aneuploidy effects.

Very little is known about the effects of mutagens on chromosomal activities at meiotic prophase. The relatively decondensed state of chromosomes at this stage makes damage difficult to resolve by direct cytogenetic analysis. However, SC analysis using electron microscopy methods provides a promising approach for such studies. SCs are proteinaceous axial correlates of meiotic bivalents at prophase (Moses 1968, 1981). In correspondence with homolog pairing and synapsis, single axes of individual chromosomes become synapsed to form lateral elements (LEs) of the SC. Chromosomal inversions or translocations are characterized by SC configurations that reflect synaptic modifications to accommodate the specific structural rearrangements (Moses et al. 1977; Poorman et al. 1981). The anomalous SCs can often portray the aberrations with exceptional clarity and detail (e.g., breakpoints).

Recently, ionizing radiation (Cawood and Breckon 1983; Kalikinskaya et al. 1986) and chemical clastogens/antimitotic agents (Moses et al. 1985; Allen et al. 1987, 1988; Backer et al. 1988) have been shown to induce various types of aberrations in SCs of treated mice and hamsters or their offspring. The types of damage thus far observed to be produced by different agents, and current questions concerning their significance and relationship to conventional chromosome damage, are discussed below. The methods used in these studies are described briefly as follows: Male C57BL/6J mice and/or Armenian hamsters were treated with a single intraperitoneal (i.p.) or intratesticular (i.t.) injection of the test chemical, or they were exposed to whole body gamma irradiation from a ^{60}Co source. SCs from pachytene spermatocytes were evaluated 1–5 days (mice) or 1–6 days (hamsters) after treatment to assess damage induced at times consistent with premeiotic S phase or at substages (leptotene, zygotene, or pachytene) of meiotic prophase. SC complements were prepared by microspreading procedures, stained with silver, and analyzed by electron microscopy. These methods have been reported in detail elsewhere (Moses 1977; Dresser and Moses 1979).

RESULTS

The normal appearance of SCs in the mouse is indicated in Figure 1. Axes of homologous chromosomes are continuous and synapsed throughout their lengths (except the XY pair, which undergoes a partial synapsis). In contrast to this regularity of SC configuration, treatment by various chemicals induces structural and/or synaptic abnormalities in SCs as indicated in Figure 2. Structural aberrations include breakage of unpaired axial elements, or breakage of one or both lateral elements of the SC, and also rearrangements, e.g., translocations or insertions (or deletions) interpreted as such by their synaptic configurations. Synaptic errors include asynapsis along regions of, or the entire length of, SCs; foldbacks of single axes that appear to align with another region of the same axis; and mispairing in which regions between two or more different SCs may become associated by abnormal (presumably nonhomologous) synapsis.

Different classes of chemicals (distinguished by different mechanisms of biological interaction) have been observed to preferentially induce different types of damage. Figure 3 describes the spectrum of damage, and highlights the primary effects, for single representative chemicals (cyclophosphamide, colchicine, m-AMSA, or bleomycin) from several mechanistically distinctive classes: alkylating agents, anti-tubulin alkaloids, topoisomerase II inhibitors, and radiomimetic agents, respectively. Limited studies indicate that the patterns of damage tend to be generally characteristic of the class. For example, another alkylating agent, mitomycin C, causes a similar pattern of damage as cyclophosphamide, although at much lower doses. Vinblastine sulfate is another anti-tubulin alkaloid that is remarkably similar to colchicine in the types of damage it produces, although relatively higher doses are required. Alkylating agents tend to be especially proficient at inducing breakage and mispairing; anti-tubulin alkaloids for asynapsis and foldbacks; the topoisomerase II inhibitor for breakage and mispairing; and the

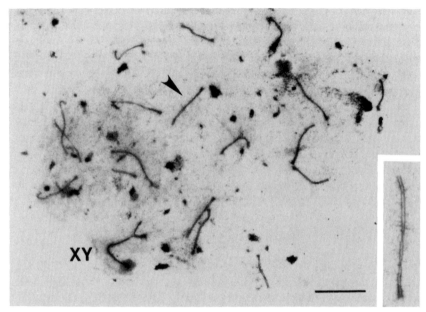

Figure 1
A normal early-to-mid-pachytene stage complement of SCs from an untreated mouse. Each SC is composed of continuous, homologous axes (the LEs) that are completely synapsed, except the XY, which is only partially homologous and partially synapsed. The SC marked by an arrowhead is enlarged three times as an insert to reveal greater detail. Bar, 10 µm.

radiomimetic agent for rearrangements. However, these effects need to be qualified; they are specific for particular times of exposure in relation to cell harvest. The extent of effects noted for cyclophosphamide and m-AMSA pertain only to exposures 4–5 days earlier, presumably when cells were at premeiotic S phase. The pronounced effects of colchicine are only evident after exposures 2–3 days earlier, when cells were in early prophase (leptotene, zygotene). Bleomycin is unusual in that it induces similar types of damage following S-phase or prophase exposure (although the response is greater following S-phase exposure).

Quantitative effects also distinguish the individual chemicals/classes. Figure 4 illustrates some dose-response curves for percent abnormal cells (combined types of effects) over which the chemicals (from Fig. 3) are active after S-phase/prophase exposure. The relatively steeper curves of colchicine and m-AMSA (as compared with cyclophosphamide and bleomycin) reflect the sensitivities of these cells to very low doses of such chemicals when injected via the i.t. route. Systemic toxicities to colchicine and m-AMSA when administered i.p. prompted use of the i.t. experimental route of exposure for these agents. Some quantitative differences among the

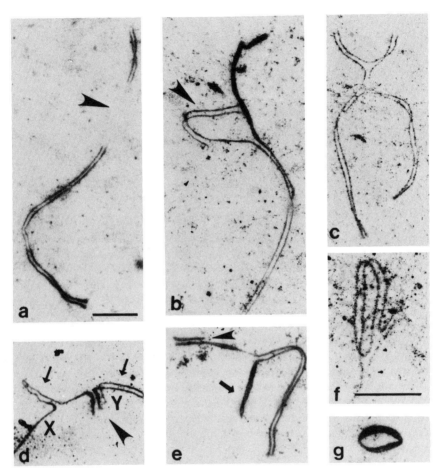

Figure 2
Various examples of aberrant SCs from treated mice and Armenian hamsters. (*a*) SC break (involving both LEs; arrowhead); from a mouse treated with bleomycin. (*b*) Asynapsis and mispairing of nonhomologous axes (arrowhead); from a hamster treated with gamma radiation. (*c*) Quadrivalent SC configuration suggestive of a reciprocal translocation; from a mouse treated with gamma radiation. (*d*) Foldback (arrow) in X and also in Y to result in triaxial pairing (arrowhead); from a mouse treated with cyclophosphamide. (*e*) Asynapsis (arrow) and LE break (arrowhead); from a hamster treated with gamma radiation. (*f*) Rearranged SC possibly deriving from an insertion or deletion; from a hamster treated with gamma radiation. (*g*) SC ring; from a hamster treated with gamma radiation. Bars, 2 μm: *a–e* and *g* are the same magnification. (*c* and *f* reprinted, with permission, from Backer et al. 1990.)

	Breakage	Mispairing	Rearrange- ments	Asynapsis	Foldbacks
Cyclophosphamide (Alkylator)	+++	+++	+	++	+
Colchicine (Anti–tubulin Alkaloid)	+	+	–*	+++	+++
m–AMSA (Topoisom. Inhibitor)	++	++	+	+	+
Bleomycin (Radiomimetic)	++	++	+++	+	+

*None observed

Figure 3
Patterns of SC damage induced by different clastogenic/antimitotic chemicals. The one-, two-, or three-plus notations indicate relative activities for a given effect. Cross-hatched boxes highlight those effects that appear to distinguish the individual chemicals and the primary mutagenic mechanisms of action by which they are presumed to operate.

chemicals are not apparent in these curves. For example, high doses (at or near cytotoxicity) of cyclophosphamide or colchicine cause multiple aberrations per cell (often nearly all SCs are damaged), whereas high doses of bleomycin or m-AMSA seldom elicit more than one or two aberrations per recovered cell.

In our studies with mice and hamsters exposed to gamma radiation at premeiotic S phase or at prophase, the types of SC damage resulting were rather similar to those described for bleomycin. Breakage, mispairing, and asynapsis were more prominent as radiation effects in the hamsters. In mice, rearrangements were especially common, although dose-dependent increases were observed for all categories of effects, except foldbacks, which roughly doubled in frequency regardless of dose. At the highest dose (4 Gy) tested as an S-phase exposure, the frequency of SC breakage was nearly 20 times the control level (0.5/control cell; Backer et al. 1990). A particularly interesting type of damage observed from radiation treatment that had not previously been described was the appearance of "bridges" of material between lateral elements (Fig. 5) (Backer et al. 1990). The bridges appeared to represent "sublateral element" components and to connect different SCs, or alternatively, different regions within an SC. An additional observation concerning the distribution of these radiation-induced sublateral element bridges in Armenian hamsters is that they appear to occur with especially

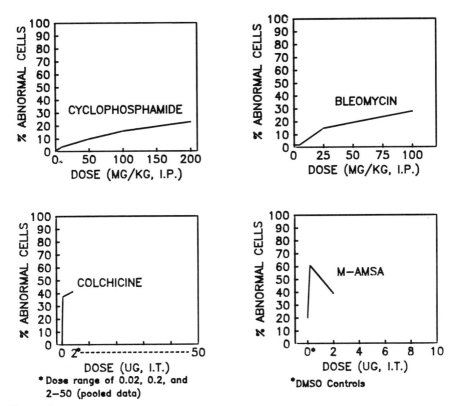

Figure 4
Dose-response curves for SC damage (percent abnormal cells) comprising all the different types of effects following exposure to the chemicals described in Fig. 3.

high frequencies in regions that contain crossovers. Data to support such associations are very limited, since only the sex chromosomes have been analyzed in this regard. However, roughly half of such bridges (6 of 12 especially clear examples) examined in the sex-bivalent occurred in regions of the short arms that have been identified with BrdUrd-differential staining as regular sites of crossing-over (Allen 1979). Most remaining bridges appeared to occur at or near tips of the chromosome arms.

Several of the agents discussed above have been assessed for quantitative and qualitative relationships among various types of induced SC damage at prophase, and chromosome aberrations at diakinesis-metaphase I. Moses et al. (this volume) have discussed colchicine in this regard. In studies with cyclophosphamide (Backer et al. 1988), mice were injected i.p. at premeiotic S phase with 100 mg/kg cyclophosphamide and also tritiated thymidine so that prophase and metaphase cell

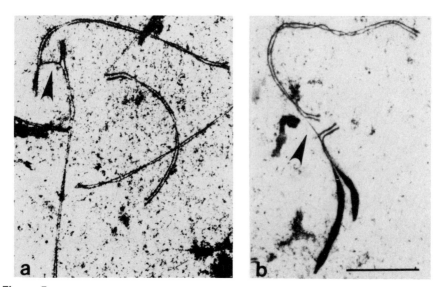

Figure 5
Sublateral element bridges (arrowheads) from a mouse (*a*) and a hamster (*b*), following exposure to gamma radiation. Bar, 4 µm.

stages sequentially harvested from the same animals could be identified by autoradiography as belonging to the same treated cell population. It was determined that prophase SCs reveal more than eight times the control level of damage events per nucleus, as compared with metaphase chromosome aberration frequencies of three times the control level. In particular, breakage events were much higher in SCs than in metaphase chromosomes, e.g., approximately nine times control levels of SC and/or LE breaks and fragments as compared with two to three times control levels of isochromatid and/or chromatid breaks and fragments. Although these forms of breakage observed at prophase and metaphase appeared highly correlated, the nature of their relationship is not clear. Ionizing radiation also induces damage that generally is detectable at much higher frequencies in SCs than in subsequent metaphase chromosomes. Cawood and Breckon (1983) first made this observation with radiation. In our studies of mice exposed at S phase to 2 or 4 Gy, rearrangements were subsequently much more prevalent in SCs than at meiotic metaphase I (Backer et al. 1990).

DISCUSSION

That various aberrations are inducible in SCs by mutagens has not been widely recognized or studied. However, several points may be concluded from the results summarized above: (1) SCs are subject to alterations by diverse agents and express a variety of modifications to integrity; (2) some types of aberrations observed in SCs may be unique in structure/function, whereas others appear related to aberrations evident at metaphase; (3) SCs tend to reveal much higher frequencies of aberrations than meiotic metaphase chromosomes (when exposure precedes or coincides with S phase); (4) different patterns of induced SC damage characterize different chemicals and irradiation; and (5) from very limited studies, some forms of SC damage appear to be nonrandom in distribution, possibly as a result of crossover sites constituting favored sites for aberration induction.

It is not known whether the induction pathways that give rise to SC damage are the same as those that are operative in chromosome aberrations. In general, SC and/or LE breaks and rearrangements have shown maximum inducibility following exposure at S phase, as would be expected if the damage corresponded to clastogen effects on DNA. In addition, the significant correlations that have been found between breakage aberrations in prophase SCs and meiotic metaphase chromosomes are consistent with the idea that these events represent different manifestations of the same damage. However, since the chemical composition of the SC and its orientative relationship to the chromosome remain uncertain, there may also be instances in which breakage occurs independently in these two structures. Perhaps damage to one could mediate damage to the other. Efforts to define the equatability of the two forms of damage are limited by other quantitative and qualitative considerations: Some cells with SC breakage may "drop out" and never reach metaphase (Moses et al., this volume), and the detection/interpretation of chromatid-type breaks in SCs is obscured by the difficulty in resolving axial counterparts of chromatids.

The unique metabolism of meiotic prophase that functions in the regulation of synapsis may afford special targets for disruption of synaptic processes and SC formation. Stern and Hotta (1985) have described specific DNA sequences that are normally suppressed in their replication at premeiotic S phase and are instead replicated and transcribed at prophase, apparently in association with chromosome pairing (zygotene or ZYG DNA), or with crossing-over and repair (pachytene or P DNA). Interference with DNA replication at zygotene is reported to interfere with SC formation (Roth and Ito 1967). Mutagens and agents with other specialized chemical properties, e.g., colchicine, may interact with such molecular targets (e.g., ZYG DNA, SC proteins) essential to SC development. Even those chemicals administered at S phase may have additional activity at prophase causing synaptic irregularities, e.g., asynapsis. Induced mispairing may be driven by homologous attraction of LE regions among different SCs or by some physical force that "takes over" when normal homologous attraction fails (as proposed by McClintock 1933; Moses et al. 1985) due to structural or metabolic impairment.

The preliminary observations of radiation-induced sub-LE bridges preferentially occurring in regions of crossing-over (in the XY of Armenian hamsters) raise some interesting questions. Are the sub-LE bridges remnants of adhesions induced during

S phase between chromosomal regions when they were in regular, close proximity, or do these bridges represent chromatid translocations? Although the LE typically appears as a singular morphological structure interpreted to represent a chromosome axis, a clear doubleness of its nature (two strands) has occasionally been observed (see, e.g., Dresser and Moses 1980). Wahrman (1981) has proposed that half-LEs may exchange and produce chiasmata. If the LE is actually two juxtaposed axes (one axis per chromatid), then sub-LE bridges may reflect the breakage and abnormal reunion of chromatids occurring either within or between bivalents. In addition to radiation, certain chemicals that can induce chromatid breaks in an S-independent manner, e.g., bleomycin, m-AMSA, and camptothecin, have also been observed to induce sub-LE bridges (at lower frequencies). Studies are under way to determine which clastogenic mechanisms appear to be associated with the production of these unusual effects. Are the bridges occurring specifically at sites of recombination? The preferential localization of these bridges may be a function of the timing of replication characterizing relatively large segments of chromatin. Crossovers in the XY short arms of Armenian hamsters give the cytological appearance of occurring in a region that differentially stains as if it were early replicating chromatin (Allen 1979). Alternatively, perhaps certain DNA sequences, conformation, or timing of replication/repair (as mentioned above) specifically associated with crossing-over may serve as preferential sites for induced damage. Further investigation is needed to determine if a sub-LE bridge may in some way manifest, or otherwise contribute to, aberrant crossing-over.

SC analysis can provide a high level of sensitivity for revealing induced damage. In the few studies wherein aberration frequencies have been compared in meiotic prophase SCs and metaphase chromosomes, the SCs have typically exhibited much higher levels of damage. There may be several explanations for this: It may be due to the broader spectrum of effects measured in SCs; e.g., some forms of mispairing and also foldbacks may have no counterpart evident at metaphase. There may be extensive loss of damaged cells during the interval immediately following that in which SCs are analyzed. Consistent with this, meiotic cells exposed to mitomycin C at premeiotic S phase have shown extensive SC damage at early to mid-pachytene, and then a significant reduction of damage recovered in late pachytene cells when the affected cohort should have progressed to that point (Allen et al. 1987). Repair of damage evident in SCs may also have played a role in leading to reduced levels of damage subsequently detected in metaphase chromosomes. However, fertility studies in mutant mouse stocks carrying certain translocations have indicated that the aberrant cells often die at, or around, pachytene (Burgoyne and Baker 1984; Handel 1987). It may be that normal synapsis is a general requisite for continued gamete maturation (Burgoyne and Baker 1984; Handel 1987), and whenever exogenous agents alter this process, at least in sensitive regions, there is a risk of cell death.

To conclude, there is increasing evidence to suggest that homologous chromosome synapsis and recombination in meiosis are crucial to maintaining normal developmental and genetic states of the mammalian gamete. Aberration of such meiotic chromosomal processes may lead to cell death or heritable mutation. In the past, SC analysis has provided considerable detail of normal chromosomal synaptic processes. In this paper, we have described recent studies in which aberrant SC formation has been used as a genotoxicity endpoint. Clearly, induced perturbations of the usually regular and precise formation of the SC are easily detected and provide a very sensitive indication of damage. Our results, in conjunction with those obtained by several other investigators using this approach, suggest that meiotic chromosomes are subject to types and levels of mutagen insult that have not generally been recognized or sufficiently appreciated with regard to their potential implications for gamete toxicity or heritable mutation. Further studies of SC damage will be aimed at defining its molecular mechanisms of induction and its biological significance for the cell.

ACKNOWLEDGMENTS

The authors gratefully acknowledge the technical assistance of Deborah Howard and James Campbell, and the early contributions of Gary DeWeese and James Gibson to this work.

REFERENCES

Allen, J.W. 1979. BrdU-dye characterization of late replication and meiotic recombination in Armenian hamster germ cells. *Chromosoma* **74:** 189.

Allen, J.W., G.K. DeWeese, J.B. Gibson, P.A. Poorman, and M.J. Moses. 1987. Synaptonemal complex damage as a measure of chemical mutagen effects on mammalian germ cells. *Mutat. Res.* **190:** 19.

Allen, J.W., J.B. Gibson, P.A. Poorman, L.C. Backer, and M.J. Moses. 1988. Synaptonemal complex damage induced by clastogenic and anti-mitotic chemicals: Implications for nondisjunction and aneuploidy. *Mutat. Res.* **201:** 313.

Backer, L.C., M.J. Moses, and J.W. Allen. 1990. Synaptonemal complex analysis in genetic toxicology. In *Mutation and the environment* part B: *Metabolism, testing methods and chromosomes* (ed. M. Mendelsohn and R. Albertini). A.R. Liss, New York. (In press.)

Backer, L.C., J.B. Gibson, M.J. Moses, and J.W. Allen. 1988. Synaptonemal complex damage in relation to meiotic chromosome aberrations after exposure of male mice to cyclophosphamide. *Mutat. Res.* **203:** 317.

Baker, B.S. and J.C. Hall. 1976. Meiotic mutants: Genic control of mitotic recombinational chromosome segregation. In *The genetics and biology of* Drosophila (ed. M. Ashburner and E. Novitski), vol. 1A, p. 352. Academic Press, London.

Burgoyne, P.S. and T.G. Baker. 1984. Meiotic pairing and gametogenic failure. *Symp. Soc. Exp. Biol.* **38:** 349.

Cawood, A.H. and G. Breckon. 1983. Synaptonemal complexes as indicators of induced structural change in chromosomes after irradiation of spermatogonia. *Mutat. Res.* **122:** 149.

Dresser, M.E. and M.J. Moses. 1979. Silver staining of synaptonemal complexes in surface spreads for light and electron microscopy. *Exp. Cell Res.* **121:** 416.

———. 1980. Synaptonemal complex karyotyping in spermatocytes of the Chinese hamster (*Cricetulus griseus*). IV. Light and electron microscopy of synapsis and nucleolar development by silver staining. *Chromosoma* **76:** 1.

Handel, M.A. 1987. Genetic control of spermatogenesis in mice. *Results Probl. Cell Differ.* **15:** 1.

Henderson, S.A. and R.G. Edwards. 1968. Chiasma frequency and maternal age in mammals. *Nature* **218:** 22.

Holliday, R. 1984. The biological significance of meiosis. *Symp. Soc. Exp. Biol.* **38:** 381.

Hotta, Y., S. Tabata, L. Stubbs, and H. Stern. 1985. Meiosis-specific transcripts of a DNA component replicated during chromosome pairing: Homology across the phylogenetic spectrum. *Cell* **40:** 785.

Juberg, R.C. and P.N. Mowrey. 1983. Origin of nondisjunction in trisomy 21 syndrome: All studies compiled, parental age analysis, and international comparisons. *Am. J. Med. Genet.* **16:** 111.

Kalikinskaya, E.I., O.L. Kolomiets, V.A. Shevchenko, and Y.F. Bogdanov. 1986. Chromosome aberrations in F_1 from irradiated male mice studied by their synaptonemal complexes. *Mutat. Res.* **174:** 59.

McClintock, B. 1933. The association of non-homologous parts of chromosomes in the midprophase of meiosis in *Zea mays*. *Z. Zellforschung Mikrosk. Anat.* **19:** 191.

Moses, M.J. 1968. Synaptonemal complex. *Annu. Rev. Genet.* **2:** 363.

———. 1977. Synaptonemal complex karyotyping in spermatocytes of the Chinese hamster (*Cricetulus griseus*). I. Morphology of the autosomal complement in spread preparations. *Chromosoma* **60:** 99.

———. 1981. Meiosis, synaptonemal complex and cytogenetic analysis. In *Bioregulators of reproduction* (ed. G. Jagiello and H. Vogel), p. 187. Academic Press, New York.

Moses, M.J., L.B. Russell, and N.L.A. Cacheiro. 1977. Mouse translocations: Visualization and analysis by electron microscopy of the synaptonemal complex. *Science* **196:** 892.

Moses, M.J., P.A. Poorman, M.E. Dresser, G.K. DeWeese, and J.B. Gibson. 1985. The synaptonemal complex in meiosis: Significance of induced perturbations. In *Aneuploidy, etiology and mechanisms* (ed. V.L. Dellarco et al.), p. 337. Plenum Press, New York.

Poorman, P.A., M.J. Moses, M.T. Davisson, and T.H. Roderick. 1981. Synaptonemal complex analysis of mouse chromosomal rearrangements. III. Cytogenetic observations on two paracentric inversions. *Chromosoma* **83:** 419.

Roth, T.F. and M. Ito. 1967. DNA-dependent formation of the synaptonemal complex at meiotic prophase. *J. Cell Biol.* **35:** 247.

Stern, H. and Y. Hotta. 1985. Molecular biology of meiosis: Synapsis-associated phenomena. In *Aneuploidy, etiology and mechanisms* (ed. V.L. Dellarco et al.), p. 305. Plenum Press, New York.

U.S. Congress, Office of Technology Assessment. 1986. Technologies for detecting heritable mutations in human beings, OTA-H-298, p. 40. U.S. Government Printing Office, Washington, D.C.

Wahrman, J. 1981. Synaptonemal complexes—Origin and fate. *Chromosomes Today* **7**: 105.

Warren, A.C., A. Chakravarti, C. Wong, S.A. Slaugenhaupt, S.L. Halloran, P.C. Watkins, C. Metaxotou, and S.E. Antonarakis. 1987. Evidence for reduced recombination on the nondisjoined chromosomes 21 in Down syndrome. *Science* **237**: 652.

COMMENTS

Bishop: You showed one slide of aberration frequencies in metaphases 1 and 2, as well as SCs, and the latter was very steep and the other two were very shallow. Considering all the chemicals that you have looked at now in this, it appeared like a lot of what you were measuring was SC damage that does not transmit through. What do you see in terms of the SC damage relative to what we know about transmissible damage for those same chemicals, and same kinds of dose levels?

Allen: That's really hard to get at because about all you can do is make correlations and that can only take you so far. There are at least three possible reasons why there were such differences in those curves. It could be that some repair occurred. More likely, there is a lot of cell loss along the way. But also, we were measuring some kinds of synaptic aberrations which are irrelevant at metaphase 1 and metaphase 2, so what is measured sometimes is different between prophase and metaphase. We do have some frequency distributions that go along with these data which tell us the amount of dropout that is occurring, some of which is after this pachytene stage, and I think Monte [Moses] may refer to some of this. We can also see the amount of dropout in terms of blockage of cells from entering meiosis. However, to be able to determine precisely what is happening in between those two stages, pachytene and metaphase, we really don't know for any specific type of aberration.

Handel: This question is about SC structure. Why do the unpaired axes have so much of a greater affinity for the stain? That was seen particularly dramatically in your micrographs of the Armenian hamster X/Y.

Moses: I don't know why, but unpaired axes tend to be thicker and more pronounced, and sometimes after certain treatments, a shortened axis that is nonhomologously paired with another axis may also appear thicker. In plants, there are nodules along the lateral elements. I'm not sure what they mean.

Handel: Do you think it's thicker because the unpaired chromatin is more condensed, less extended, or do you think there are different kinds of, or quantitative differences in, associated proteins?

Moses: We have no way of really knowing that.

Allen: I don't know either, but it clearly happens very quickly. It's not only in the normal single axes, but if you induce a break in the lateral element of a complex, regions near the break can flare out and become single and very thick. The same applies to the abnormal loop configurations; if you have a lateral element that loops off by itself, it thickens in the hamster to about 2–3 times the width size of the other one. It is curious that the other lateral element does not get thick even though it is also single. It could be something like it is just more taut, something is pulling on it—I don't know what it is—but when you do have a loop the lateral element that is looping out thickens and the other one does not.

Generoso: What I am going to say does not apply to radiation or to those chemicals you were talking about, but probably applies to alkylating agents. Triethylenemelamine is a cross-linking agent. If you expose pachytene spermatocytes, you will induce dominant lethals, and you will induce heritable translocations. But if you look a few days later in metaphase 1 cytogenetically, you would not see the aberrations there. The point I am trying to make is that for many of these alkylating chemicals, you need an intervening S for the conversion of the aberrations.

Also, we have sterile male translocation carriers where apparently spermatogenesis was disrupted early in meiosis. When we looked at the synaptonemal complex of the sterile translocations, the synaptonemal complex is just abnormal. Is it possible that some of the configurations you see may in fact be secondary effects because many of these agents that you use kill spermatocytes?

Allen: I can imagine one scenario in which some of those observations would be consistent. Suppose the alkylating agent was inducing a lesion that led to a single strand break, perhaps in association with meiotic prophase replication; then that could survive through meiosis and you wouldn't see any sort of breakage in diakinesis or metaphase chromosomes, but it may have another type of an effect on the synaptonemal complex, some kind of mispairing or whatever. Certainly, an alkylating agent is going to need S, but remember, there is a small amount of S that is occurring in meiotic prophase, and possibly some damage being mediated through that can survive and not be detectable in metaphase and yet may be implicated in the heritable effects.

Generoso: We don't see it in metaphase 1, that's the thing, and it's not only us. Brewen earlier on looked at metaphase 1 from treated pachytene spermatocytes and could not see anything either.

Allen: Well, if there was single strand breakage you wouldn't expect to see anything anyway. You would have to have both strands broken. That's why these agents, irradiation and bleomycin, given in either S phase or prophase, cause lots of obvious damage, whereas the alkylating agents given at prophase do not. Really, you would have to treat prior to S, or at early S, with these alkylating agents. If you hit at prophase, even if those cells survive to metaphase, I don't think you would see the damage.

Adler: Relating to that very point, you don't see anything if you treat pachytene and look at metaphase. What you do recover is translocations that are preleptotene- or leptotene-induced. Because of the timing of mating, you can't differentiate whether you have treated pachytenes or leptotenes. It's just too much of a smear if you mate the animals. The lesions you induce in pachytene, if you've got cross-linking lesions, are going to be repaired. They are not going to stay around until replication and translation in oocytes. That's just too long a period of time.

Generoso: What chemical are you talking about that you use to induce?

Adler: TEM, mitomycin C.

Generoso: Leptotene?

Adler: Preleptotene induced.

Generoso: During S?

Adler: Yes, during S. That's where you get recovery of translocations as well, not from treated pachytenes.

Meiosis and Experimentation in the Yeast *Saccharomyces cerevisiae*

MICHAEL E. DRESSER
Program in Molecular and Cell Biology
Oklahoma Medical Research Foundation
Oklahoma City, Oklahoma 73104

OVERVIEW

More is known about the molecular biology of meiosis in the yeast *Saccharomyces cerevisiae* than in any other organism, and the body of salient information is growing rapidly. Evolutionary conservation in critical steps of meiotic DNA metabolism and chromosome behavior indicates that an awareness of the information and insights gained from yeast can contribute substantially to ongoing efforts to understand how heritable damage arises in germ cells of mammals. A brief sketch of the yeast model for meiosis is presented here to suggest areas that may be of interest to those unfamiliar with the unique capabilities that the model currently provides.

INTRODUCTION

Meiosis is required for development of genetically diverse haploid gametes. Heritable damage that arises before or during meiosis occurs in the midst of mechanisms that normally produce considerable genetic variation by metabolizing the DNA in unique ways. In this setting, exogenous agents could effect deleterious results by causing DNA damage that is processed in meiosis-specific ways or by interfering with the structures and functions that carry out meiosis-specific DNA replication, organization, recombination, and repair. Clearly, information at the molecular level concerning meiotic structures and functions will be required to understand the contributions of meiosis per se to the events occurring in germ cells that lead to genetic abnormalities in offspring.

The complexities of DNA metabolism in eukaryotes are nowhere more evident than in meiotic prophase. Meiotic chromosomes condense, pair, synapse, exchange, and form stable associations that orient the reduction division while DNA recombination is regulated, initiated, processed, and resolved. It is apparent that a combination of cytology, genetics, and biochemistry will be required to understand the molecular mechanisms underlying these activities and to determine the relationships between the chromosomal and DNA activities.

Currently, there is no mammalian model system that supports the ease of manipulation and experimentation required for an efficient, combined analysis of meiotic chromosome behavior. Fortunately, the major events of meiosis are well-conserved, elevated genetic recombination and haploidization being hallmarks of meiosis throughout eukaryotes. More is currently known about the molecular biology of meiosis in one of the simplest eukaryotes, the yeast *S. cerevisiae*, than in any other organism, and it has recently been shown cytologically that the pattern of meiotic chromosome condensation, pairing, and synapsis seen in mammals is followed in yeast. The purpose of this paper is to present a brief overview of meiosis in yeast and to provide examples of its utility for examining the mechanisms that lead to heritable genetic damage.

Life Cycle

The life habits of yeast provide its greatest strengths as an experimental organism (Fig. 1) (Esposito and Klapholz 1981; Botstein and Fink 1988), and its manipulability allows such rigorous testing of hypotheses that the standards of proof in the field are high. Strains grow and negotiate meiosis as single cells in suspension culture, so that there are no cell-cell interactions to contend with as in mammalian testes and ovaries. Cells can be grown into clonal populations on solid medium, allowing replica-plating for rapid genetic analysis, and wild-type cells can grow in widely varying culture conditions, affording numerous possibilities for isolating conditional mutants. Haploids can easily be made into homozygous diploids, for example, to test the effects of a recessive mutation in a normal diploid meiosis, by using cells that switch mating types and mate with sisters. This can be done either by using homothallics, cells that normally switch mating type until successfully mated, or by transforming in the gene for the endonuclease responsible for the mating-type switch (Jensen et al. 1983). Healthy strains grow quickly, doubling every 100–120 minutes in presporulation medium, and it requires only 3–4 days to get from single-cell isolates to a sizable clonal population in meiosis. Diploids, triploids, tetraploids, and most sorts of aneuploids are viable and capable of negotiating meiosis successfully, i.e., sustaining elevated levels of recombination between homologous chromosomes, undergoing the two meiotic divisions, and forming viable spores (if sometimes rare and aneuploid). Even haploid strains can be constructed that enter meiosis and produce viable spores when genetically restricted to a single meiotic division (Wagstaff et al. 1982).

Cells that contain both the a and α mating type genes (as in normal diploids) or contain mutations that circumvent mating-type control can be induced to enter meiosis by transfer to medium lacking a nitrogen source and containing only a nonfermentable carbon source (Olempska-Beer 1987; Smith and Mitchell 1989). After 3 or more hours (depending on the strain) recombination increases and reaches two to three orders of magnitude over that seen spontaneously in cells growing in vegetative medium, and final levels of meiotic recombination are extremely high—on average, approximately 1 cM per 3 kb of DNA (Mortimer et al. 1989) in a diploid genome of only 2 x 14,000 kb pairs. Each of the four products of

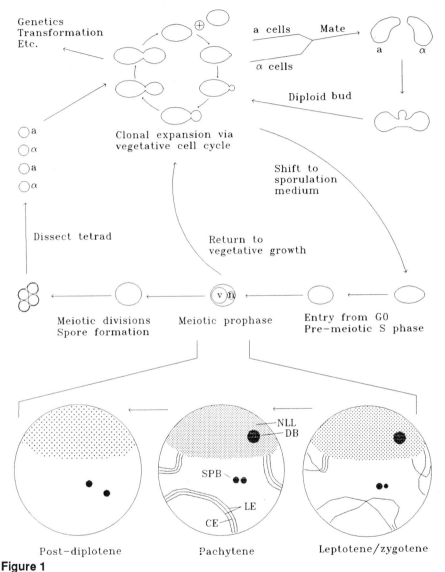

Figure 1
The yeast life cycle, with an emphasis on meiotic prophase. The lower row depicts parts of nuclei as seen in spread preparations with the electron microscope; the earlier two stages show two bivalents, one of which is divided by the nucleolus organizer region. (a and α) Mating types, (v) vacuole, (n) nucleus, (NLL) nucleolus, (DB) dense body, (SPB) spindle pole bodies, (LE) lateral elements of the synaptonemal complex, (CE) central element.

a normal meiosis can be recovered as the spores in a single ascus, allowing tetrad analysis of recombination and segregation. Furthermore, meiotic recombination can be assessed in the absence of the meiotic divisions by transferring cells in meiotic prophase back to vegetative medium ("Return to vegetative growth" in Fig. 1) before they become committed to the meiotic divisions (Sherman and Roman 1963).

Molecular Genetics

Genetic manipulation is greatly facilitated by the efficiency with which genes can be moved into and out of the cell, either on one of the multitudes of available plasmids or by integration into the genome via targeted gene replacement. For example, in order to make a haploid strain that enters meiosis, the a mating-type gene can be introduced into an α mating-type cell or vice versa (Gottlieb and Esposito 1989), or a mutation can be introduced directly into one of the genes responsible for mating-type control of meiosis (Rine and Herskowitz 1987). New genes can be isolated, processed, and often identified quickly. For example, a clone carrying a gene of interest can be isolated from a library, hybridized to chromosomes separated electrophoretically to narrow the position of the gene in the genome, reinserted in the genome in one step (Rothstein 1983) along with an easily scored genetic marker, and then mapped on the fine scale using classic genetics guided by the burgeoning yeast genetic map (Mortimer et al. 1989). Targeted deletion of the entire coding region of sequenced genes provides a reasonable starting point for genetic analysis of function that avoids problems of leaky or partially functioning alleles, and replacement of the intact gene into a strain that carries the deletion provides a convincing control. A particularly important aspect of the manipulability of yeast genetically is that analysis of multigene families is feasible even when there is considerable overlap in the functions of the genes (for a good example, see Werner-Washburne et al. 1987). Another feature is the relative ease and efficacy with which operationally useful phenotypes can be combined in ongoing analyses. For example, by combining an increase in permeability to exogenous agents and a removal of the putative target, it has been shown that topoisomerase I is required, and is likely to be the sole target, for killing of cells by camptothecin (Nitiss and Wang 1988). Treatment of mouse spermatocytes in vivo with camptothecin causes breaks in the synaptonemal complexes (J. Allen, pers. comm.), but the effects on yeast meiocytes are not yet known.

Meiotic DNA Metabolism

Which DNA lesions have the most potential to be destructive and how they act to cause heritable damage or get repaired or screened out of developing germ cells are questions that can be addressed by examination of DNA repair in meiosis. Many elements of the DNA repair systems are used in both meiotic and vegetative cells (Game et al. 1980; Resnick 1987), and a number of genes that are involved have been cloned. Analyses of the DNA double-strand break and mismatch repair systems are particularly well developed and provide good examples of the information available in yeast.

Genetic, biochemical, and cytological analyses of mutant phenotypes of two double-strand break repair genes required for meiosis have shown that RAD52 acts after RAD50, in that a rad52 mutant is rescued from lethality by a mutation in rad50 (Malone 1983), rad52 but not rad50 makes recombined DNA molecules (Borts et al. 1986), and rad52 but not rad50 makes synaptonemal complexes (Farnet et al. 1988; Giroux et al. 1989). These results and others indicate an interesting role for the double-strand break repair pathway in meiotic recombination, and the possibility that double-strand breaks per se initiate meiotic recombination is currently of much interest (Nicolas et al. 1989; Sun et al. 1989).

Mismatches arise in heteroduplexes that form as part of meiotic recombination (discussed in Hastings 1987). Certain pairs of heteroalleles in yeast give rise to elevated levels of postmeiotic segregation, suggesting that some properties of these particular mismatches, their structures or locations, abrogate repair during meiosis. A possible analogy in mammals is the damage sustained before or during meiosis following treatment with triethylenemelamine, which is not seen cytologically in chromosomes in the meiotic divisions but nevertheless results in translocations in the offspring (W. Generoso, pers. comm.). Premeiotic lesions that lead to mosaic offspring are another possibility, and would be a closer analog to postmeiotic segregation in yeast. Mutations in PMS1 in yeast, a gene with sequence similarity to mismatch repair genes from prokaryotes, result in a mutator phenotype in the cell cycle and in postmeiotic segregation at a number of loci (Kramer et al. 1989; also, see Bishop et al. 1989). The involvement of mismatch repair in recombination may have complex genetic consequences. From the results of a particularly detailed molecular analysis, it seems that the outcome of recombination is affected by the level of heterozygosity in a restricted region, leading to the suggestion that mismatch repair of heteroduplexes can lead to a second round of recombination (Borts and Haber 1987; Hastings 1987). Analysis of the same region in a pms1-1 mutant background gives results consistent with this interpretation (referred to in Borts and Haber 1989). It has been suggested that interference with mismatch repair per se, even if not coupled with damage, could have particularly unfavorable results in or following meiosis (Rayssiguier et al. 1989). Testing these ideas in mammalian meiosis will be difficult, but the issues are clearly important.

Meiotic recombination in yeast occurs at sufficiently high levels to encourage a biochemical approach to isolating the molecules involved in meiotic DNA metabolism (e.g., see Sugino et al. 1988). Even so, regulation of reciprocal events parallels that in larger eukaryotes, in that crossing-over shows positive interference (see Hastings 1987; Jones 1987). Curiously, this seems not to be the case for another yeast, *Schizosaccharomyces pombe*, a finding that may be related to the apparent absence of synaptonemal complexes in this species (Olson et al. 1978).

A number of elegant, detailed analyses have demonstrated that the level of recombination differs across the genome, as is true for larger eukaryotes (discussed in Jones 1987). This has been shown in yeast by comparing genetic with molecular maps (Kaback et al. 1989) and by determining the levels of recombination between alleles moved to different locations in the genome (Klar and Strathern 1986), in a number of adjacent intervals set off by restriction site heterologies (Symington and Petes 1988), between the multiple copies of retrovirus-like Ty elements (Kupiec and Petes 1988), and between alleles before and after moving an adjacent centromere (Lambie and Roeder 1986, 1988). Surprisingly, reciprocal recombination and gene conversion occur at a relatively high level between alleles placed at nonhomologous locations (Jinks-Robertson and Petes 1986; Klar and Strathern 1986). Whether yeast differs from mammals in this respect, quantitatively and/or qualitatively, remains to be seen. Translocations, which could result from such activities, certainly do appear in the products of mammalian meioses.

Meiotic Chromosome Behavior

Early in meiotic prophase, when chromosomes begin to condense and pair, distinct proteinaceous axes form along each chromosome. These axes reflect the behavior of the chromosomes and come together in pairs. As they align, they form a well-ordered structure termed the synaptonemal complex, which is composed of the two axes (lateral elements) connected by transverse filaments to the central element (see Fig. 1 and J. Allen et al.; M. Moses et al., both this volume). Synaptonemal complex formation and chromosome condensation during meiosis appear similar in yeast and mammals (Byers and Goetsch 1975; Zickler and Olson 1975; Dresser and Giroux 1988). The cytological phenotypes for mutations in two cloned genes required specifically for normal levels of meiotic recombination (SPO11, Giroux et al. 1989; HOP1, Hollingsworth and Byers 1989) demonstrate that these genes are required for normal synaptonemal complex formation, as is RAD50 (Farnet et al. 1988). Also, as with rad50, no recombined DNA molecules have been seen in a spo11-1 mutant strain (Borts et al. 1986). As intriguing as these results are, too little is known of the functions of the gene products to indicate in molecular terms the relationship between recombination and synaptonemal complex formation. Whether any of the cloned genes that are required for synaptonemal complex formation encode structural components is being examined using well-defined antibody probes and immunocytology. A related approach is under way in mammals (Dresser 1987; Moens et al. 1987; Heyting et al. 1989), where antibodies that label the synaptonemal complexes will be used to isolate the structural genes, and the same approach is being applied to yeast (Fig. 2A,B). Conservation of epitopes should provide another means of getting from yeast to mammalian proteins, and vice versa.

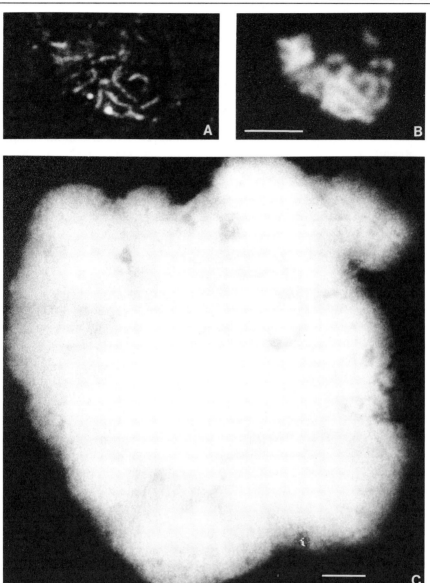

Figure 2
(*A,B*) Yeast pachytene nucleus stained in *A* with antibodies from a human autoimmune serum that labels, in part, the synaptonemal complexes (FITC secondary antibody) and in *B* with the fluorescent DNA stain DAPI. Bar in *B*, 4 microns. (*C*) Human spermatocyte pachytene nucleus (prepared in M. J. Moses' laboratory) labeled with the DNA stain DAPI. A single bivalent lies at upper right outside the main mass of the nucleus; the axial "shadow" is the synaptonemal complex. Bar, 4 microns.

FUTURE DIRECTIONS

The successes in identifying and isolating genes required for meiotic recombination and chromosome behavior in yeast represent only the beginning of what can be accomplished using the approaches that are now in place, and more of the same is clearly indicated. Two further avenues seem timely as well.

A number of agents have been identified that disrupt formation of the synaptonemal complex in mammals (see J. Allen et al.; M. Moses et al., both this volume), and some of these are known to affect yeast cells in meiosis (Kelly et al. 1983; Sora et al. 1983), although their effects on yeast synaptonemal complexes are unknown. For those with conserved targets, as seems likely for camptothecin, analysis in yeast may provide the most efficient means of understanding their mechanisms of action. It is possible that these agents affect chromosome structure primarily, and synaptonemal complex structure only secondarily. In either case, experimental manipulation of lateral element formation along the chromosomal axes may provide a means of examining the principles responsible for the linear organization of chromosomes generally as well as providing information on meiotic chromosomes per se. The relatively small size of the yeast genome simplifies the cytological as well as the genetic and biochemical approaches to this question, since the loops of chromatin that emanate from the chromosomal axes are shorter than in, for example, human spermatocytes (compare Fig. 2, B and C), and it may be that a larger fraction of the DNA is associated with the axes. Whether meiotic chromosome organization depends on specific DNA sequences to anchor the DNA loops is a fundamental and open question.

Yeast artificial chromosomes (YACs) can be made from exogenous DNA (Burke et al. 1987). The YACs can be the size of the endogenous chromosomes and are relatively stable yet are completely dispensible in nonselective medium. This should allow studies of chromosome behavior without apparent selective pressures on maintenance of the chromosomes that might influence experimental outcomes. For example, meiotic chromosome loss in mutant strains or following challenge with exogenous agents may provide more easily interpreted results with this system. Obviously, the meaning of such experiments depends on the manner in which the yeast mechanisms handle foreign DNA, and clearly the YACs bring in sequences that have not been selected for function in yeast. However, the differences themselves may also be informative; for example, a simple question (perhaps with a complicated answer) is whether the loops of chromatin formed from human DNA in a yeast meiosis are the same lengths as those in the endogenous chromosomes. Finally, the YACs bring in different overall base compositions and high levels of repeat sequences (see, e.g., Little et al. 1989) that may challenge the machinery responsible for meiotic chromosome behavior in unexpected ways, emphasizing known processes or even revealing new ones.

ACKNOWLEDGMENTS

The author was supported by a National Research Council Research Associate Award (while at the National Institute for Environmental Health Sciences) and by the Oklahoma Medical Research Foundation.

REFERENCES

Bishop, D.K., J. Andersen, and R.D. Kolodner. 1989. Specificity of mismatch repair following transformation of *Saccharomyces cerevisiae* with heteroduplex plasmid DNA. *Proc. Natl. Acad. Sci.* **86:** 3713.

Borts, R.H. and J.E. Haber. 1987. Meiotic recombination in yeast: Alteration by multiple heterozygosities. *Science* **237:** 1459.

―――. 1989. Length and distribution of meiotic gene conversion tracts and crossovers in *Saccharomyces cerevisiae*. *Genetics* **123:** 69.

Borts, R.H., M. Lichten, and J.E. Haber. 1986. Analysis of meiosis-defective mutations in yeast by physical monitoring of recombination. *Genetics* **113:** 551.

Botstein, D. and G.R. Fink. 1988. Yeast: An experimental organism for modern biology. *Science* **240:** 1439.

Burke, D.T., G.F. Carle, and M.V. Olson. 1987. Cloning of large segments of exogenous DNA into yeast by means of artificial chromosome vectors. *Science* **236:** 806.

Byers, B. and L. Goetsch. 1975. Electron microscopic observations on the meiotic karyotype of diploid and tetraploid *Saccharomyces cerevisiae*. *Proc. Natl. Acad. Sci.* **72:** 5056.

Dresser, M.D. 1987. The synaptonemal complex and meiosis: An immunocytochemical approach. In *Meiosis* (ed. P.B. Moens), p. 245. Academic Press, Orlando, Florida.

Dresser, M.E. and C.N. Giroux. 1988. Meiotic chromosome behavior in spread preparations of yeast. *J. Cell Biol.* **106:** 567.

Esposito, R.E. and S. Klapholz. 1981. Meiosis and ascospore development. In *Molecular biology of the yeast Saccharomyces: Life cycle and inheritance*, (ed. J.N. Strathern et al.), p. 211. Cold Spring Harbor Laboratory, Cold Spring Harbor, New York.

Farnet, C., R. Padmore, L. Cao, E. Alani, and N. Kleckner. 1988. The *RAD50* gene of *S. cerevisiae*. In *Mechanisms and consequences of DNA damage processing* (ed. E. Friedberg et al.), p. 201. A.R. Liss, New York.

Game, J.C., T.J. Zamb, R.J. Braun, M. Resnick, and R.M. Roth. 1980. The role of radiation (*rad*) genes in meiotic recombination in yeast. *Genetics* **94:** 51.

Giroux, C.N., M.E. Dresser, and H.F. Tiano. 1989. Genetic control of chromosome synapsis in yeast meiosis. *Genome* **31:** 88.

Gottlieb, S. and R.E. Esposito. 1989. A new role for a yeast transcriptional silencer gene, *SIR2*, in regulation of recombination in ribosomal DNA. *Cell* **56:** 771.

Hastings, P.J. 1987. Meiotic recombination interpreted as heteroduplex correction. In *Meiosis* (ed. P.B. Moens), p. 107. Academic Press, Orlando, Florida.

Heyting, C., J.J. Dietrich, P.B. Moens, R.J. Dettmers, H.H. Offenberg, E.J.W. Redeker, and A.C.G. Vink. 1989. Synaptonemal complex proteins. *Genome* **31:** 81.

Hollingsworth, N.M. and B. Byers. 1989. *HOP1:* A yeast meiotic pairing gene. *Genetics* **121:** 445.

Jensen, R., G.F. Sprague, Jr., and I. Herskowitz. 1983. Regulation of yeast mating-type interconversion: Feedback control of *HO* gene expression by the mating-type locus. *Proc. Natl. Acad. Sci.* **80:** 3035.

Jinks-Robertson, S. and T.D. Petes. 1986. Chromosomal translocations generated by high-frequency meiotic recombination between repeated yeast genes. *Genetics* **114:** 731.

Jones, G.H. 1987. Chiasmata. In *Meiosis* (ed. P.B. Moens), p. 213. Academic Press, Orlando, Florida.

Kaback, D.B., H.Y. Steensma, and P. de Jonge. 1989. Enhanced meiotic recombination on the smallest chromosome of *Saccharomyces cerevisiae*. *Proc. Natl. Acad. Sci.* **86:** 3694.

Kelly, S.L., C. Merrill, and J.M. Parry. 1983. Cyclic variations in sensitivity to X-irradiation during meiosis in *Saccharomyces cerevisiae*. *Mol. Gen. Genet.* **191:** 314.

Klar, A.J.S. and J.N. Strathern, eds. 1986. *Current communications in molecular biology: Mechanisms of yeast recombination*, p. 123. Cold Spring Harbor Laboratory, Cold Spring Harbor, New York.

Kramer, W., B. Kramer, M.S. Williamson, and S. Fogel. 1989. Cloning and nucleotide sequence of DNA mismatch repair gene *PMS1* from *Saccharomyces cerevisiae:* Homology of PMS1 to procaryotic MutL and HexB. *J. Bacteriol.* **171:** 5339.

Kupiec, M. and T.D. Petes. 1988. Meiotic recombination between repeated transposable elements in *Saccharomyces cerevisiae*. *Mol. Cell. Biol.* **8:** 2942.

Lambie, E.J. and G.S. Roeder. 1986. Repression of meiotic crossing over by a centromere (*CEN3*) in *Saccharomyces cerevisiae*. *Genetics* **114:** 769.

———. 1988. A yeast centromere acts in *cis* to inhibit meiotic gene conversion of adjacent sequences. *Cell* **52:** 863.

Little, R.D., G. Porta, G.F. Carle, D. Schlessinger, and M. D'Urso. 1989. Yeast artificial chromosomes with 200- to 800-kilobase inserts of human DNA containing HLA, V_K, 5S, and Xq24-Xq28 sequences. *Proc. Natl. Acad. Sci.* **86:** 1598.

Malone, R.E. 1983. Multiple mutant analysis of recombination in yeast. *Mol. Gen. Genet.* **189:** 405.

Moens, P.B., C. Heyting, A.J.J. Dietrich, W. van Raamsdonk, and Q. Chen. 1987. Synaptonemal complex antigen location and conservation. *J. Cell Biol.* **105:** 93.

Mortimer, R., D. Schild, C.R. Contopoulou, and J.A. Kans. 1989. Genetic map of *Saccharomyces cerevisiae*, edition 10. *Yeast* **5:** 321.

Nicolas, A., D. Treco, N.P. Schultes, and J.W. Szostak. 1989. An initiation site for meiotic gene conversion in the yeast *Saccharomyces cerevisiae*. *Nature* **338:** 35.

Nitiss, J. and J.C. Wang. 1988. DNA topoisomerase-targeting antitumor drugs can be studied in yeast. *Proc. Natl. Acad. Sci.* **85:** 7501.

Olempska-Beer, Z. 1987. Current methods for *Saccharomyces cerevisiae*. II. Sporulation. *Anal. Biochem.* **164:** 278.

Olson, L.W., U. Eden, M. Egel-Mitani, and R. Egel. 1978. Asynaptic meiosis in fission yeast? *Hereditas* **89:** 189.

Rayssiguier, C., D.S. Thaler, and M. Radman. 1989. The barrier to recombination between *Escherichia coli* and *Salmonella typhimurium* is disrupted in mismatch-repair mutants. *Nature* **342:** 396.

Resnick, M.A. 1987. Investigating the genetic control of biochemical events in meiotic recombination. In *Meiosis* (ed. P.B. Moens), p. 157. Academic Press, Orlando, Florida.

Rine, J. and I. Herskowitz. 1987. Four genes responsible for a position effect on expression from *HML* and *HMR* in *Saccharomyces cerevisiae*. *Genetics* **116:** 9.

Rothstein, R. 1983. One-step gene disruption in yeast. *Methods Enzymol.* **101:** 202.

Sherman, F. and H. Roman. 1963. Evidence for two types of allelic recombination in yeast. *Genetics* **48:** 225.

Smith, H.E. and A.P. Mitchell. 1989. A transcriptional cascade governs entry into meiosis in *Saccharomyces cerevisiae*. *Mol. Cell. Biol.* **9:** 2142.

Sora, S., M. Crippa, and G. Lucchini. 1983. Disomic and diploid meiotic products in *Saccharomyces cerevisiae:* Effect of vincristine, vinblastine, adriamycin, bleomycin, mitomycin C and cyclophosphamide. *Mutat. Res.* **107:** 249.

Sugino, A., J. Nitiss, and M.A. Resnick. 1988. ATP-independent DNA strand transfer catalyzed by protein(s) from meiotic cells of the yeast *Saccharomyces cerevisiae*. *Proc. Natl. Acad. Sci.* **85:** 3683.

Sun, H., D. Treco, N.P. Schultes, and J.W. Szostak. 1989. Double-strand breaks at an initiation site for meiotic gene conversion. *Nature* **338:** 87.

Symington, L.S. and T.D. Petes. 1988. Expansions and contractions of the genetic map relative to the physical map of yeast chromosome III. *Mol. Cell. Biol.* **8:** 595.

Wagstaff, J.E., S. Klapholz, and R.E. Esposito. 1982. Meiosis in haploid yeast. *Proc. Natl. Acad. Sci.* **79:** 2986.

Werner-Washburne, M., D.E. Stone, and E.Z. Craig. 1987. Complex interactions among members of an essential subfamily of *hsp70* genes in *Saccharomyces cerevisiae*. *Mol. Cell. Biol.* **7:** 2568.

Zickler, D. and L.W. Olson. 1975. The synaptonemal complex and the spindle plaque during meiosis in yeast. *Chromosoma* **50:** 1.

COMMENTS

Arnheim: It has been speculated that HOP1 is a synaptonemal complex gene. Is anything more known about that?

Dresser: Currently, the most direct evidence that a gene encodes a structural component of the SC is that antibodies raised against the gene product label the SC. The single micrograph that I have seen, of one yeast cell labeled with anti-HOP1 antibodies, was not convincing; but there may well be better evidence by now.

With regard to the search for SC genes in yeast, mutations in four different genes, which were chosen for analysis because they reduce or eliminate meiotic recombination, in every case affect SC formation and are, for now, candidate SC genes. I'll be very surprised if each gene encodes an SC component, although it is comforting that the tie between recombination and the SC is holding up as well as it is. An important question is whether formation of the SC fosters meiotic recombination or whether recombination, or its initiation, leads to SC formation.

Hecht: Are there SPO11 or SPO11-like sequences in the mouse?

Dresser: Preliminary results from Craig Giroux's lab suggest that sequences similar to the SPO11 coding sequence can be found in other species, but whether such a sequence exists in mouse is not yet clear.

Human Sperm Cytogenetics and the One-cell Zygote

BRIGITTE F. BRANDRIFF AND LAURIE A. GORDON
Biomedical Sciences Division
Lawrence Livermore National Laboratory
University of California
Livermore, California 94550

OVERVIEW

Human reproductive wastage is known to be a common event. It is important to take this high background frequency into consideration when attempting to identify detrimental effects from environmental agents. One major cause of embryonic and fetal losses is chromosomal aberrations, identified by karyotyping spontaneous abortion material and in-vitro-fertilized human embryos. Karyotyping of human gametes has made it possible to document types and frequencies of chromosomal aberrations directly in eggs and sperm themselves. Our studies with human sperm from normal, healthy men support the view that chromosome-specific aneuploidy does in fact occur and that frequencies of structural chromosomal aberrations appear to be person-specific and stable over time. The types of structural aberrations identified suggest that normal human spermiogenesis may be vulnerable to breakage events or precursor lesions leading to such breakage events. After entry into egg cytoplasm and preceding the formation of first-cleavage mitotic chromosomes, the male and the female genomes replicate their DNA in a pattern qualitatively similar to that in somatic cells. However, it is not known what relationship exists between spontaneous chromosome breaks seen at first cleavage and DNA replication activities. Limited data on survivors of radiotherapy lend support to the view that long-term effects on sperm chromosomal integrity can be identified. Studies on sperm cytogenetics thus have the potential for identifying adverse environmental effects on human spermatogenesis as monitored by this well-defined endpoint.

INTRODUCTION

The high frequencies of spontaneous fetal wastage in human beings must be considered as a confounding factor when addressing questions on potential effects from environmental agents. About 15–20% of clinically recognized pregnancies are lost as spontaneous abortions. Earlier embryonic losses, between implantation and the recognition of pregnancy around 4–5 weeks after conception, can be determined through sensitive immunological tests for human chorionic gonadotropin and have been estimated to occur with similar frequencies (Warburton 1987).

From cytogenetic studies on spontaneous abortion material, it has become clear that chromosomal abnormalities are one major cause of these losses (Hassold and Jacobs 1984; Hassold 1986). In particular, trisomies contribute over 90% of abnormalities encountered, with only three autosomal trisomies, those involving chromosomes 21, 18, and 13, occurring in the liveborn. In contrast, in spontaneous abortion material, trisomies for all autosomes except chromosome 1 have been identified. However, their relative frequencies vary dramatically. Trisomy 16 contributes fully one third of these cases, whereas others, e.g., 17 and 11, are extremely rare (Hassold and Jacobs 1984). Another, apparently nonrandom effect is seen in the paternal contribution to these nondisjunctional events, documented by fluorescent polymorphisms (Hassold 1986) and more recently by restriction fragment length polymorphism (RFLP) analysis (Jacobs et al. 1988). The paternal contribution to all trisomics combined amounts to less than 10%, in contrast to trisomy 21, for which it is about 20% (Hassold 1986).

To account for these widely varying frequencies, two mechanisms have been proposed. These are either differential survival of the trisomic conceptions as seen in trisomies 13, 18, and 21, the only ones surviving to birth, or differential primary events, specifically chromosome-specific nondisjunction, for which no direct evidence exists.

Structural aberrations, even though less commonly observed than trisomies, nevertheless add to the human mutational load. In spontaneous abortion material, they account for approximately 4% of abnormalities seen (Boué and Boué 1976). Among the liveborn, de novo structural aberrations have been estimated to occur at a rate of 45.1 per 100,000 gametes (Hook et al. 1989). Accounting for such structural anomalies, a preferential paternal origin has been postulated. For instance, out of 41 de novo structural aberrations, 84% were transmitted to the offspring by the male parent (for summary, see Magenis 1988). Similarly, in Prader-Willi syndrome children with structurally defective chromosomes 15, 21 out of 21 cases were transmitted by the fathers (Butler et al. 1986). Furthermore, de novo dominant mutations are associated with increasing paternal age (for review, see Carothers et al. 1986; Martínez-Frías et al. 1988).

Pushing this kind of analysis back to an earlier timepoint, closer to the actual time of conception, cytogenetic studies on human oocytes and in-vitro-fertilized preimplantation embryos (Rudak et al. 1985; Angell et al. 1986) have confirmed the occurrence of high frequencies of chromosomal abnormalities. However, because of the scarcity of this material, these data bases are small and are likely to remain so in the foreseeable future. A more extensive data base has been accumulated in recent years through analysis of the chromosomal constitution of individual human sperm. These analyses have been obtained by the technique of fusing human sperm with eggs from the golden hamster (Rudak et al. 1978). The fusion event activates the eggs such that after some hours in culture, they display the haploid chromosomal complement of the fused sperm, after which they divide once and then die. These chromosomes can be analyzed by standard cytogenetic techniques for numerical and structural aberrations, for sex chromosome constitution, and for effects from potentially detrimental environmental agents. This technique offers the advantage of a well-defined endpoint providing direct information on gametes themselves instead of postconception events and their attendant uncertainties.

RESULTS

Our work using this approach initially focused on obtaining baseline information on types and frequencies of chromosomal anomalies in sperm from normal, healthy individuals as a basis for future studies on potential environmental effects on sperm chromosomal integrity.

Numerical Aberrations

We addressed the question of chromosome-specific aneuploidy by direct inspection of the chromosomal constitution of individual human sperm cells. The results of 5997 such analyses are shown in Table 1. From the significant excess of disomy 1 complements, an almost fourfold incidence observed compared to expected frequency, it appears that chromosome-specific aneuploidy is indeed a reality in male meiosis. Ironically, trisomy 1, unlike the other autosomal trisomies, has never been seen in spontaneous abortion material, leading to the assumption that it is embryolethal at the very earliest stages of gestation. Only a single report exists of an in-vitro-fertilized 8-day-old embryo trisomic for chromosome 1, in which the extra chromosome was paternally derived (Watt et al. 1987). In our data set, no excess of extra chromosomes 21 or 16 was seen. Therefore, these results do not offer an explanation for the differential observations on these two chromosomes seen in spontaneous abortion material described above. However, to detect a more modest increase such as a doubling, a larger data set will be required. The overall frequency of hyperploid sperm is around 1% for all chromosomes, including X and Y. Hypohaploidy occurs more frequently (Table 1). However, since chromosome loss can be the result of slide preparation, it is difficult to interpret this finding.

Structural Aberrations

The presence of substantial numbers of structural aberrations in human sperm chromosomes has been confirmed by a number of laboratories employing human sperm/hamster egg techniques, with overall frequencies reported to be around 8% (for review, see Brandriff et al. 1990; Zenzes 1987). We have found in a longitudinal study that frequencies in individual men are stable over time and thus appear to be person-specific, although not correlated to frequencies of aberrations in lym-

Table 1
Tally of Aneuploid Chromosomes

Chromosome	No. Hyperhaploid	No. Hypohaploid
1	10[a]	0
2	4	2
3	1	1
4	3	2
5	2	2
6	1	1
7	2	1
8	0	1
9	4	3
10	4	2
11	0	1
12	1	1
C	1	0
13	0	3
14	1	4
15	1	5
16	4	2
17	2	2
18	3	5
E	0	1
19	3	5
20	2	5
F	0	1
21	5	9[b]
22	0	9[b]
X	0	n.a.
Y	2	n.a.
X or Y	3	11
Total	59	79

Total number of complements analyzed: 5159 from 19 normal men and 838 from 5 genetic counseling cases (aneuploidy reported was not related to reason for genetic counseling). n.a. indicates not applicable.
[a] $p < 0.001$, chi-square test.
[b] $p < 0.005$, chi-square test.

phocyte chromosomes from the same men. The largest category of structural aberrations reported by us, and confirmed by other laboratories, consisted of chromosome breaks, i.e., unrepaired lesions (Brandriff et al. 1988b). Lesions giving rise to these breaks in human sperm chromosomes must in most cases arise postmeiotically because of the presence of both pieces in the same cell, as seen in Figure 1. These in vitro data, together with those obtained in vivo described in the Introduction, suggest the possibility of an inherent tendency toward mutational lesions arising during human spermiogenesis. Such a tendency might increase the susceptibility of human spermiogenic stages to potential damage from environmental agents. In this connection, it is of note that the chemotherapeutic agent chlorambucil induced large-lesion mutations, including one that was cytogenetically identifiable, in early spermatids and spermatozoa of mice, whereas stem cells were unaffected (Russell et al. 1989). This observation also raises the interesting possibility of a continuum from lesions seen at the molecular level to those that are cytogenetically identifiable.

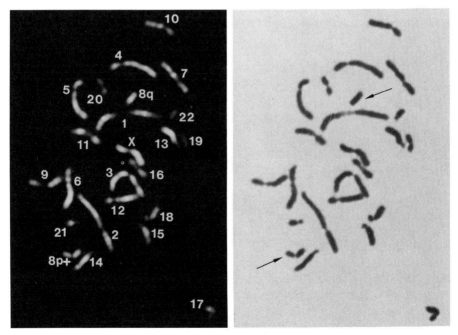

Figure 1
Human sperm chromosome complement showing unrejoined break in chromosome 8. Karyotype is 23,X,csb(8q1). (*Left* panel) Quinacrine banding to identify chromosomes; (*right* panel) giemsa staining to confirm abnormality. Arrows point to two unrejoined pieces of chromosome 8.

DNA Replication

Chromosomes at metaphase are static, and chromosomal aberrations are a relatively nonspecific endpoint in general. Their formation must be preceded by a sequence of steps, the nature of which is not well known, first in the male germ cells, and then in the egg following sperm-egg fusion. As a first step in an inquiry into the nature of the observed abnormalities, we have undertaken studies of pronuclear function and interphase nuclear architecture as expressed in two parameters, DNA replication and the spatial distribution of chromatin.

In a timed study with bromodeoxyuridine pulses lasting from 0.5 to 5 minutes, the earliest incorporation sites, detectable at 2 minutes, were distributed throughout the pronuclei and became brighter but not noticeably more numerous or distributed differently during 5-minute pulses (Fig. 2) or 30-minute pulses (Brandriff and Gordon 1989). These results indicate that even after short pulses, large numbers of sites are replicating at what appear to be similar rates, resembling replication activities described for somatic cells, in which such "replication granules" may represent tandemly arranged replicon clusters coordinately replicating DNA (Nakayasu and Berezney 1989). It thus appears that at least in a qualitative sense, replication in eggs follows the general pattern in eukaryotic cells.

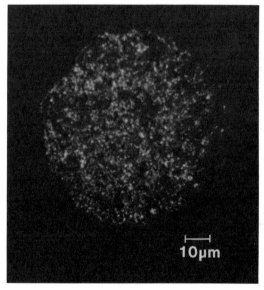

Figure 2
Zygote pronucleus showing bromodeoxyuridine (BU) incorporation sites after 5-min pulse. BU was detected by a monoclonal antibody to BU followed by a fluorescein-conjugated secondary antibody.

Caution is indicated, however, for direct extrapolations from somatic cells to eggs relating DNA synthetic activities, repair activities, and the formation of chromosomal aberrations. For instance, in a study of chromosomal aberrations induced by in vitro irradiation of sperm and lymphocytes from the same man, even though the total aberration frequency was the same in the two cell types, there were many more unrejoined lesions in the sperm compared to the lymphocytes, which had many more rejoined lesions than the sperm (Brandriff et al. 1988a). In addition, it must be remembered that the aberrations seen in human sperm chromosomes can be assumed to be spontaneous rather than induced, whereas most studies of chromosomal aberration formation have been conducted with induced abnormalities. Some of the confounding factors in interpreting germ-cell chromosomal anomalies have been reviewed by Adler and Brewen (1982).

In Situ Hybridization

The arrangement of various constituents within the interphase nucleus is of interest because it is during that phase that critical physiological activities take place. In the currently accepted model of interphase nuclear organization, "beads on a string," 10-nm nucleosomes are wound into 30-nm fibers attached in loops to the nuclear matrix (Pienta et al. 1989). In the special case of zygote nuclei, examination of pronuclear architecture would be expected to provide information relating to the reorganization of the male genome after gamete fusion, when it expands from a sperm nuclear volume of a few cubic microns into a pronuclear volume of several thousand cubic microns. We have, therefore, begun to examine the spatial distribution of chromatin in the pronuclei of these hybrid cells employing in situ hybridization techniques (Pinkel et al. 1988) with DNA probes for repetitive elements in the human genome. From preliminary results, it appears that chromatin fibers can be extended at least tenfold greater than equivalent regions in interphase somatic nuclei (B. Brandriff et al., unpubl.). We expect that studies of this kind will eventually show that nuclear volume and chromatin distribution in zygotes are parameters, in addition to others, special to germ cells (Adler and Brewen 1982), needing to be taken into consideration when first-cleavage chromosomal analysis is the endpoint chosen for monitoring adverse environmental effects on male germ-cell stages. For instance, for two damaged chromosomes to form a translocation, the likelihood for being spatially close enough to do so would be greater in a compact, diploid nucleus than in an expanded, haploid zygote pronucleus.

DISCUSSION

Since a relatively extensive data base on chromosomal abnormalities in sperm from normal, healthy men has been accumulated by a number of laboratories (for review, see Zenzes 1987), cytogenetic analysis of human sperm promises to be a useful endpoint in mutagenesis studies of individual men, although too labor-intensive at this time to be applied to large groups. Some beginnings have been made in this direction. Persistent elevated frequencies of structural and numerical abnormalities have been observed in some individuals for several years following radiation treatment (Martin et al. 1986; Brandriff et al. 1987). In addition, a few reports on small numbers of analyses from chemotherapy patients (Jenderny and Röhrborn 1987; Templado et al. 1988) have appeared. Both of these sets of data need to be expanded so that generalizations can be drawn from them on long-term effects. One of the difficulties inherent in expanding these data bases is that in the hamster egg assay, sperm have to be sufficiently functional to be able to become capacitated, undergo an acrosome reaction, fuse with the egg plasma membrane, decondense, and undergo the transformations required for chromosome formation. In men previously subjected to radio- and chemotherapy, sperm counts and/or function may be compromised, reducing yields of analyzable chromosome complements. This difficulty is reflected in the small number of analyses published to date. However, these studies are worth pursuing.

One question suggested by information on spontaneous abortion material and data on sperm chromosomes is whether *Homo sapiens* has in general a higher incidence of germ-cell chromosomal abnormalities than other species. Considering only the male contribution, frequencies of structural aberrations ranged from 2% to 15% among 20 individuals in our studies (Brandriff et al. 1988b), and from 0.0% to 23% among 30 men in the study of Martin et al. (1987). In contrast, in Chinese hamster embryos, for instance, analyzed for spontaneous aberrations at the first cleavage division, the combined incidence for both male and female was around 6%, excluding triploids (Mikamo and Kamiguchi 1983). As pointed out by Brent (1987), laboratory animals may not be the best reference group for comparison, because they are usually selected for good breeding success and therefore may not be really representative of the species as a whole. Wild populations might present a very different picture. For instance, in a natural population of South American marsh rats, 26 differing karyotypes were observed among 42 individuals (Nachman and Myers 1989). What seems to be the much greater variability among human populations compared to other animal species, whether seen at the molecular level or in epidemiological studies, may be more apparent than real, the result of our selection of otherwise invaluable laboratory animal models for mutagenesis studies.

ACKNOWLEDGMENTS

We thank A.V. Carrano for his collaboration on sperm cytogenetics. This work was performed under the auspices of the U.S. Department of Energy by the Lawrence Livermore National Laboratory under contract W-7405-ENG-48.

REFERENCES

Adler, I.-D. and J.G. Brewen. 1982. Effects of chemicals on chromosome-aberration production in male and female germ cells. In *Chemical mutagens: Principles and methods for their detection* (ed. F.J. de Serres and A. Hollaender), vol. 7, p. 1. Plenum Press, New York.

Angell, R.R., A.A. Templeton, and R.J. Aitken. 1986. Chromosome studies in human in vitro fertilization. *Hum. Genet.* **72:** 333.

Boué, J.G. and A. Boué. 1976. Chromosomal anomalies in early spontaneous abortion. (Their consequences on early embryogenesis and in vitro growth of embryonic cells.) *Curr. Top. Pathol.* **62:** 193.

Brandriff, B.F. and L.A. Gordon. 1989. Analysis of the first cell cycle in the cross between hamster eggs and human sperm. *Gamete Res.* **23:** 299.

Brandriff, B.F., L.A. Gordon, and A.V. Carrano. 1990. Cytogenics of human sperm: Structural aberrations and DNA replication. In *Mutation and the environment* part B: *Metabolism, testing methods, and chromosomes* (ed. M.L. Mendelsohn and R.J. Albertini). A.R. Liss, New York. (In press.)

Brandriff, B.F., L.A. Gordon, L.K. Ashworth, and A.V. Carrano. 1988a. Chromosomal aberrations induced by in vitro irradiation: Comparisons between human sperm and lymphocytes. *Environ. Mol. Mutagen.* **12:** 167.

Brandriff, B.F., L.A. Gordon, D. Moore II, and A.V. Carrano. 1988b. An analysis of structural aberrations in human sperm chromosomes. *Cytogenet. Cell Genet.* **47:** 29.

Brandriff, B.F., L.A. Gordon, I. Sharlip, and A.V. Carrano. 1987. Sperm chromosomal analysis in a survivor of seminoma and associated radiotherapy. *Environ. Mutagen.* (suppl. 8) **9:** 19.

Brent, R.L. 1987. Etiology of human birth defects: What are the causes of the large group of birth defects of unknown etiology? *Banbury Rep.* **26:** 287.

Butler, M.G., F.J. Meaney, and C.G. Palmer. 1986. Clinical and cytogenetic survey of 39 individuals with Prader-Labhart-Willi syndrome. *Am. J. Med. Genet.* **23:** 793.

Carothers, A.D., S.J. McAllion, and C.R. Paterson. 1986. Risk of dominant mutation in older fathers: Evidence from osteogenesis imperfecta. *J. Med. Genet.* **23:** 227.

Hassold, T.J. 1986. Chromosome abnormalities in human reproductive wastage. *Trends Genet.* **2:** 105.

Hassold, T.J. and P.A. Jacobs. 1984. Trisomy in man. *Annu. Rev. Genet.* **18:** 69.

Hook, E.B., N.P. Healy, and A.M. Willey. 1989. How much difference does chromosome banding make? Adjustments in prevalence and mutation rates of human structural cytogenetic abnormalities. *Ann. Hum. Genet.* **53:** 237.

Jacobs, P.A., T.J. Hassold, E. Whittington, G. Butler, S. Collyer, M. Keston, and M. Lee. 1988. Klinefelter's syndrome: An analysis of the origin of the additional sex chromosome using molecular probes. *Ann. Hum. Genet.* **52:** 93.

Jenderny, J. and G. Röhrborn. 1987. Chromosome analysis of human sperm I. First results with a modified method. *Hum. Genet.* **76:** 385.

Magenis, R.E. 1988. Invited editorial: On the origin of chromosome anomaly. *Am. J. Hum. Genet.* **42:** 529.

Martin, R.H., K. Hildebrand, J. Yamamoto, A. Rademaker, M. Barnes, G. Douglas, K.

Arthur, T. Ringrose, and I.S. Brown. 1986. An increased frequency of human sperm chromosomal abnormalities after radiotherapy. *Mutat. Res.* **174:** 219.

Martin, R.H., A.W. Rademaker, K. Hildebrand, L. Long-Simpson, D. Peterson, and J. Yamamoto. 1987. Variation in the frequency and type of sperm chromosomal abnormalities among normal men. *Hum. Genet.* **77:** 108.

Martínez-Frías, M.L., I. Herranz, J. Salvador, L. Prieto, M.A. Ramos-Arroyo, E. Rodríguez-Pinilla, and J.F. Cordero. 1988. Prevalence of dominant mutations in Spain: Effect of changes in maternal age distribution. *Am. J. Med. Genet.* **31:** 845.

Mikamo, M.W. and Y. Kamiguchi. 1983. Primary incidences of spontaneous chromosomal anomalies and their origins and causal mechanisms in the Chinese hamster. *Mutat. Res.* **108:** 265.

Nachman, M.W. and P. Myers. 1989. Exceptional chromosomal mutations in a rodent population are not strongly underdominant. *Proc. Natl. Acad. Sci.* **86:** 6666.

Nakayasu, H. and R. Berezney. 1989. Mapping replicational sites in the eucaryotic cell nucleus. *J. Cell Biol.* **108:** 1.

Pienta, K.J., A.W. Partin, and D.S. Coffey. 1989. Cancer as a disease of DNA organization and dynamic cell structure. *Cancer Res.* **49:** 2525.

Pinkel, D., J. Landegent, C. Collins, J. Fuscoe, R. Segraves, J. Lucas, and J. Gray. 1988. Fluorescence *in situ* hybridization with human chromosome-specific libraries: Detection of trisomy 21 and translocations of chromosome 4. *Proc. Natl. Acad. Sci.* **85:** 9138.

Rudak, E., P.A. Jacobs, and R. Yanagamachi. 1978. Direct analysis of the chromosome constitution of human spermatozoa. *Nature* **274:** 911.

Rudak, E., J. Dor, S. Mashiach, L. Nebel, and B. Goldman. 1985. Chromosome analysis of human oocytes and embryos fertilized *in vitro*. *Ann. N.Y. Acad. Sci.* **442:** 476.

Russell, L.B., P.R. Hunsicker, N.L.A. Cacheiro, J.W. Bangham, W.L. Russell, and M.D. Shelby. 1989. Chlorambucil effectively induces deletion mutations in mouse germ cells. *Proc. Natl. Acad. Sci.* **86:** 3704.

Templado, C., J. Benet, A. Genescà, J. Navarro, M.R. Caballin, R. Miró, and J. Egozcue. 1988. Human sperm chromosomes. *Hum. Reprod.* **3:** 133.

Warburton, D. 1987. Reproductive loss: How much is preventable? *N. Engl. J. Med.* **316:** 158.

Watt, J.L., A.A. Templeton, I. Messinis, L. Bell, P. Cunningham, and R.O. Duncan. 1987. Trisomy 1 in an eight cell human pre-embryo. *J. Med. Genet.* **24:** 60.

Zenzes, M.T. 1987. Chromosome analyses using human spermatozoa. In *Cytogenetics: Basic and applied aspects* (ed. G. Obe and A. Basler), p. 198. Springer-Verlag, Berlin.

COMMENTS

Bishop: Have you determined whether the sperm being in the hamster egg contributes somehow to some of the damage? For instance, if you did this with mouse sperm in which you know a set amount of damage is going to occur.

Brandriff: We have tried to address that, but for some reason or other, hamster eggs are willing to make human sperm chromosomes, but they are very unwilling to make any other kinds of sperm chromosomes, even though they are perfectly capable of fusing with sperm. We have tried mouse and rabbit but gave up because we could not get it to work. After many years of trying, other people have finally made it with bull sperm and pig sperm, but I have seen no papers that show anything except maybe 20 analyses or something like that. So that experiment is not doable. We have shown something by doing these longitudinal studies, by following individual men and showing that these frequencies do not change all over the place for an individual man, but that they stay approximately the same over many years.

Bishop: That is all measured by one measurement, not multiple measurements?

Brandriff: Oh, no. For this person that we followed for five years, we did our first set, say three or four experiments, and then maybe a year later we got three or four more samples, then another year later we got four or five samples, and in that way we accumulated those data from individual men.

Bishop: But I mean all of the data are accumulated from hamster egg and not by any other confirmatory method?

Brandriff: Yes. Unfortunately, I guess these frequencies are not high enough to do any kind of alkaline elution.

Favor: On those longevity studies, is there any chromosomal specificity to the aberrations that you see?

Brandriff: We were hopeful at first when we started out doing that. We saw repeat breaks at specific bands, but that has not held up over time. Also, these breaks do not correlate with, for instance, fragile sites.

Favor: Could it be the lifestyle of the individual? Could they be exposed to something like that, or do you think it's endogenous?

Brandriff: That's another problem you have to put up with when you work with people, because you cannot ask people. We are doing a study with chemotherapy patients, but we do not have enough data yet to be able to say anything about dose.

Wyrobek: Along those lines, maybe the urine analysis, maybe Ames-positive urine might be interesting to look at, to see if there is a lifestyle issue there.

Brandriff: I have not been willing to make that request of the donors.

Wyrobek: Urine is different.

Brandriff: Yes, that's perfectly true. But then, you really have to ask these people, "Are you willing to tell us what you are doing?"

Bishop: But you measured somatic chromosomes in these same people, right?

Brandriff: Yes.

Bishop: And you got nothing?

Brandriff: Nothing.

Bishop: That was disturbing.

Brandriff: Well, you can see that there is no selection pressure for making 100 million perfect sperm a day, but there is a lot of selection pressure for making perfect lymphocytes.

Bridges: About 10% of the sulfhydryl bridges in the protamines were not joined in the sperm. Is it known whether that is general over the whole sperm population or whether there is 8% of the sperm where that was not joined? In other words, is this a sperm-specific thing? If you look at sperm morphology, that's the sort of proportion of "junk" sperm that you get. I was wondering if there was any connection.

Wyrobek: As far as I know, we do not know. Just like the percent of histone among sperm, it's based on pooled analysis. We do not know.

Chromosome Aberrations Associated with Induced Mutations: Effect on Mapping New Mutations

MURIEL T. DAVISSON[1] AND SUSAN E. LEWIS[2]
[1]The Jackson Laboratory, Bar Harbor, Maine 04609
[2]Research Triangle Institute, Research Triangle Park, North Carolina 27709

OVERVIEW

A potential problem in mapping induced mutations or using induced mutations to map other genes is the possibility that they are tightly associated with a chromosome aberration that may interfere with recombination and alter estimates of recombination frequency. We have karyotyped mice carrying mutations that have arisen in mutagenesis experiments to determine how frequently this occurs. First, we karyotyped from The Jackson Laboratory breeding colonies mice carrying preexisting mutations that may have been induced by mutagens. Second, we karyotyped all mice carrying presumptive new induced mutations from a mutagenesis experiment ongoing at Research Triangle Institute (RTI). One tightly linked chromosome aberration has been identified among the 17 preexisting mutations screened to date, and one among the new mutations induced in the RTI program. Our results suggest that karyotyping should be part of the analysis of new induced mutations. We also present here a list of the mutations screened at The Jackson Laboratory and shown to have no cytologically detectable chromosomal rearrangements.

INTRODUCTION

In this paper, we discuss a potential problem in mapping induced mutations or using induced mutations to map other genes. Many mutations causing changes in visible phenotype or in protein-encoding loci have been induced or have arisen in mutagenesis experiments. Mice carrying such mutations are not always examined for chromosome aberrations, depending largely on the laboratory where the mutation was induced or whether the laboratory was also screening for chromosome aberrations. These induced mutations are described in the literature and are retained as useful mutations, but the information that they were induced may be forgotten over time. Mutations carrying undetected chromosome aberrations can drastically alter recombination results in subsequent mapping studies.

Two mutations recovered in mutagenesis experiments at the Harwell Medical Research Council (MRC) unit in England, agouti-suppressor (A^s) and bare patches (Bpa), were subsequently shown to be associated with paracentric chromosomal inversions (Evans and Phillips 1975, 1978). We encountered this problem at The Jackson Laboratory a few years ago when we discovered that the hairy ears mutation (*Eh*) appeared to inhibit recombination in the distal half of chromosome (Chr) 15. In the case of A^s and *Bpa*, recombination suppression in three-point crosses and production of XO females, respectively, alerted the investigators analyzing them that chromosomal rearrangements might be involved. In the case of *Eh*, we were unaware it suppressed recombination until we discovered in mapping studies that it did not recombine with several different genes on Chr 15 (M.T. Davisson, in prep.). This led us to wonder how much of a problem this is.

Two studies are in progress to assess how often induced mutations may be associated with chromosome aberrations. First, Dr. Davisson has identified all such mutations presently maintained at The Jackson Laboratory and is examining the chromosomes carrying the mutations for abnormalities. Second, we are karyotyping mice carrying mutations induced in a large mutagenesis project under Dr. Lewis's direction at RTI. In this paper, we summarize the results of these two studies.

RESULTS

Karyotyping Established Induced Mutations

In the Mouse Mutant Resource at The Jackson Laboratory, we were simultaneously searching for the chromosomal location of three new mutations when we discovered that none of the mutations recombined with the *Eh* locus on Chr 15, yet subsequent mapping studies clearly showed that the three loci mapped to different regions of the chromosome. Subsequently, analysis of G-banded mitotic chromosomes and synaptonemal complexes from *Eh*/+ heterozygotes and +/+ littermate controls showed that the *Eh* mutation is associated with a small paracentric inversion in the distal end of Chr 15 (M.T. Davisson, in prep.). We are now completing genetic analysis to determine the genetic extent of the inversion.

To assess how frequently induced mutations may be associated with detectable chromosome rearrangements, we decided to identify and analyze karyotypically as many induced mutations as possible. Margaret Green's catalog (Green 1989) identified over 100 mutations that had been induced or had arisen in mutagenesis experiments (Table 1), 38 of which are maintained at The Jackson Laboratory. We are systematically screening them for chromosome aberrations. While searching the literature, we also identified 10 translocations or inversions known to be associated with phenotypic changes (Table 2).

G-banded chromosomes were prepared from peripheral lymphocyte cultures (Davisson and Akeson 1987) of heterozygotes for each of the mutations listed in Table 3. The two homologs of the chromosome carrying the mutation were compared in at least 10 well-banded cells. To date, we have screened 29 mutations

Table 1
Mutations That Have Possibly, Probably, or Surely Been Induced

1. X-irradiation

a^e	ec	mi	p^d	sla	Ul
a^l	Gy	my	p^s	Slf	Ve
an	kr	ol	pu	Str	W^{19H}
Bhd	Kw	Os	s	sy	W^a
Bpa	ld	oto	se (many)	T^c	
Dmm	lz	p^{6H}	sh-2	T^{Or} (6)	

2. Gamma irradiation

Acc	Apyc	Idc	Nuca	Sl^m
Alm	Ccd	Mi^{or}	Nzc	Vlm
Anc	Er	Mo^{dp}	se (many)	
Apoc	Iac			

3. Neutron irradiation

bl	Eh	Re^{wc}	Sl^{cg}	Tht
Br	p^{11}	Rw	Sl^{con}	Xt^{bph}
c^m	p^{cp}	Sig	thd	
Cm	Ps			

4. Ethylnitrosourea

a^{16H}	$G6pd^{low}$	Lop-4	l(17)-4	pge
Bru	$Gdc\text{-}1^e$	l(17)-1	mi^{di}	Pk-1
Bsk	Hba^{g2}	l(17)-2	Nan	Tsk-2
Chy	hph-1	l(17)-3	Pep-3 (several)	

5. Irradiation (type unspecified)

am	A^s	fl	p^{25H}	sa	shm
a^{da}	bg	Ht	p^{bs}	se^1 (8)	Swl
a^m	Dfp	Ie	p^{m1}	se^x	tl
a^u	dv	Mp	p^{m2}	se^{sv}	wh
a^x	Eo	Och	pf	Sey^H	Xcat

6. Other

fro	tris (1-aziridinyl) phosphine sulfide	$Ldh\text{-}1^c$	procarbazide hydrochloride
hpc	thiotepa	xn	DDT
lst	methylcholanthrene	pdn	EMS
Pt	methylcholanthrene	bt-2	MNNG
W^s	methylcholanthrene	cr	nitrogen mustard
Tal	X-ray + hydroxyurea	di	nitrogen mustard
Pep-3	procarbazine	tu	nitrogen mustard

Table 2
Chromosome Aberrations Associated with Visible Mutations

Gene symbol	Phenotype or gene name	Chromosome aberration	Induced by	Reference
—	"diver"	T(2;14)1Gso	TEM[a]	Rutledge et al. (1986)
—	"steeloid"	T(5;10)9Rl T(10;12)10Rl T(10;18)10Rl	fission neutron X-irradiation neutron	Cachiero and Russell (1975)
—	T/A = "agouti-umbrous" T/T = dark agouti	T(2;8)26H	fission neutron	Batchelor et al. (1966)
se^{5H}	T/se = short ear	T(1;9)27H	fission neutron	Batchelor et al. (1966) Searle (1981)
—	T/a = non-agouti	T(2;11)30H	TEM	Cattanach (1966)
—	light coat color	T(In1;5)44H	X-irradiation	Beechey et al. (1987)
pg^J	pygmy	In(10)17Rk	TEM	Roderick (1983)
Tim	translocation-induced mutation	T(4;17)2Lws	ethylene oxide	Lewis et al. (1990)

[a](TEM) Triethylene melamine.

that either were reported to be induced or arose in mutagenesis experiments. Of these, in addition to the *Eh* mutation, we have identified a second mutation, rump white (*Rw*) on Chr 5, that may be associated with a chromosomal rearrangement that appears to be an inversion. Meiotic studies are in progress to determine whether this rearrangement is an inversion.

In the Mutant Resource at The Jackson Laboratory, we have now begun to include chromosomal analysis, in addition to genetic mapping, as a routine part of characterizing new mutations arising apparently spontaneously in our breeding colonies. All new mutants are karyotyped using G-banded metaphase chromosomes from peripheral lymphocyte cultures to determine whether they carry an associated chromosome aberration. Although no chromosomal rearrangements have been identified by karyotyping mice with spontaneous mutations, we have one spontaneous

Table 3
Mutations at The Jackson Laboratory Examined for Chromosome Aberrations

Locus	Mutant name	Chr	Mutagen	Result
a^e	extreme non-agouti	2	X-irradiation	normal
an	Hertwig's anemia	4	X-irradiation	normal
Bpa	bare paatches	X	X-irradiation	normal
bg	beige	13	irradiation	normal
Bsk	bare skin	11	ENU	normal
Cm	coloboma	2	neutron irradiation	normal
cr	crinkled	13	nitrogen mustard	normal
Eh	hairy ears	15	neutron irradiation	inversion
Er	repeated epilation	4	probably gamma irradiation	normal
Gs	greasy	X	possibly irradiation	normal
Gy	gyro	X	X-irradiation	normal
mi	microphthalmia	6	X-irradiation	normal
Och	ochre	4	irradiation	normal
Os	oligosyndactylism	8	X-irradiation	normal
oto	otocephaly	1	irradiation	normal[a]
Ps	polysyndactyly	4	neutron irradiation	normal
Pt	pintail	4	methylcholanthrene	normal
pu	pudgy	7	X-irradiation	normal
Rw	rump white	5	neutron irradiation	inversion?
sa	satin	13	probably irradiation	normal
sh-2	shaker-2	11	probably irradiation	normal
shm	shambling	11	possibly irradiation	normal
Sig	sightless	6	neutron irradiation	normal
Sl^{con}	contrasted	10	neutron irradiation	normal
sla	sex-linked anemia	X	X-irradiation	normal
spf	sparse fur	X	possibly irradiation	normal
T^{hp}	T-hair pin	17	X-irradiation	normal
Ul	ulnaless	2	X-irradiation	normal
Ve	velvet ears	15	X-irradiation	normal

[a] *oto* was induced in a chromosome already carrying In(1)1Rk. No abnormality other than In1Rk was detected.

mutation, patchy fur (*Paf*), that may involve a small deletion or rearrangement of the X chromosome. *Paf* causes delayed disjunction of the X and Y chromosomes in *Paf*/Y males (Lane and Davisson 1990). No obvious rearrangement can be detected in G-banded X chromosomes, and further cytogenetic studies are in progress to resolve whether a rearrangement is associated with the *Paf* mutation.

Table 4
Karyotypes of Mice Carrying Mutations Induced in the RTI/Lewis Project

Mouse No.	Expt.	Mutation	Treatment	Karyotype result
30,272	P[6]	Car-2 null	ENU	normal
30,813[a]		Car-2 null		normal
30,985	B[3]	coat color variant	ENU	normal
31,324[a]		coat color variant		normal
29,005	A[4]	coat color variant	EtO	normal
30,645	W[5]	neurological; bilateral cataracts	EtO	T(4A2;17E4)2Lws
29,298	Z[7]	Pgd mobility variant	EtO	mosaic for T(12;14), Chrs 4 normal
29,582[a]		Pgd mobility variant		normal
30,949	D[7]	coat color variant	600R	T(X;14), X may be deleted
30,294	K[7]	Mod-1 null	600R	normal

[a]Two different mice carrying the same mutation were karyotyped.

Screening Mutations Induced at RTI

We have begun karyotyping all presumptive new mutations induced in Dr. Lewis's mutagenesis program at RTI. We have screened a total of seven mutations induced in the mutagenesis project: three mutations that cause loss of mobility or mobility shifts in proteins (29298, 30272, 30294); three mutations that cause coat color changes (29005, 30949, 30985); and one mutation that causes neurological defects and bilateral cataracts (30645) (Table 4). One mouse (29298) carrying a mutation that causes a mobility shift in 6-phosphoglucocate dehydrogenase (PGD) was mosaic for a reciprocal translocation involving Chrs 12 and 14. Since the *Pgd* locus is on Chr 4, presumably this translocation will be separable from the mutation in subsequent generations. Indeed, another mutant mouse (29582) from the same line lacked the translocation. One coat color mutant (30949) had a translocation between Chr 14 and the X chromosome. It was probably not X-linked, but the mutation was lost before the preliminary breeding analysis could be confirmed.

Most interestingly, a neurological mutant (30645) had a translocation between Chrs 4 and 17, T(4A2;17E4)1Lws, that has not yet been separated in breeding tests and stock maintenance from the mutation causing the neurological and eye defects (Lewis et al. 1990). The mutation has been named translocation induced circling mutation (*Tim*). Thus, of seven mutants karyotyped, at least one had a chromosome rearrangement that is closely associated with the mutation and may alter recombination in Chrs 4 and 17.

DISCUSSION

We have examined two populations of mice with induced mutations to determine how frequently an induced mutation is likely to be inseparably associated with a chromosome aberration that drastically alters recombination. The discussion of preexisting mutations will be restricted to those that have been retained in gene lists, and excludes repeat mutations at specific loci induced in specific locus mutagenesis studies. Close association between a mutation and a chromosome rearrangement can occur in two ways: (1) One of the chromosome aberration breakpoints interrupts or deletes a gene or (2) a mutation is induced simultaneously with the chromosome aberration elsewhere in the affected chromosome. There are examples of both. The mutation at the agouti locus on Chr 2, agouti-suppressor (A^s), has been inseparable from the inversion In(2)2H (Evans and Phillips 1978), and the translocation-induced mutation, Tim, has been inseparable from T(4;17)1Lws (Lewis et al. 1990). Bare patches, Bpa, on the X chromosome, on the other hand, was induced simultaneously with In(X)1H but has since been separated from the inversion by recombination (Evans and Phillips 1975).

There is no strong evidence that chromosome aberrations themselves cause mutant phenotypes. Perhaps because chromosome aberrations in human beings are often associated with mental retardation or morphological abnormalities, the aberrations themselves are often spoken of as causing the abnormal phenotypes. In those cases where chromosome aberrations associated with diseases have been studied at the molecular level, small deletions or gene interruptions have been discovered. For example, there are numerous studies showing that translocations associated with various kinds of cancer interrupt oncogenes or immunoglobulin loci (for review, see Croce 1987), and translocations associated with Duchenne muscular dystrophy have been shown to cause mutations or deletions in the X-linked dystrophin locus (Love and Davies 1989). The chromosomal break that occurs prior to rearrangement interrupts a specific locus causing a mutation that in turn causes the aberrant phenotype.

There are now at least 15 mutations in the mouse that are known to be closely associated with cytologically detectable chromosome aberrations, including the aberrant phenotypes associated with identified translocations (Table 1) and the A^s, Bpa, Eh, Rw, and Tim mutations. Other mutations are known to be associated with chromosomal deletions that are not detectable cytologically. Examples of mutations known to have small deletions include two alleles at the brachyury locus, T-hairpin (T^{hp}) and T-Oak Ridge-1 (T^{1Or}) (Bennett et al. 1975; Green 1989). Additional induced mutations that produce lethality when homozygous, such as repeated epilation (Er) and velvet coat (Ve), quite possibly also have small deletions that are not detectable cytologically. Such small deletions are not likely to interfere with recombination estimates as drastically as chromosomal rearrangements such as translocations and inversions, although they will obviously alter recombination in the small regions in which they occur.

The potential for mapping error resulting from undetected chromosome aberrations associated with induced mutations is real, but the likelihood of such aberrations going undetected appears to be minimal. In two of the four mutations for which the mutant phenotype was detected for some time before the chromosome aberration was detected (A^s, Bpa, Eh, Rw), subsequent breeding tests signaled the presence of the associated aberration. The inversion associated with A^s was detected by its suppression of recombination in three-point linkage studies to map it. The inversion associated with Bpa was identified because it caused abnormal sex chromosome segregation. For the remaining mutations or deviant phenotypes listed in Tables 2 and 4, the mutants were screened cytologically as well as phenotypically as part of mutagenesis programs. Only Eh and Rw were described without knowledge of their associated inversions, and only Eh has been used to map other loci.

Our results indicate that an important part of characterizing induced mutations is looking for associated chromosome aberrations. By screening for chromosome aberrations associated with the induced mutations available at The Jackson Laboratory, we have also verified that the mutants so far tested do not carry associated chromosome aberrations that can alter recombination estimates. The remaining mutations maintained now only as frozen embryos at The Jackson Laboratory will be screened when available, and the results will be reported in Mouse News Letter.

Screening known mutants for chromosome aberrations has been a useful aspect of the characterization of mutants identified in the RTI program, either by electrophoretic or visual screening. The Car-2 null mutation, identified by electrophoretic screening analysis, has provided a potentially useful animal model of human disease (Lewis et al. 1988). Elimination of the possibility of a chromosomal defect associated with the mutant gene constituted an important aspect of the characterization of this mutant.

The cytogenetic studies reported here complete (as far as possible) the analysis of four visually identified and two electrophoretically identified mutants induced by ethylene oxide (EtO) (Lewis et al. 1986). The karyotype of mouse 29005, which was heterozygous for one of the EtO-induced mutants, was normal. One of the visible mutations induced by this agent could not be carried beyond the progeny of the original mutant and could not be recovered for analysis. The two other visible mutations recovered from this study, the neurological mutant described here (Lewis et al. 1990) and a runted dysmorphic F_1 mutant (Lewis et al. 1986), were associated with chromosome aberrations.

Of the electrophoretically detected mutants found in the EtO study, one, a null at Car-2, has already been shown to be associated with a translocation involving the relevant portion of Chr 3 (Lewis et al. 1986). The other, a mutation specifying altered mobility at Pgd, shown in the study reported here, was not associated with a chromosomal defect in Chr 4, although a translocation unrelated to the mutation at Pgd was found in one carrier animal.

Finally, it should be pointed out that mutations closely associated with chromosome aberrations are useful tools. Although using the *Eh* mutation to map genes on Chr 15 will give erroneous recombination estimates, *Eh* is a good locus to screen for linkage of new mutations to Chr 15. It tests for linkage over the entire length of Chr 15 because it inhibits recombination almost completely in the distal half of the chromosome and reduces recombination in the proximal half (M.T. Davisson in prep.). The "steeloid" translocations provided the clue that linkage group X was on Chr 10 of the mouse because they involved Chr 10, and the phenotype produced by the presumed mutations associated with them resembled that produced by the steel (*Sl*) mutation in linkage group X (Cacheiro and Russell 1975).

ACKNOWLEDGMENTS

We thank Ellen C. Akeson, Cecilia Schmidt, Belinda S. Harris, and Lois B. Barnett for technical assistance with various parts of this project, and Margaret C. Green and Eva M. Eicher for helpful comments on the manuscript. The research described here was supported by National Science Foundation grant BSR-8418828, National Institutes of Health grant RR-01183, a gift from the Eleanor Naylor Dana Charitable Trust, and National Institute of Environmental Health Science contract N0-1-ES-55078. The National Institutes of Health are not responsible for the contents of this paper nor do the comments necessarily represent their official views. The Jackson Laboratory is fully accredited by the American Association for Accreditation of Laboratory Animal Care.

REFERENCES

Batchelor, A.L., R.J.S. Phillips, and A.G. Searle. 1966. A comparison of the mutagenic effectiveness of chronic neutron- and γ-irradiation of mouse spermatogonia. *Mutat. Res.* **3:** 291.

Beechey, C.V., A.G. Searle, M.D. Burtenshaw, and E.P. Evans. 1987. T44H: A combined translocation and inversion with visible effect. *Mouse News Lett.* **77:** 126.

Bennett, D., L.C. Dunn, M. Spiegelman, K. Artzt, J. Cookingham, and E. Schermerhorn. 1975. Observations on a set of radiation-induced dominant *T*-like mutations in the mouse. *Genet. Res.* **26:** 95.

Cacheiro, N.L.A. and L.B. Russell. 1975. Evidence that linkage group IV as well as linkage group X of the mouse are in chromosome 10. *Genet. Res.* **25:** 193.

Cattanach, B.M. 1966. Chemically induced mutations in mice. *Mutat. Res.* **3:** 346.

Croce, C.M. 1987. Role of chromosome translocations in human neoplasia. *Cell* **49:** 155.

Davisson, M.T. and E.C. Akeson. 1987. An improved method for preparing G-banded chromosomes from mouse peripheral blood. *Cytogenet. Cell Genet.* **45:** 70.

Evans, E.P. and R.J.S. Phillips. 1975. Inversion heterozygosity and the origin of XO daughters of *Bpa*/+ female mice. *Nature* **256:** 40.

———. 1978. A phenotypically marked inversion (In(2)2H). *Mouse News Lett.* **58:** 44.

Green, M.C. 1989. Catalog of mutant genes and polymorphic loci. In *Genetic variants and strains of the laboratory mouse*, 2nd edition (ed. M.F. Lyon and A.G. Searle), p. 12. Oxford University Press.

Lane, P.W. and M.T. Davisson. 1990. Patchy fur, *(Paf)*, a semi-dominant X-linked gene, associated with a high level of X-Y non-disjunction in male mice. *J. Hered.* **81:** 43.

Lewis, S.E., L.B. Barnett, E.C. Akeson, and M.T. Davisson. 1990. A new dominant neurological mutant induced in the mouse by ethylene oxide. *Mutat. Res.* (in press).

Lewis, S.E., R.P. Erickson, L.B. Barnett, P.J. Venta, and R.E. Tashian. 1988. N-ethyl-N-nitrosourea-induced null mutation at the mouse *Car-2* locus: An animal model for human carbonic anhydrase II deficiency syndrome. *Proc. Natl. Acad. Sci.* **85:** 1962.

Lewis, S.E., L.B. Barnett, C. Felton, F.M. Johnson, L.C. Skow, N. Cacheiro, and M.D. Shelby. 1986. Dominant visible and electrophoretically-expressed mutations induced in male mice exposed to ethylene oxide by inhalation. *Environ. Mutagen.* **8:** 867.

Love, D.R. and W.E. Davies. 1989. Duchenne muscular dystrophy: The gene and the protein. *Mol. Biol. Med.* **6:** 7.

Roderick, T.H. 1983. Using inversions to detect and study recessive lethals and detrimentals in mice. In *Utilization of mammalian specific locus studies in hazard evaluation and estimation of genetic risk* (ed. F.J. de Serres and W. Sheridan), p. 135. Plenum Publishing, New York.

Rutledge, J.C., H.T. Cain, N.L.A. Cacheiro, C.V. Cornett, C.G. Wright, and W.M. Generoso. 1986. A balanced translocation in mice with a neurological defect. *Science* **231:** 395.

Searle, A.G. 1981. Numerical variants and structural rearrangements. In *Genetic variants and strains of the laboratory mouse* (ed. M.C. Green), p. 324. Fischer, Stuttgart.

COMMENTS

Cattanach: You did not include any of the inversions or the deletions like W^{19H}, $S2^{12H}$, or $S2^{18H}$. There is also the Ta^{25H} deletion of the X chromosome which, curiously, seems to enhance crossing-over. If you remember, way back I had a mutation called Snaker which looked like it had a two-gene basis but probably was an unbalanced product of a translocation.

Davisson: Many of you have been at this a lot longer than I have. Our sample just started out with screening induced mutations that have been published and are in the literature—those that were available to us at The Jackson Laboratory and those that were not necessarily already known to be small deletions. We didn't look at multiple series like all the mutations that have been induced in specific locus tests. Some mutations, like the tabby one you mentioned, are not available to us. I didn't really do an extensive survey of all the literature.

Cattanach: I wasn't suggesting you had missed them. I just meant there were other things.

Davisson: I know I probably missed some, or I haven't been able to get my hands on some. One of the most frustrating things is that we have about six at the lab in the freezer; I'm not going to spend $200 a mouse to get them out just to look, but when they come out we'll look.

Liane Russell: I wasn't quite sure what the Chr 17 translocation was, what the phenotype was; and the segmental trisomy, what was the phenotype of that?

Davisson: The translocation mouse has a neurological behavioral phenotype which is very similar to the circular mutations that have inner ear defects or cerebellar defects; it also has bilateral cataracts. One segmental trisomy we recovered has the identical phenotype. So, whatever mutation is causing it, the mutated gene is in that little translocation product.

Selby: I think it's quite likely that there are many dominant mutations that are associated with translocations. When I did my large breeding test experiment for dominant skeletal mutations, out of 31 transmitted mutations, at least 3 of them were translocations. Ilse-Dore Adler helped me with that. It is also interesting that when committees make risk estimates for translocations, they never have assumed any risk or included any numerical risk for the balanced individuals; it's always for the aneuploid segregants, whereas, indeed, there almost certainly is an important component of damage for balanced individuals.

Davisson: Sometimes people talk about chromosome aberrations causing a mutant phenotype. I don't know how you sort that out from a mutation in a gene that was affected by the breakpoint, but I think it's more likely that the breakpoint and not the translocation or the inversion itself causes the mutant phenotype. That is important because that breakpoint could allow you to get into the gene and clone it.

Lewis: I just wanted to add something that might interest Liane [Russell], since she asked about the phenotype. They do circle, but they do not have quite the same swimming defect as most of the mutants like shaker-1. They seem to be able to orient themselves in the water; it's the coordination that drags them down. It doesn't seem to be vestibular. They're not typical circlers, in other words.

Wyrobek: Some years ago Jack Bishop and I looked at the offspring of some triethylenemelamine-treated male mice and looked at their sperm abnormalities. We found some that had very high sperm abnormalities, up to 100%, and we found in about 15 animals that at least 10 of them were translocation carriers. So, we followed this dominant effect for a small handful of these further up to 5 generations, and they behave like dominant effects each time with the translocation mice having the high sperm abnormality. Those offspring in generations F2, F3, F4 that didn't have the translocation did not have the sperm abnormality.

Davisson: Did you hang on to them? Those would be interesting to look at.

Wyrobek: Unfortunately, we didn't. But I think it could be reproduced. It was a very striking proportion.

Lewis: The T2 Lws carriers are sub-fertile, but the original mutant was a male and his fertility may have been decreased because of problems during meiosis or whatever. Fertility is also decreased as far as litter size goes, because we have dissected a few matings of such males to +/+ females, and it looks like the embryos with the unbalanced component are dead early on.

Wyrobek: The translocation carriers did have reduced litter size.

Lewis: Probably for two reasons: problems with meiosis itself, and the fact that some of the unbalanced progeny die.

Handel: Are you suggesting that, in addition, the translocation breakpoints were perhaps in genes affecting sperm morphology?

Wyrobek: That's the interesting thing that is coming out of this talk. That's why I thought I would raise the issue, because it just was so clear. There was high concordance in subsequent generations between the presence of the translocation in meiosis and the high sperm abnormality.

Bishop: Two of them involved Chromosome 18, at least two of the ones that did get karyotyped.

Davisson: Probably with translocations you wouldn't miss induced aberrations because you would see the reduced litter size and you would begin to wonder about it. With inversions, however, there really isn't any effect. With some of them there is not an effect on litter size that you would pick up just by breeding studies, and so they can well be hidden and you wouldn't pick them up unless you looked.

Variables Affecting the Rate and Nature of Mutations

Factors Affecting Mutation Induction by X-rays in the Spermatogonial Stem Cells of Mice of Strain 101/H

BRUCE M. CATTANACH, CAROL RASBERRY, AND COLIN BEECHEY
MRC Radiobiology Unit
Chilton, Didcot, OX11 ORD
United Kingdom

OVERVIEW

Investigation of the factors underlying the differences in radiation response of the spermatogonial stem cells of certain inbred mouse strains could enhance understanding of the basis of the species differences that have been observed and thereby assist valid extrapolation of mouse mutation data to man. Data are presented that show that the high radiosensitivity of 101/H strain stem cells is not evident in other male germ-cell stages and that the 101/H mouse oocytes have a normal capacity to repair paternally induced genetic damage. Moreover, the low yield of translocations from irradiated 101/H stem cells that is associated with the high sensitivity to cell killing was not found when the heterogeneity in radiosensitivity of the stem cell populations was reduced. An intrinsic difference in radiosensitivity does not therefore appear to be responsible for the strain difference but, instead, 101/H mice may have a higher proportion of radiosensitive cells in their heterogeneous stem cell populations than most other strains.

INTRODUCTION

It is clear that the genetic response to ionizing radiation of the spermatogonial stem cells of different mammalian species varies considerably. Thus, with reciprocal translocations as the genetic endpoint, all species studied have shown or given clear indications of the humped dose-response curves first demonstrated in the mouse, but with peak yields being lower and occurring at lower doses (van Buul 1980). Such findings clearly hamper valid extrapolation to man of the wealth of mutation data obtained in the mouse. An understanding of the factors underlying the species differences in radiation response is needed.

One approach to this problem is to investigate the bases of strain differences in the mouse, the species most amenable to genetic and mutation research. Thus, Cattanach and Kirk (1987) have investigated the mutation and cell killing responses to X-rays of the spermatogonial stem cells of the 101/H inbred mouse strain. On the basis of duration-of-sterile-period and recovered-testis-weight data, they concluded that the stem cell populations of this strain were more sensitive to radiation killing than those of the C3H/HeH x 101/HF$_1$ hybrid (hereafter, 3H1) that is routinely used in mutation experiments in Harwell, Neuherberg, and Oak Ridge. Moreover, the translocation dose-response curve also differed, more closely resembling the curves obtained with other rodent species. Analysis of the data, using parameters defined by Leenhouts and Chadwick (1981), showed that the novel genetic response of 101/H mice could be accounted for either by a higher proportion of radiosensitive cells in the heterogeneous stem cell populations or by a higher ratio of cell killing to recoverable genetic damage. The latter could imply a reduced repair capacity.

In this paper, we present the results of a series of experiments designed to distinguish, first, if 101/H germ cell types other than spermatogonial stem cells have an enhanced sensitivity/reduced repair capacity relative to those of 3H1 animals and, second, if the stem cell difference is dependent on the heterogeneity in radiosensitivity of these cells.

Experiment 1 assayed dominant lethal induction by X-rays in 101/H and 3H1 postmeiotic male germ cells, after mating with 101/H and 3H1 females. The males were sequentially mated with females of both types, vaginal plugs being taken as evidence of mating. A 4-Gy dose was used for investigating the spermatozoal response (week 1) and a 3-Gy dose for late and early spermatids (weeks 2 and 3). In each treatment group, 16 males were used, and for each group, 16 unirradiated males provided control data. Experiment 2 assessed killing of spermatocytes and differentiated spermatogonia on the basis of testis weights and sperm counts 4, 5, 6, 7, and 8 weeks after a 3-Gy X-ray dose. At each dose, 7 or 8 males of each genotype were used, and 15 or 16 provided the unirradiated controls. Finally, Experiment 3 looked for a 101/H-3H1 difference in genetic response of spermatogonial stem cells in which the heterogeneity in radiosensitivity had been reduced by giving a small (1 Gy) priming dose 24 hours prior to the challenging dose of 5 Gy X-rays. A single 6-Gy dose given to males of each type provided positive control data. Each treatment group comprised ten males. These were killed at about 110 days postirradiation when their testes were weighed and their spermatocytes, prepared according to the method of Evans et al. (1964), were screened for translocations (100 cells/male). In all experiments, irradiations (250 kV; 14 mA; filter 0.25 mm Cu; HVT 1.2 mm Cu; 0.76 Gy/min) were limited to the posterior third of the body.

RESULTS

Experiment 1: Dominant lethal mutations, expressed as early postimplantation losses, were detected following irradiation of the postmeiotic germ cells of both 101/H and 3H1 males. The induced frequencies, calculated according to the formula of Kirk and Lyon (1984),

$$DL\% = 1 - \frac{[\text{live embryos/total implants (treated)}]}{[\text{live embryos /total implants (control)}]} \times 100$$

are shown together with the basic data in Table 1.

Several conclusions can be drawn. First, when realized in 3H1 females, dominant lethality was no higher following irradiation of 101/H, than of 3H1, male postmeiotic cells. Second, dominant lethal induction following 3H1 male irradiation clearly did not differ when realized in 101/H and 3H1 females. There was, therefore, no evidence of a reduced repair capacity in the oocytes of 101/H females. However, third, when dominant lethality, induced in 101/H and 3H1 males, was assayed in 101/H females, a significant difference was found in week 1 ($\chi^2_1 = 6.86$; $p = 0.00088$) and, with a consistent trend in the same direction in weeks 2 and 3, a highly significant difference overall ($\chi^2_1 = 7.25$; $p = 0.00071$) was indicated. Finally, when the dominant lethality induced in 101/H males was assessed in 101/H and 3H1 females, the data appeared to indicate that 101/H oocytes might indeed have a reduced repair capacity. Thus, the week-1 data from the 101/H male/101/H female cross were significantly different ($\chi^2_1 = 13.6$; $p = 0.00022$) from those of the corresponding 101/H male/3H1 female control; and over all germ cell stages, a highly significant difference was indicated ($\chi^2_1 = 12.7$; $p = 0.00036$). Because both groups of males were equally tested with both groups of females and the results at each mating interval were consistent, a technical artifact cannot be responsible for the high dominant lethal yields obtained in the pure-strain matings. The cause appears to lie within the 101/H strain itself; but, rather than indicating an enhanced sensitivity of 101/H germ cells, the data suggest that pure 101/H strain embryos are more likely to express potential dominant lethal damage than those of mixed genotype (101/H female x 3H1 male; 3H1 female x 101/H male; 3H1 female x 3H1 male). In view of this, the pure 101/H data can justifiably be excluded from Experiment 1, and it can be concluded from the remainder of the results that the postmeiotic germ cells of 101/H males are no more sensitive to chromosome damage than those of 3H1 males, and that the repair capacities of 101/H and 3H1 oocytes also do not differ.

Experiment 2. Figure 1 plots the absolute mean values for testis weight and sperm count obtained with 101/H and 3H1 males 4–8 weeks after 3-Gy X-irradiation. Table 2 presents the results as percentages of the control values. Both the 101/H and 3H1 testis weights were about 40% of normal at 4 weeks postirradiation; this suggests that there was a similar level of killing of spermatocytes and premeiotic germ cells in the two groups. At later times, the testis weights of both groups increased, as would be expected with repopulation of the germinal epithelium, but now the 101/H testis weights were proportionately smaller (Table 2). This could be interpreted to mean either that there is more extensive killing of the early meiotic and premeiotic cells in 101/H than in 3H1 mice, or that the reestablishment of normal stem cell numbers occurs later, or both. A later reestablishment of the stem

Table 1
Induction and Detection of Dominant Lethal Mutations in 101/H and 3H1 Mice

Week tested (germ cell)	Strain of male	X-ray dose	Strain of female	No. females mated	No. fertile females	No. live embryos	No. early deaths	Dominant lethality (% ± S.E.)
1 Spermatozoa	101/H	4 Gy	101/H	14	9	27	43	51.9 ± 7.5
		–		23	21	130	32	
		4 Gy	3H1	18	17	108	34	18.3 ± 4.2
		–		22	19	162	12	
	3H1	4 Gy	101/H	23	20	94	56	27.6 ± 5.0
		–		29	27	173	27	
		4 Gy	3H1	25	24	153	64	26.3 ± 3.4
		–		34	34	307	14	
2 Late spermatids	101/H	3 Gy	101/H	15	11	31	36	39.4 ± 8.5
		–		21	18	103	32	
		3 Gy	3H1	18	18	91	40	28.2 ± 4.3
		–		22	22	180	6	
	3H1	3 Gy	101/H	26	23	106	59	26.1 ± 4.8
		–		26	23	167	25	
		3 Gy	3H1	12	12	65	30	29.3 ± 5.0
		–		19	18	147	5	
3 Early spermatids	101/H	3 Gy	101/H	18	11	20	41	60.9 ± 7.4
		–		19	11	67	13	
		3 Gy	3H1	17	16	44	39	45.8 ± 5.6
		–		22	21	182	4	
	3H1	3 Gy	101/H	20	13	33	47	53.9 ± 6.3
		–		20	17	118	14	
		3 Gy	3H1	22	22	63	86	56.2 ± 4.2
		–		24	24	216	8	

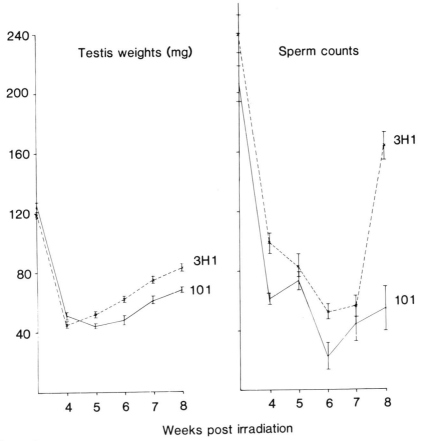

Figure 1
Change in testis weights and sperm counts of 101/H (shown as 101) and 3H1 mice following 3-Gy X-irradiation.

cell population would be consistent with the known facts on 101/H mice (Cattanach and Kirk 1987).

Sperm counts should not be influenced by repopulation of the germinal epithelium until at least 7–8 weeks postirradiation. They therefore provide a better measure of spermatocyte and differentiated-spermatogonial killing in weeks 4–5 and 6, respectively. On this basis, no consistent 101/H-3H1 difference in spermatocyte killing was indicated. There was a significant difference for differentiated spermatogonia at week 6, but the major difference was found with stem cells at 8 weeks (Table 2). The latter, again, is in accord with the earlier results on 101/H mice

Table 2
Measures of Meiotic and Premeiotic Cell Killing in 101/H and 3H1 Mice

Weeks postirrad.	Testis weight (% of control)					Sperm count (% of control)				
	101/H	3H1	t	d.f.	p	101/H	3H1	t	d.f.	p
4	41.5 ± 1.4	37.9 ± 0.9	2.15	17.9	0.46×10^{-1}	29.7 ± 2.7	41.2 ± 3.5	2.37	24.5	0.26×10^{-1}
5	35.9 ± 0.8	43.7 ± 0.9	6.45	33.3	0.25×10^{-6}	35.5 ± 3.8	34.0 ± 2.9	0.28	25.5	0.78
6	38.9 ± 2.6	52.3 ± 1.2	4.24	8.0	0.28×10^{-2}	10.8 ± 4.7	21.8 ± 2.4	2.56	6.5	0.40×10^{-1}
7	49.4 ± 2.7	63.0 ± 2.3	3.74	12.5	0.27×10^{-2}	21.6 ± 6.2	23.5 ± 3.4	1.03	6.6	0.34
8	54.6 ± 1.6	69.7 ± 2.2	5.50	17.3	0.30×10^{-4}	26.3 ± 7.4	68.4 ± 5.4	3.28	5.9	0.17×10^{-1}

Statistical analysis is based on a comparison of the logarithmically weighted means, using Satterthwaite's (1946) method of allowing differences between the within-sample variances.

(Cattanach and Kirk 1987). Interpretation of the week 6 and 7 findings is complicated by interanimal variation (range of counts, 4–80).

Experiment 3. Table 3 presents the recovered-testis-weight and translocation-yield data obtained following X irradiation of 101/H and 3H1 males with single, acute 6-Gy and 24-hour fractionated 1 + 5-Gy doses. The 6-Gy testis weights of both groups of males were lower than those found in the earlier study (Cattanach and Kirk 1987). Nevertheless, a bigger reduction was seen with the 101/H than with 3H1 mice ($t = 4.61$; $p = 0.000044$), demonstrating the greater sensitivity of 101H spermatogonial stem cell populations to radiation killing. The fractionated dose could also be so deduced to cause more stem cell killing in 101/H than 3H1 mice ($t = 5.12$; $p = 0.011$), but the data do not allow the distinction of whether the greater effect is due to the first dose, the second, or both, and therefore do not help to resolve the basis of the higher sensitivity of 101/H mouse stem cell populations. It is known that 1-Gy X-rays bring about a sterile period in 101/H mice of a duration similar to that caused by 4–5 Gy in 3H1 animals (Cattanach and Kirk 1987). Therefore, a significant proportion of the enhanced stem cell killing response of 101/H mice to the 1 + 5-Gy fractionated dose could be attributed to the first treatment.

The translocation data are potentially more informative. The results presented in Table 3 show that with the split 1 + 5-Gy X-ray dose the translocation yield from 101/H males was no lower than that from 3H1 males. This generally supports the results of the earlier study, which gave a similar type of response but could not be properly evaluated because of the lack of a concurrent 3H1 control (Cattanach and Kirk 1987). The validity of the similar 101/H and 3H1 responses to the split dose in the present experiment, however, is somewhat reduced by the fact that the 6-Gy responses of the two groups of males did not differ ($\chi^2_1 = 0.21$; $p = 0.65$). Clearly, this experiment will need to be repeated, but, given the consistency of the lower 101/H translocation yields following single X-ray doses in the earlier study (Cattanach and Kirk 1987), together with the fact that strain differences in response

Table 3
Response of 101/H and 3H1 Males to Fractionated and Single X-ray Doses

Treatment	101/H males		3H1 males	
	mean testis weight (mg)	cells with translocations (% ± S.E.M.)	mean testis weight (mg)	cells with translocations (% ± S.E.M.)
1 ± 5 Gy (24-hr interval)	69.6	20.8 ± 1.3	87.0	17.9 ± 1.2
6 Gy	77.9	8.4 ± 0.9	87.9	9.0 ± 0.9

should be more evident when translocation yields are high, it is probable that the present split-dose result is valid. If correct, this could mean that the differences between 101/H and 3H1 males in stem cell killing and translocation yield following X-irradiation are due not to a different sensitivity within the cells themselves, but rather to a difference in the proportion of sensitive and resistant cells within the heterogeneous stem cell populations.

DISCUSSION

The 101/H mouse has been reported to have a higher spontaneous incidence of chromosome aberrations in bone marrow cells than a number of other strains and to give higher frequencies of such changes following chemical mutagen treatments (Surkova and Malashenko 1975a,b). In addition, females of a similarly sensitive strain (C57BL/6JY) have permitted a higher realization of dominant lethals caused by paternal mutagen treatment than have females of several other strains (Malashenko and Surkova 1975). On this basis, it might be expected that the novel translocation and cell killing responses of 101/H spermatogonial stem cells following radiation exposure would be attributable to a generalized intrinsic sensitivity or reduced repair capacity.

The data presented here give little to support this expectation. Dominant lethality was no greater following irradiation of 101/H than of 3H1 postmeiotic cells; 101/H females gave no indication of reduced repair capacities compared with 3H1 females; there was no evidence that 101/H spermatocytes were more sensitive to killing than those of 3H1 males. Only with differentiated spermatogonia was there evidence of a difference, and this was complicated by high interanimal variation. The strain sensitivity differences may therefore be confined to the stem cell population and, here again, the similar 101/H and 3H1 responses to the 24-hour split dose tend to refute the concept that some intrinsic difference in radiosensitivity of the stem cells is responsible for the altered translocation and cell killing responses found with single X-ray doses. Therefore, it now seems more likely that the cause of the novel radiosensitivity of 101/H stem cell populations is that they contain a higher proportion of radiosensitive cells and a smaller radioresistant component. Further work could include the use of a chemical mutagen, which kills stem cells but does not induce recoverable chromosome damage, as a priming dose in translocation experiments. The possibility of investigating specific stages of the stem cell cycle through the use of combined hydroxyurea-X-ray experiments could also prove informative. Beyond this, it is evident from the literature (Meistrich et al. 1984; Bianchi et al. 1985; Generoso et al. 1985) that other strains could be used to investigate other factors that modify the genetic response of spermatogonial stem cells to radiation (and other environmental agents) and so facilitate valid extrapolation of mouse data to man for risk assessment.

There is also the additional observation made in this study that genetic damage induced by paternal radiation exposure is more likely to result in dominant lethality in inbred 101/H embryos than in those of some other strains and crosses. A strain with a high sensitivity to dominant lethal induction would be valuable for mutagen screening. The basic observation, however, also raises a risk issue; in more resistant strains a proportion of potentially lethal damage may be transmitted to later generations and cause abnormality or deleterious effects. Remarkably, essentially the same finding was made over 20 years ago in A/J mice (Storer 1967) but appears to have been ignored.

ACKNOWLEDGMENTS

I thank Mr. D. Papworth for his help with the statistical analysis of the data. This work was funded in part by the CEC Radiation Protection Programme Contract B16-E-143-UK.

REFERENCES

Bianchi, M., J.I. Delic, G. Hurtado-de-Catalfo, and J.H. Hendry. 1985. Strain differences in the radiosensitivity of mouse spermatogonia. *Int. J. Radiat. Biol.* **48**: 579.

Cattanach, B.M. and M.J. Kirk. 1987. Enhanced spermatogonial stem cell killing and reduced translocation yield from X-irradiated 101/H mice. *Mutat. Res.* **176**: 69.

Evans, E.P., G. Breckon, and C.E. Ford. 1964. An air-drying method for meiotic preparations from mammalian testes. *Cytogenetics* **3**: 289.

Generoso, W.M., K.T. Cain, C.V. Cornett, N.L.A. Cachiero, L.A. Hughes, and P.W. Braden. 1985. Difference in the response of two hybrid stocks of mice to X-ray induction of chromosome aberrations in spermatogonial stem cells. *Mutat. Res.* **152**: 217.

Kirk, M. and M.F. Lyon. 1984. Induction of congenital malformations in the offspring of male mice treated with X-rays at pre-meiotic and post-meiotic stages. *Mutat. Res.* **125**: 75.

Leenhouts, H.P. and K.H. Chadwick. 1981. An analytical approach to the induction of translocations in the spermatogonia of the mouse. *Mutat. Res.* **82**: 305.

Malashenko, A.M. and N.I. Surkova. 1975. The mutagenic effect of thio-TEPA in laboratory mice. V. Influence of the genotype of females on the realization of dominant lethal mutations induced in spermatids of males. *Soviet Genet.* **11**: 210.

Meistrich, M.L., M. Finch, C.C. Lu, A.K. de Ruiter-Bootsma, and D.C. de Rooij. 1984. Strain differences in the response of mouse testicular stem cells to fractionated radiation. *Radiat. Res.* **97**: 478.

Satterthwaite, F.E. 1946. An approximate distribution of estimates of variance components. *Biometrics Bull.* **2**: 110.

Storer, J.B. 1967. On the relationship between genetic and somatic sensitivity to radiation damage in inbred mouse strains. *Radiat. Res.* **31**: 699.

Surkova, N.I. and A.M. Malashenko. 1975a. Mutagenic effect of thio-TEPA in laboratory mice. IV. Influence of the genotype and sex on the frequency of induced chromosome aberrations in bone marrow. *Soviet Genet.* **11**: 49.

———. 1975b. Mutagenic effect of mitomycin C and cytosine arabinoside in mice of different genotypes. *Soviet Genet.* **11:** 1106.

van Buul, P. 1980. Dose-response relationships for X-ray-induced reciprocal translocations in stem-cell spermatogonia of the Rhesus monkey *(Macaca mulatta). Mutat. Res.* **73:** 363.

COMMENTS

Lewis: Just a comment. We have, somewhat to our frustration, found that when we have used both B6 and DBA in our system, B6 has more trouble recovering from mutagens—less fertile and all. However, it seems to be, as far as mutation induction, the better strain. But we have ended up using DBA exclusively because it just isn't practical to do the other cross.

Cattanach: Yes, and there are Russian workers who, as I have indicated, looked at B6 as well and indicated that it showed 101-type response. Then there is the radiation work, by Bianchi and others, looking at the population indexes with B6, that has shown that strain is particularly sensitive.

Generoso: The length of the sterile period is at best an indirect measure of stem cell killing. One can imagine there are several components that can go into the response. One is the resistance to irradiation and the other is the ability for recovery. Those two may not be necessarily independent of one another.

Cattanach: Well, that's true for any measure.

Generoso: So why can't we have a direct measure of the stem cells for the A_s counts before and after? It seems to me we need to have a baseline idea of the number of those cells before irradiation in two different strains.

Cattanach: Yes. There are lots of problems in here, of course. As you, I'm sure, are very well aware, there are a whole lot of different methods being used, from sterility, testis weights, LDH assays, sperm counts, the population index method, and others that don't come to mind immediately. They all have advantages and disadvantages. Sterile period has the advantage, as far as I'm concerned, of being easy to measure. It's there, it's obtainable; all you've got to do is to make a note of it. In addition, you can get measures at low radiation doses. Whereas, when you start looking at population indexes you have to go to higher doses before you can assess effects. So, I think each of the methods has difficulties.

I agree, it would be nice to know exactly what the situation is, but there is a problem, it seems to me, of distinguishing exactly what is happening. I don't mean so much distinguishing stem cells from the other kinds of spermatogonial cells that are there, which is some degree of a problem. Rather that once you have depleted the population and got rid of most of the differentiated cells, you have an abnormal testis. Recognizing then, what you have in the way of stem cells, they are then going to behave differently at different times postirradiation; and how are you going to recognize those? This is a question that Dirk de Rooij would answer better than I. There are a great many complications in obtaining really good measures of what has happened.

de Rooij: I can add to that. You can count the A_s spermatogonia before irradiation. After irradiation, it takes 8 days at least before all the lethally irradiated stem cells have died. During that period, the ones that did survive and started to proliferate have already formed new A_s spermatogonia, and maybe even paired. Therefore, you cannot directly count the A_s spermatogonia after irradiation, but you can take the number of differentiated spermatogonia present at day 10 as a measure of the number of surviving stem cells, and then you can go with low doses of X-irradiation.

Lonnie Russell: It seems to me that there are literally hundreds of studies on this particular topic in the literature, but not one that addresses what else could happen in the testis besides stem cell killing, especially at a high resolution level morphology. Is that correct?

Cattanach: I think it's largely true, but not totally true. There certainly are other effects besides just the stem cell killing.

Generoso: Let me give a more important point. You are comparing a hybrid with an inbred. Being an inbred devotee, it seems to me that they may be on the threshold already but their baseline may be different.

Cattanach: I don't think that's true. The 101/H male is really quite a good breeding animal, in our hands at least. The one other strain I mentioned, JU, which has a dramatic response to radiation, is one of our most fertile stocks.

Generoso: It might be in terms of their breeding ability, but what do we know about their baseline? What do we know about the level that will make them fertile again? The distances that the two strains must recover to become fertile again may be quite different.

Cattanach: All that you are saying is probably perfectly correct, but if it is going to be true between one mouse strain and another, it is surely also going to be true between one species and another; and this is, after all, what one is seeking to investigate.

Favor: Coming back to the dominant lethal data, where you had either 101s or hybrids, it almost looks as if there is a recessive mutation in strain 101. One could interpret it that way. In other words, you only get the effect when you treat 101 and mate to 101?

Cattanach: I find it hard to imagine how a single gene could bring about a general enhancement of probabilities of induced genetic damage resulting in dominant lethality. It may just be that the more vigorous hybrid embryos have a better chance of surviving deleterious changes than those of the inbred. The inbred has a high postimplantation loss, about 20%. I find it amazing that the radiation damage can be detected with this high background loss. The situation is complex, however. About half of the observed loss is attributable to a maternal effect and half to the embryos themselves.

Toward an Understanding of the Mechanisms of Mutation Induction by Ethylnitrosourea in Germ Cells of the Mouse

JACK FAVOR
GSF-Institut für Säugetiergenetik
D-8042 Neuherberg
Federal Republic of Germany

OVERVIEW

Specific-locus mutation results for ethylnitrosourea (ENU) represent the most extensive data base for a chemical mutagen in germ cells of the mouse and allow an examination into the mechanisms of mutation induction in vivo in mammals. Both a dose-response analysis of mutation induction by ENU and a characterization of ENU-induced mutations are presented. Analyses indicate a thresholded dose response due to a saturable repair system in germ cells of the mouse. A high frequency of induced mutation mosaics suggests that the ENU-induced lesions are confined to a single DNA strand and may persist a long period even in repair-competent cells. Finally, the molecular characterization of ENU-induced mutations in germ cells of the mouse is inconsistent with a mechanism of mutagenesis involving O^6-ethylguanine. This is in contrast to results in bacterial systems and emphasizes the complexity of in vivo mammalian mutagenesis.

INTRODUCTION

ENU has been shown to be a potent mutagen in spermatogonia of the mouse for the induction of recessive specific-locus mutations (Russell et al. 1979; Ehling et al. 1982). Due to the potency of ENU, mutagenesis experiments in germ cells of the mouse could be realistically undertaken to explore factors affecting the mutational process. The initial ENU spermatogonial treatment experiments at 250 mg/kg were extended to additional doses (Russell et al. 1982b; Ehling and Neuhäuser-Klaus 1984), and the effects of dose fractionation were studied (Russell et al. 1982a; Ehling and Neuhäuser-Klaus 1984; Hitotsumachi et al. 1985; Favor et al. 1988). Furthermore, the ENU-induced mutation rate was estimated in postspermatogonial stages (Favor et al. 1990a), oocytes (Ehling and Neuhäuser-Klaus 1988), and early postfertilization zygotic stages (Russell et al. 1988). Finally, the mutagenic effectiveness of ENU has been studied for alternative genetic endpoints such as dominant cataract mutations and enzyme-charge or -activity mutations (Johnson and

Lewis 1981; Ehling et al. 1982, 1985; Favor 1983, 1986a,b; Peters et al. 1986; Charles and Pretsch 1987). Together these results represent the most extensive data base for a chemical mutagen in germ cells of the mouse.

To provide insight toward an understanding of the mechanisms of mutation induction by ENU in mammalian germ cells, an analysis to establish the relationship between the level of mutagen exposure and the extent of mutational response as well as a characterization of recovered mutations was undertaken.

RESULTS

Specific-locus mutagenicity data available to analyze the relationship between the level of ENU exposure and the extent of mutational response are presented in Table 1. Where clustering occurred, the number of independent mutational events is also presented. However, all calculations were based on the total number of recovered mutations, since this provides an unbiased estimate of the mutation rate (Searle 1974; Engels 1979; Russell et al. 1982b; Cattanach et al. 1985). The mutation-rate data represent a quantal event that may be best described by a binomial distribution. The binomial parameter, p, is the probability of a mutational event and is related to the level of ENU exposure. The following alternative models were chosen for regression analysis:

Linear	$p_i = a + b * D_i$	
Linear-quadratic	$p_i = a + b * D_i + c * D^2_i$	
Power	$p_i = a + b * D^c_i$	
Threshold	$p_i = a$	for $D_i < c$
	$ = a + b * (D_i - c)$	for $D_i \geq c$

where D_i is the dose of ENU in the ith group.

The first three models represent the low-dose behavior of commonly used dose-response models for quantal response toxicity data (Armitage 1982; Van Ryzin 1982). Thus, the linear model corresponds to the one-hit model, the linear-quadratic model corresponds to the multistage model, and the power model corresponds to the multihit model as well as to the various tolerance distribution models, Probit, Logit, and Weibull. The threshold model is a modification of the linear model to allow for a dose region of no effect, also known as the no-effect level model (NOEL). The threshold model implies a saturable process intervening between mutagen exposure and the fixation of DNA adducts to mutations, such as mutagen detoxification or repair of DNA damage. The nonlinear regression program BMDPAR (Dixon 1983) was employed to obtain maximum-likelihood estimates of the model parameters by maximizing the binomial likelihood function with an iteratively weighted regression technique using the usual binomial weights of the data points, the reciprocal of the

Table 1
ENU-induced Specific-locus Mutations in Spermatogonia of Mice

Dose (mg/kg)	Mutants[a]	Offspring
0	19 (13)	227,805
40	3	11,410
80	20	13,274
160	35 (32)	8,658
250	64 (58)	9,766

[a]The number of independent mutational events is given in parentheses. (Reprinted, with permission, from Ehling and Neuhäuser-Klaus 1984.)

variance. To evaluate the alternative models, the goodness-of-fit likelihood-ratio statistic for generalized linear models was calculated, often designated the deviance statistic (McCullagh and Nelder 1983), which may be compared to the chi-square distribution.

Table 2 gives the observed per-locus mutation rates following ENU spermatogonial treatment, as well as the mutation rates predicted by the models fitted. In addition, the estimated parameters and the standard error of the estimates are presented. The linear model was shown to be inadequate in describing the relationship between mutagen dose and mutation rate. In comparison to the linear model, which estimates two parameters, all remaining models estimate three parameters

Table 2
Regression Analyses of the Frequency of ENU-induced Specific-locus Mutations in Spermatogonia of Mice

Dose (mg/kg)	Observed MR x 10^5	Predicted MR			
		linear[a]	lin-quad[b]	power[c]	threshold[d]
0	1.2	1.1	1.2	1.2	1.2
40	3.8	13.9	8.1	7.4	3.8
80	21.5	26.6	18.4	18.8	21.3
160	57.8	52.1	49.4	51.2	56.1
250	93.6	80.8	100.5	99.3	95.3
Deviance statistic		11.03	3.93	2.79	0.05

(MR) Mutation rate per locus. (Reprinted, with permission, from Favor et al. 1990b.)
[a]Predicted MR x 10^{-5} = $(1.1 \pm 0.3) + (0.3 \pm 0.03) \times D$
[b]Predicted MR x 10^{-5} = $(1.2 \pm 0.3) + (0.1 \pm 0.06) \times D + ([1.1 \pm 0.4] \times 10^{-3}) \times D^2$
[c]Predicted MR x 10^{-5} = $(1.2 \pm 0.3) + (0.02 \pm 0.02) \times D^{(1.5 \pm 0.2)}$
[d]Predicted MR x 10^{-5} = (1.2 ± 0.3) : for D < 34
  = $(1.2 \pm 0.3) + (0.4 \pm 0.05) \times (D-[34\pm5.1])$: : for D ≥ 34

and, as expected, provide a better fit of the data than the two-parameter linear model. The linear-quadratic and power function models both yielded similar deviance statistic values, 3.93 and 2.79, respectively. The fit for the threshold model resulted in a deviance statistic value of 0.05, indicating this model to be the far superior model. Indeed, a comparison of the observed and predicted mutation rates indicates minimal deviations.

Similar results for specific-locus mutagenicity experiments following ENU treatment of spermatogonia obtained at Oak Ridge (Russell et al. 1982b) were also subjected to regression analyses (Favor et al. 1990b). A thresholded model also proved to be the best fit. More important, the threshold parameter (39 mg/kg ENU) was similar to the value fitted for the Neuherberg data (34 mg/kg ENU).

The concept of a thresholded response to mutation induction implies two biologically significant parameters: (1) The threshold dose is that dose below which induced DNA lesions are repaired or the mutagen is detoxified. (2) The recovery time is that time required for the cells to reestablish the functional repair or detoxification system to constitutive levels. Pertinent information from ENU fractionation treatments are given in Figure 1, along with the threshold curve obtained in Table 2 for single acute ENU treatment, extrapolated to a dose of 400 mg/kg. From fractionation experiments in which the multiply applied dose was within the linear portion of the dose-response curve and the fraction interval was long (3 x 100 mg/kg or 4 x 100 mg/kg with an interval between treatments of 1 week), a simple additive effect was observed as expected. When the fractionation interval for a multiply applied dose of ENU was short (2 x 80 or 4 x 40 mg/kg with a 24-hour interval between treatments), the observed mutation rate was similar to the application of an unfractionated total dose. Finally, when the multiply applied dose was lower than the threshold dose and the fractionation interval was long (10 x 10 mg/kg with a 1-week interval between treatments), the mutagenic response was lower than that observed for an unfractionated application of the total dose. Together, these observations indicate the recovery period to be longer than 24 hours but shorter than 1 week. However, the observed mutation rate following 10 x 10 mg/kg treatment with a 1-week interval between exposures was higher than the control rate, although the multiply applied dose of ENU was below the threshold dose and the fractionation interval was long enough to allow recovery of the saturable process. These results suggest that not every induced DNA lesion that may ultimately be fixed as a mutation immediately induces a repair response, and that the threshold model is probably an oversimplification of the process of mutation induction in spermatogonia of mice.

Stem-cell spermatogonia are mitotically active, such that the processes of DNA replication and cell proliferation also intervene between mutagen exposure and mutation fixation. A more accurate understanding of mutation induction may be obtained from results of ENU treatment of postspermatogonial stages, since the complications of treating a mitotically active cell population are eliminated. Table 3

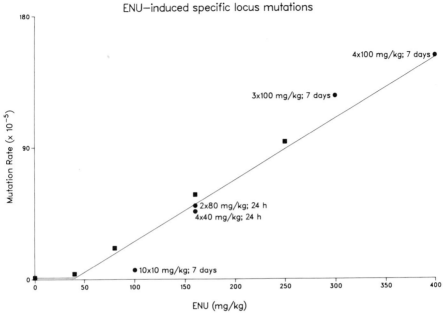

Figure 1
Dose-response curve of specific-locus-mutation induction by acute, unfractionated exposure of spermatogonia to ENU in the mouse (boxes). The fitted curve has been extrapolated to a total dose of 400 mg/kg, and experimental results for fractionated ENU exposure have been plotted (circles). Original data are from acute, unfractionated dose points as well as 4 × 40 mg/kg, Ehling and Neuhäuser-Klaus (1984); 2 × 80 mg/kg, Favor et al. (1988); 10 × 10 mg/kg, Russell et al. (1982a); 3 × 100 and 4 × 100 mg/kg, Hitotsumachi et al. (1985).

summarizes the results of extensive experiments to determine the mutagenic effects of ENU treatment in postspermatogonial stages. Of importance is that the mutations mainly occurred as mosaics, which suggests that the DNA lesions ultimately fixed as a mutation were confined to a single DNA strand. Figure 2 indicates the stages of spermatogenesis in which mutation mosaics were induced along with the activity of the various cell stages for DNA synthesis and DNA repair. It is evident that ENU-induced mutation mosaics occurred in all postspermatogonial cell stages. It is important to note that some mutation mosaics occurred in cell stages that are repair competent. This observation supports the suggestion from low-dose fractionation experiments that ENU-induced DNA lesions may not immediately invoke a DNA repair response and that the DNA lesions may persist before being fixed as a mutation. Indeed, some mutation mosaics were shown to be carried in only a small fraction of the germ line. Considering the probabilities of inclusion of mutation-bearing cells following fertilization, cell proliferation and the differentiation of the

Table 3
ENU-induced Whole-body and Mosaic Mutants Derived from Treated Postspermatogonial Stages of the Mouse

Dose (mg/kg)	Offspring	Mutants			
		dominant cataract		specific locus	
		whole-body	mosaic	whole-body	mosaics
0	22,594	1	0	19	1
2 × 80	4,215	0	1	1	2
160	7,983	4	1	0	4
250	3,344	1	1	0	3

(Reprinted, with permission, from Favor et al. 1990a.)

germ line would suggest that the ENU-induced DNA lesions may persist through a number of cleavage divisions before eventually being fixed as a mutation.

DISCUSSION

The decision whether a threshold exists usually precipitates controversy and is most critical when assessing the toxic, mutagenic, teratogenic, or carcinogenic risks associated with human exposure at low doses (see, e.g., Freese 1973; Gehring 1977; Rall 1978; Park and Snee 1983; Gaylor 1985; Cox 1987). Given the variability inherent in animal experimentation, an alternative explanation to a thresholded response of the observed dose effects often invoked is that at lower doses, experiments of the power necessary to demonstrate an effect could not be or were not carried out. For the presently discussed ENU-induced specific-locus mutation-rate data in spermatogonia of the mouse, the situation is different. Two laboratories have independently shown ENU to be relatively less effective in the low dose range tested, but there was a positive effect (Russell et al. 1982b; Ehling and Neuhäuser-Klaus 1984). Furthermore, a low dose of ENU repeatedly applied also resulted in a reduced but positive effect (Russell et al. 1982b). Thus, the power of the experiments performed was great enough to demonstrate an effect, albeit a reduced effect. Of the dose-response models employed in regression analysis of the data, a threshold model proved to be far superior to the others. The level of DNA ethylation in mouse testis was shown to increase following ENU exposure even at a low dose (Sega et al. 1986), which leads to the conclusion that the saturable process responsible for the thresholded mutagenicity response is repair rather than detoxification. Considering the occurrence of ENU-induced mutation mosaics, the ENU-induced DNA lesion ultimately fixed as a mutation is confined to a single strand and may persist for a relatively long period even in repair-competent cell stages. The high frequency of intermediate alleles and the low frequency of homo-

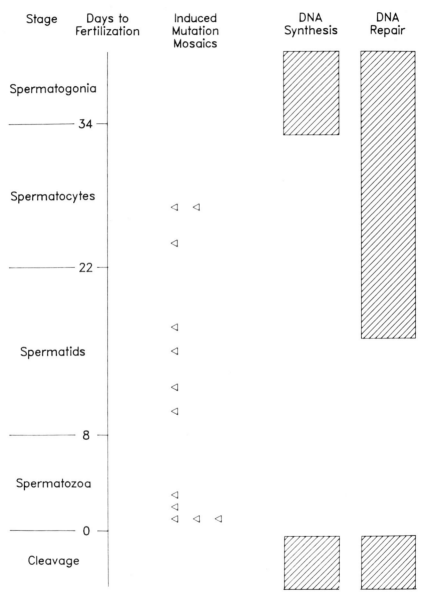

Figure 2
Occurrence of ENU-induced specific-locus mutation mosaics in postspermatogonial germ-cell stages of the mouse. The stages of spermatogenesis as well as of DNA replication and DNA repair competency are given. Adapted from Oakberg (1956), Monesi (1962), and Sega (1974).

zygous lethal alleles recovered in specific-locus experiments in which mutations were induced in spermatogonia of mice with ENU (Russell et al. 1979; Ehling and Favor 1984; Favor et al. 1989) support the hypothesis that in spermatogonia a similar mechanism of ENU mutation induction exists.

The O^6-ethylguanine DNA adduct has been suggested to be the relevant adduct that eventually leads to a mutation (Loveless 1969; Van Zeeland et al. 1985). The repair of the O^6-ethylguanine adduct has been shown to be accomplished by a saturable alkyltransferase (Karran et al. 1979), and indeed the level of O^6-ethylguanine DNA adduct formation following ENU treatment in the mouse testis was shown to be less efficient at lower doses due to this saturable repair system (Van Zeeland et al. 1985; Sega et al. 1986). It has also been shown that the O^6-ethylguanine DNA adduct has a long persistence in mammalian cells (Goth and Rajewsky 1974a,b; Singer et al. 1981; Sega et al. 1986). Thus, the formation, repair, and persistence of O^6-ethylguanine DNA adduct as well as the correlation of the levels of O^6-ethylguanine adduct and the observed mutation rate are consistent with a threshold model for ENU-induced mutation and with the observation of mutations occurring as mosaics in the mouse.

Six independent ENU-induced germ-cell mutations in the mouse have been characterized, and all were shown to be transitions or transversions of an AT base pair (Popp et al. 1983; Lewis et al. 1985; Peters et al. 1985; E. Zdarsky et al., in

Table 4
Extent and Consequences of DNA Alkylation by ENU

Site	DNA alkylated (% of total DNA)	Mispairing	Base Substitution
7-Gua	12.5	Neg	
O^6-Gua	8.7	T	GC→AT
3-Gua	0.6	Neg	
3-Ade	3.9	Neg	
1-Ade	0.3	A, G, C	AT→TA
			AT→CG
			AT→GC
7-Ade	0.4	Neg	
O^2-Thy	7.4	C	AT→CG
O^4-Thy	2.6	G, C	AT→GC
			AT→CG
3-Thy	n.d.	Neg	
O^2-Cyt	3.4	A	CG→TA
3-Cyt	0.2	A, T, C	CG→TA
			CG→AT
			CG→GC

Adapted from Montesano (1981) and Singer et al. (1978). n.d. indicates not determined.

prep.; S. Lewis; J. Peters; both pers. comm.). This is in contrast to the mutational events characterized in *Escherichia coli* (Richardson et al. 1987) and *Salmonella typhimurium* (Zielenska et al. 1988) and indicates differences in the mechanism of ENU mutagenesis in bacteria and mammalian germ cells. Table 4 indicates the extent of base ethylation following ENU treatment and the consequences of the various ethylated bases. Although the O^6-ethylguanine adduct is the most prevalent of the DNA lesions that lead to a base substitution, a number of other DNA adducts are induced by ENU, some of which also result in base-pair substitution. In particular, 1-ethyladenine and O^4-ethylthymine adducts result in a substitution of an AT base pair. Although the O^6-ethylguanine base alkylation product has been shown to be the mutagenically significant DNA adduct resulting in a GC→AT transition in *S. typhimurium*, Zielenska et al. (1988) have shown that both the exposure level to ENU and the presence or absence of the pKM101 plasmid, which imparts error-prone DNA repair competence, affect the spectrum of ENU-induced mutational events. In the presence of error-prone repair, a significant fraction of the recovered mutations were base substitutions involving an AT base pair. This situation may be compared to a mitotically active population of spermatogonia in which error-prone repair would be expected if ENU-induced DNA lesions were repaired during a round of DNA synthesis. The question remains, however, why no ENU-induced mutations recovered to date in mouse germ cells involve a GC→AT transition. A characterization of additional ENU-induced germ-cell mutations in the mouse is of considerable importance to determine if the spectra of ENU-induced mutants are indeed as widely variant between bacteria and mammals as the present data indicate.

If DNA alkylation products other than O^6-ethylguanine should be the relevant DNA adducts for ENU mutagenesis, then a relatively long persistence and a saturable repair system for the DNA adducts would be predicted from the germ-cell mutagenicity results in the mouse. Indeed, the persistence time in vivo for three alternative base ethylations, O^2-ethylthymine, O^4-ethylthymine, and O^2-ethylcytosine, which are induced at significant levels by ENU and which lead to base mispairing, is relatively long (Singer et al. 1981). However, to date, studies to demonstrate a repair process for these adducts in mammalian cells have been negative (Brent et al. 1988).

ACKNOWLEDGMENTS

This research was supported in part by research contract EV4V-00640D (B) of the Commission of the European Communities.

REFERENCES

Armitage, P. 1982. The assessment of low-dose carcinogenicity. *Biometrics* (suppl. *Current topics in biostatistics and epidemiology*):119.

Brent, T.P., M.E. Dolan, H. Fraenkel-Conrat, J. Hall, P. Karran, F. Laval, G.P. Margison, R. Montesano, A.E. Pegg, P.M. Potter, B. Singer, J.A. Swenberg, and D.B. Yarosh. 1988. Repair of O-alkylpyrimidines in mammalian cells: A present consensus. *Proc. Natl. Acad. Sci.* **85**: 1759.

Cattanach, B.M. and C. Jones. 1985. Specific-locus mutation response to unequal, 1+9 Gy X-ray fractionations at 24-h and 4-day fraction intervals. (With Appendix by D.G. Papworth.) *Mutat. Res.* **149**: 105.

Charles, D.J. and W. Pretsch. 1987. Linear dose-response relationship of erythrocyte enzyme-activity mutations in offspring of ethylnitrosourea-treated mice. *Mutat. Res.* **176**: 81.

Cox, C. 1987. Threshold dose-response models in toxicology. *Biometrics* **43**: 511.

Dixon, W.J. 1983. *BMDP statistical software.* University of California Press, Berkeley.

Ehling, U.H. and J. Favor. 1984. Recessive and dominant mutations in mice. In *Mutation, cancer, and malformation* (ed. E.H.Y. Chu and W.M. Generoso), p. 389. Plenum Press, New York.

Ehling, U.H. and A. Neuhäuser-Klaus. 1984. Dose-effect relationships of germ-cell mutations in mice. In *Problems of threshold in chemical mutagenesis* (ed. Y. Tazima et al.), p. 15. Kokusai-bunken, Tokyo.

———. 1988. Induction of specific-locus mutations in female mice by 1-ethyl-1-nitrosourea and procarbazine. *Mutat. Res.* **202**: 139.

Ehling, U.H., J. Favor, J. Kratochvilova, and A. Neuhäuser-Klaus. 1982. Dominant cataract mutations and specific-locus mutations in mice induced by radiation or ethylnitrosourea. *Mutat. Res.* **92**: 181.

Ehling, U.H., D.J. Charles, J. Favor, J. Graw, J. Kratochvilova, A. Neuhäuser-Klaus, and W. Pretsch. 1985. Induction of gene mutations in mice: The multiple endpoint approach. *Mutat. Res.* **150**: 393.

Engels, W.R. 1979. The estimation of mutation rates when premeiotic events are involved. *Environ. Mutagen.* **1**: 37.

Favor, J. 1983. A comparison of the dominant cataract and recessive specific-locus mutation rates induced by treatment of male mice with ethylnitrosourea. *Mutat. Res.* **110**: 367.

———. 1986a. The frequency of dominant cataract and recessive specific-locus mutations in mice derived from 80 or 160 mg ethylnitrosourea per kg body weight treated spermatogonia. *Mutat. Res.* **162**: 69.

———. 1986b. A comparison of the mutation rates to dominant and recessive alleles in germ cells of the mouse. *Prog. Clin. Biol. Res.* **209B**: 519.

Favor, J., A. Neuhäuser-Klaus, and U.H. Ehling. 1988. The effect of dose fractionation on the frequency of ethylnitrosourea-induced dominant cataract and recessive specific locus mutations in germ cells of the mouse. *Mutat. Res.* **198**: 269.

———. 1990a. The frequency of dominant cataract and recessive specific locus mutations and mutation mosaics in F_1 mice derived from post-spermatogonial treatment with ethylnitrosourea. *Mutat. Res.* (in press).

Favor, J., A. Neuhäuser-Klaus, J. Kratochvilova, and W. Pretsch. 1989. Towards an understanding of the nature and fitness of induced mutations in germ cells of mice: Homozygous viability and heterozygous fitness effects of induced specific-locus, dominant cataract and enzyme-activity mutations. *Mutat. Res.* **212**: 67.

Favor, J., M. Sund, A. Neuhäuser-Klaus, and U.H. Ehling. 1990b. A dose response analysis of ethylnitrosourea-induced recessive specific locus mutations in treated spermatogonia of the mouse. *Mutat. Res.* (in press).

Freese, E. 1973. Thresholds in toxic, teratogenic, mutagenic and carcinogenic effects. *Environ. Health Perspect.* (exp. issue) **6:** 171.

Gaylor, D.W. 1985. The question of the existence of thresholds: Extrapolation from high to low dose. In *Mechanisms and toxicity of chemical carcinogens and mutagens* (ed. W.G. Flamm and R.J. Lorentzen), p. 249. Princeton Scientific, Princeton, New Jersey.

Gehring, P. 1977. The risk equations—The threshold controversy. *New Sci.* **75:** 426.

Goth, R. and M.F. Rajewsky. 1974a. Persistence of $O6$-ethylguanine in rat-brain DNA: Correlation with nervous system-specific carcinogenesis by ethylnitrosourea. *Proc. Natl. Acad. Sci.* **71:** 639.

———. 1974b. Molecular and cellular mechanisms associated with pulse-carcinogenesis in the rat nervous system by ethylnitrosourea: Ethylation of nucleic acids and elimination rates of ethylated bases from the DNA of different tissues. *Z. Krebsforsch.* **82:** 37.

Hitotsumachi, S., D.A. Carpenter, and W.L. Russell. 1985. Dose-repetition increases the mutagenic effectiveness of *N*-ethyl-*N*-nitrosourea. *Proc. Natl. Acad. Sci.* **82:** 6619.

Johnson, F.M. and S.E. Lewis. 1981. Electrophoretically detected germinal mutations induced in the mouse by ethylnitrosourea. *Proc. Natl. Acad. Sci.* **78:** 3138.

Karran, P., T. Lindahl, and B. Griffin. 1979. Adaptive response to alkylating agents involves alteration *in situ* of O^6-methylguanine residues in DNA. *Nature* **280:** 76.

Lewis, S.E., F.M. Johnson, L.C. Skow, D. Popp, L.B. Barnett, and R.A. Popp. 1985. A mutation in the α-globin gene detected in the progeny of a female mouse treated with ethylnitrosourea. *Proc. Natl. Acad. Sci.* **82:** 5829.

Loveless, A. 1969. Possible relevance of *O*-6 alkylation of deoxyguanosine to the mutagenicity and carcinogenicity of nitrosamines and nitrosamides. *Nature* **223:** 206.

McCullagh, P. and J.A. Nelder. 1983. *Generalized linear models.* Chapman and Hall, London.

Monesi, V. 1962. Autoradiographic study of DNA synthesis and the cell cycle in spermatogonia and spermatocytes of mouse testis using tritiated thymidine. *J. Cell Biol.* **14:** 1.

Montesano, R. 1981. Alkylation of DNA and tissue specificity in nitrosamine carcinogenesis. *J. Supramol. Struct. Cell. Biochem.* **17:** 259.

Oakberg, E.F. 1956. Duration of spermatogenesis in the mouse and timing of stages of the cycle of the seminiferous epithelium. *Am. J. Anat.* **99:** 507.

Park, C.N. and R.D. Snee. 1983. Quantitative risk assessment: State-of-the-art for carcinogenesis. *Am. Stat.* **37:** 427.

Peters, J., S.J. Andrews, J.F. Loutit, and J.B. Clegg. 1985. A mouse β-globin mutant that is an exact model of hemoglobin Rainier in man. *Genetics* **110:** 709.

Peters, J., S.T. Ball, and S.J. Andrews. 1986. The detection of gene mutations by electrophoresis, and their analysis. *Prog. Clin. Biol. Res.* **209B:** 367.

Popp, R.A., E.G. Bailiff, L.C. Skow, F.M. Johnson, and S.E. Lewis. 1983. Analysis of a mouse α-globin gene mutation induced by ethylnitrosourea. *Genetics* **105:** 157.

Rall, D.P. 1978. Thresholds? *Environ. Health Perspect.* **22:** 163.

Richardson, K.K., F.C. Richardson, R.M. Crosby, J.A. Swenberg, and T.R. Skopek. 1987. DNA base changes and alkylation following *in vivo* exposure of *Escherichia coli* to N-methyl-*N*-nitrosourea or *N*-ethyl-*N*-nitrosourea. *Proc. Natl. Acad. Sci.* **84:** 344.

Russell, L.B., J.W. Bangham, K.F. Stelzner, and P.R. Hunsicker. 1988. High frequency of mosaic mutants produced by *N*-ethyl-*N*-nitrosourea exposure of mouse zygotes. *Proc. Natl. Acad. Sci.* **85**: 9167.

Russell, W.L., P.R. Hunsicker, D.A. Carpenter, C.V. Cornett, and G.M. Guinn. 1982a. Effect of dose fractionation on the ethylnitrosourea induction of specific-locus mutations in mouse spermatogonia. *Proc. Natl. Acad. Sci.* **79**: 3592.

Russell, W.L., P.R. Hunsicker, G.D. Raymer, M.H. Steele, K.F. Stelzner, and H.M. Thompson. 1982b. Dose-response curve for ethylnitrosourea-induced specific-locus mutations in mouse spermatogonia. *Proc. Natl. Acad. Sci.* **79**: 3589.

Russell, W.L., E.M. Kelly, P.R. Hunsicker, J.W. Bangham, S.C. Maddux, and E.L. Phipps. 1979. Specific-locus test shows ethylnitrosourea to be the most potent mutagen in the mouse. *Proc. Natl. Acad. Sci.* **76**: 5818.

Searle, A.C. 1974. Mutation induction in mice. *Adv. Radiat. Biol.* **4**: 131.

Sega, G.A. 1974. Unscheduled DNA synthesis in the germ cells of male mice exposed *in vivo* to the chemical mutagen ethyl methanesulfonate. *Proc. Natl. Acad. Sci.* **71**: 4955.

Sega, G.A., C.R. Rohrer, H.R. Harvey, and A.E. Jetton. 1986. Chemical dosimetry of ethyl nitrosourea in the mouse testis. *Mutat. Res.* **159**: 65.

Singer, B., H. Fraenkel-Conrat, and J.T. Kusmierek. 1978. Preparation and template activities of polynucleotides containing O^2- and O^4-alkyluridine. *Proc. Natl. Acad. Sci.* **75**: 1722.

Singer, B., S. Spengler, and W.J. Bodell. 1981. Tissue-dependent enzyme-mediated repair or removal of *O*-ethyl pyrimidines and ethyl purines in carcinogen-treated rats. *Carcinogenesis* **2**: 1069.

Van Ryzin, J. 1982. Discussion of the paper by Armitage, "The assessment of low-dose carcinogenicity." *Biometrics* (suppl. *Current topics in biostatistics and epidemiology*):130.

Van Zeeland, A.A., G.R. Mohn, A. Neuhäuser-Klaus, and U.H. Ehling. 1985. Quantitative comparison of genetic effects of ethylating agents on the basis of DNA adduct formation. Use of O^6-ethylguanine as molecular dosimeter for extrapolation from cells in culture to the mouse. *Environ. Health Perspect.* **62**: 163.

Zielenska, M., D. Beranek, and J.B. Guttenplan. 1988. Different mutational profiles induced by *N*-nitroso-*N*-ethylurea: Effects of dose and error-prone DNA repair and correlations with DNA adducts. *Environ. Mol. Mutagen.* **11**: 473.

COMMENTS

Peters: In a recent paper, it has been indicated that 20% of ENU-induced mutations in human cells were AT-TA inversions.

Favor: Yes. I didn't review all of the literature here. There is a large jump between mouse and bacteria. There are a few steps in between. One is *Drosophila*. *Drosophila*, interestingly, is actually quite consistent with bacteria. Mutations are mostly at GC sites although there are some at AT, and they are mostly transitions. If you noticed in the mouse, there were actually a lot of transversions. In human cells, there is also a large frequency of mutations at AT sites, and a high frequency are transversions. However, just to make the provocative statement, it's really interesting that all of the germ-cell mutations in the mouse are at AT sites. In *Drosophila* and in humans, mutations are recovered affecting different bases. To date, in the mouse, we have only seen mutations affecting AT sites.

Bridges: In this context, you have said that the lesions will persist through repair-proficient stages. Could you tell me what you mean by "repair" in this context?

Favor: This is the work of Gary Sega in which he has methods to measure unscheduled DNA synthesis throughout the postspermatogonial stages. He has identified those stages which exhibit unscheduled DNA synthesis and interpreted this to be DNA repair.

Bridges: So what one must get through is that, whatever the lesion is, it is not susceptible to an excision process which leads to UDS.

Favor: Right. And just to go one step further, it may be that the lesions actually persist even beyond the first DNA replication cycle. This is based on the observation that some mutations were carried in a very small percent of the germ cells of the mutation mosaic. For example, 3-ethylthymine was indicated to be negative, although infrequently there is mispairing. This could be, for example, a candidate for a base alkylation which may persist through a number of DNA replications before being fixed as a mutation.

Bridges: Yes, but then you throw away your correlation with the...

Favor: Repair.

Bridges: ...the multilocus gene on the plasmid in the bacteria.

Favor: Right.

Bridges: Clearly, those results show that the characteristics of that lesion will be that it is a potential block to DNA synthesis, doesn't lead to mispairing...

Favor: Yes, exactly.

Bridges: ...and the cell has to actively put a base into the AT site. So the chemists have got something to look for.

Favor: I don't think it's error-prone, because if it were error-prone we should also see base substitutions at all sites, it shouldn't be confined to the AT site.

Bridges: Well, except there must be something about this lesion which is so minimal a distortion that it is not recognized as a lesion and therefore not excised.

Favor: Exactly. It doesn't normally induce a repair response.

Liane Russell: Jack, I'd like to ask you about the mosaics that you got at day 27. I believe there were 2 on day 27?

Favor: Right. And there was one, a cataract, on day 24.

Liane Russell: So those really were induced prior to meiosis-I and meiosis-II, which means that they are going to get sorted out at meiosis and you wouldn't expect them to be mosaics still after that time.

Favor: Why not? The way I interpret it, the lesion is just confined to a single strand. DNA replication has already taken place. The DNA lesion is confined to one of the strands and will be sorted out in meiosis, but it will occur in a descendent cell as a single strand lesion. The lesion will not be fixed as a mutation until there is a round—or more rounds—of DNA replication.

Liane Russell: Which will be in the zygote?

Favor: Right. So I don't think it's inconsistent. The lesion is induced, confined to a single strand, goes through the meiotic divisions, still confined to a single strand, and fixed after fertilization.

Liane Russell: Well, if that's the case, why can't you also get mosaics from exposed differentiating gonia?

Favor: I think because they go through so many rounds of DNA replication in spermatogonia.

Liane Russell: In differentiating gonia, you might only have one round, and you might expect to find some mosaics.

Favor: I'd have to break out all the data and see if we actually screened the differentiating gonia stage. I don't think we screened this stage adequately. If you could, I would say you should find mosaics.

Liane Russell: I don't think we find mosaics, I mean not in huge numbers, and we've scored fairly large numbers of offspring from differentiating gonia.

Mohrenweiser: Just a comment. Mosaicism is turning out to be a fairly common event in humans. If you look at de novo germinal mutations—Duchenne's is the classic example—a fairly substantial number of them are actually mosaics, where you see two kids with the same molecular lesion born to a mother who has a "normal" dystrophy gene. Very consistent with what you are seeing.

Favor: Yes. I think it depends on being able to identify a mosaic individual. This depends on your powers of observation, and if you are able to actually identify the mosaic phenotypes and then to genetically confirm them. I should say there were two presumed mutation mosaics that we could not genetically confirm. A second point is that the mutation rate we observed when mutations were mostly occurring as mosaics is probably an underestimate because there are certainly going to be some that we cannot pick up. Also, as I said, we may be able to identify presumed mosaics but could not genetically confirm them. Once our powers of observation become better, where we can identify presumed mosaics and go back and genetically characterize them, I think mosaics will become more important.

Liane Russell: In that connection, we pointed out several years ago that, among *spontaneous* specific-locus mutations in the mouse, there is an incredibly high frequency of mosaics. At the c locus, for example, mosaics are readily detectable, 16 of 18 spontaneous c-locus mutations were mosaics. In these 16 animals, on the average, 50% of the somatic and germinal tissue was mosaic. So, spontaneous mutations presumably occur preferentially in those stages that can give rise to 50/50 mosaicism. Then, when we looked at spontaneous mutations at all loci, even those where mosaicism was not as readily detectable, there still was a very high frequency of mosaics.

Mosaic Mutants Induced by Ethylnitrosourea in Late Germ Cells of Mice

SUSAN E. LEWIS,[1] LOIS B. BARNETT,[1] AND RAYMOND A. POPP[2]
[1]Research Triangle Institute
Research Triangle Park, North Carolina 27709
[2]Oak Ridge National Laboratory
Oak Ridge, Tennessee 37830

OVERVIEW

Electrophoretic screening of F_1(DBA/2J x C57BL/6J) mice, originating from late oocytes or spermatids treated with ethylnitrosourea (ENU), resulted in the detection of mutants that were mosaic for both the normal and mutant alleles from the treated parent. In two of the five mutants, the variant trait was transmitted to progeny of backcrosses and, hence, was present in the germ line as well as somatic tissues. These findings are discussed in terms of mutational mechanisms, developmental dynamics during embryogenesis, and considerations for human risk.

INTRODUCTION

The first mosaic mutants identified in mice were almost certainly of spontaneous origin (Russell 1964, 1979). Recently, high proportions of mosaics have been found among mutants following ENU treatment of zygotes (Russell et al. 1988), mature oocytes (Ehling and Neuhäuser-Klaus 1988), and spermatids (Favor et al. 1989).

ENU has been shown to be a particularly effective mutagen in spermatogonia (Russell et al. 1979; Johnson and Lewis 1981, 1982), but no mosaicism has been reported for the large number of mutants recovered from spermatogonia. Achieving an understanding of the mechanisms underlying the phenomenon of mosaic mutants is important for predicting human risk associated with exposure to mutagens.

The studies reported here focus on the somatic tissue distribution and genetic transmission of electrophoretically detected mutations induced in the postmeiotic germ cells of mice. Comparisons are made to mosaics constructed by fusing two embryos of different origins (Tarkowski 1961).

RESULTS

The electrophoretically detected mutants described here were induced in late germ cells of both sexes. Matings of C57BL/6J females treated with ENU were carried out for 1 week after treatment so that progeny originated from mutagenized late oocytes. Treated DBA males were mated for 3 weeks after treatment so that the mutagenized germ cells giving rise to progeny were mature sperm and mid to late spermatids. When (C57BL/6J x DBA/2J)F_1 progeny mice were screened electrophoretically, as described previously (Johnson and Lewis 1981; Johnson et al. 1981), two mutants were detected in the group derived from treated females and three in the group derived from treated males. The details of the screening experiment, including the number of progeny screened and relative frequencies, will be published elsewhere.

The locus specificities and genetic analysis of all of the mutants detected electrophoretically are presented in Table 1. Presumed mutants were tested for heritability in backcrosses of the F_1 variant animal to one of the parental strains. Following heritability studies, the F_1 mutants were dissected, and the electrophoretic phenotype was studied in a variety of tissues. Heritability studies are not yet completed on the *Pgm-2* mutant induced in male germ cells.

An analysis of the mutation at the *Hbb* locus induced in female germ cells has been published (Lewis et al. 1985). Because hemoglobin is expressed only in red blood cells, it was not possible to perform comparative studies of the tissue distribution of mutant and normal gene expression. However, a study of the expression pattern of mutant and normal hemoglobins suggests that the somatic cells of the original F_1 mutant were mosaic for the gene product of the mutant and normal alleles from the treated maternal parent. The heritability studies showed that the original F_1 animal carrying the new allele was a germ-cell mosaic for the normal and mutant alleles (see Table 1).

Because of the expression of the relevant genes in more than one tissue, the rest of the mutants in this series could be studied for expression of mutant and normal

Table 1
Heritability of Electrophoretically Detected Mutations Obtained from Late Germ Cells Exposed to Ethylnitrosourea

			Alleles recovered in backcrosses			
			mutant	normal		
Sex of treated parent	RTI exp. #	Locus	treated parent	treated parent	untreated parent	Total
F[a]	D^3	*Hbb*	3	4	4	11
F	X^4	*Idh-1*	0	23	26	49
M	C^3	*Mod-1*	2	9	22	33
M	N^4	*Pgd*	0	18[c]		18
M[b]	T^{12}	*Pgm-2*	—	—	—	—

[a]Data from Lewis et al. (1985).
[b]Heritability studies are still under way.
[c]Parental alleles are indistinguishable electrophoretically.

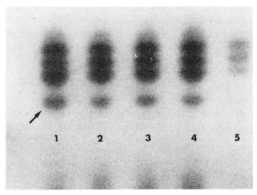

Figure 1
Electrophorogram of *IDH-1* expression in the kidney (lane *1*) and samples from two lobes of liver (lanes *2–4*) taken from the original F_1 *Idh-1* mobility mutant. The tissues in lanes *1–4* have the original F_1 three-banded pattern above a slow mutant band (indicated by arrow). The ovarian sample in lane *5* is too light to be evaluated.

alleles in various tissues. The other mutation from the female-treated group, specifying an electrophoretic mobility at *Idh-1*, was a somatic mosaic. The normal DBA/2J and C57BL/6J bands and an extra slow band were revealed on electrophoresis of tissues in which the relevant gene product is expressed (Fig. 1). The slow band was in the same position as that derived from two other previously discovered mutations at *Idh-1* induced in C57BL/6J mice (Johnson and Lewis 1981; Lewis and Johnson 1983; S. Lewis, unpubl.). The variant pattern was transmitted to none of the 49 backcrossed progeny (Table 1), but was thought to be due to a mutation because it was expressed in the somatic tissues of the F_1 carrier.

An electrophoretic variant of MOD-I was clearly demonstrated to be both a germ-line and somatic mosaic. The original F_1 carrier transmitted three types of *Mod-1* alleles to its progeny; two normal alleles, one each from the treated and untreated parent, and a mutant allele from the treated parent (Table 1). Subsequent carriers all transmitted the trait in appropriate Mendelian fashion (Cobb et al. 1990). The tissue expression of this mutant was interesting because, unlike the *Idh-1* mutant mentioned above, the proportion of the mosaic components was clearly not the same in all somatic tissues examined. In fact, in four liver samples taken from the original F_1 mutant, two were predominantly wild type and two were predominantly mutant (Fig. 2). Approximately equivalent amounts of the mutant and normal DBA/2J gene products are present in the kidneys of the original mutant.

The PGD variant, found during electrophoretic screening of progeny from treated males, failed to transmit the variant trait to backcrossed progeny, but the extra slow band manifested in the original kidney sample was present in all tissues

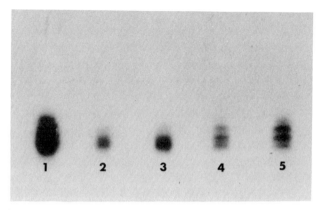

Figure 2
Electrophorogram of MOD-1 expression in tissue samples from original *Mod-1* mobility mutant. The kidney sample is in lane *1*, and four liver samples are in lanes *2–5*.

from the mutant in which PGD is expressed. Mosaicism is suggested not by the mutant pattern per se, but by the absence of mutant transmission in the germ line, whereas the expression is documented in all somatic tissues studied. A germ-line lethal is not ruled out, but is unlikely as ENU causes relatively small genetic lesions.

The last mutation identified among the progeny of treated male germ cells resulted in a mobility change in the electrophoretic pattern of *Pgm-2*. The presumed mutant band is lighter than the normal band, possibly suggestive of mosaicism. Heritability studies are still under way.

DISCUSSION

The findings reported here indicate that electrophoretically detected mutations, induced by ENU in postreplicative germ-cell stages, tend to be mosaics. Mutations induced in prereplicative germ-cell stages do not prove to be mosaic. No mosaic mutants have been found among 31 electrophoretically detected mutations from spermatogonia treated with ENU (Lewis and Johnson 1983, 1986; S. Lewis, unpubl.). They are not reported to be among ENU-induced mutations from spermatogonia in the 7-locus recessive visible system (Russell et al. 1982; Hitotsumachi et al. 1985).

In general, the induction of mosaic mutants by certain mutagens is a common phenomenon in late germ-cell stages in organisms. Many chemically induced mosaic mutations in the late stages of *Drosophila* have been identified (Lee et al. 1967, 1969). Treatment of mature spermatozoa of the zebrafish also results in the induction of mosaic mutants (Golin et al. 1982).

A number of explanations could be put forward to explain the induction of mosaics in late germ cells. If, as suggested by Russell et al. (1988) for mosaics induced by ENU in zygotes, one strand of DNA was affected by a mutagen and the other not, the fixation of the mutation in the first cleavage division could produce two cells with different genotypes. A large proportion of cells from the cleavage-stage mouse embryos is allotted to extraembryonic structures. Thus, cells carrying either the mutant or normal allele might not end up in the fetus at all if they were distributed only to the extraembryonic portions of the conceptuses. Russell et al. (1988) point out that two "whole body" mutants recovered by them from ENU-treated zygotes could have actually been mosaic. The wild-type component could have been present only in the extraembryonic portions of the conceptus and, thus, not be present in the mature animal.

Unequal tissue distribution of mutant gene expression in different tissues of the same animal has been observed so far in only one of the mosaic mutants discussed here. In most of the aggregation chimeras examined, approximately the same proportion of cells from the two embryos fused to make the chimera are found in all tissues sampled. However, exceptions have been reported (Falconer et al. 1978).

The concordance is not perfect between what may be expressed in somatic tissue and what is present in the germinal tissue. Such discrepancies between the allocation of mutant genotypes to mutant and somatic cell lineages are well demonstrated for postgonial mutations induced in *Drosophila* (Lee et al. 1967). Most of the mosaics detected by the visible system after zygotic (Russell et al. 1988) or late germ-cell (Ehling and Neuhäuser-Klaus 1988; Favor et al. 1989) treatment with ENU transmit the trait to progeny and, thus, are present in both somatic and germinal tissues (Russell et al. 1988; Favor et al. 1989). It is possible that because the gene products of all three alleles can be visualized in somatic cells where selection is made, the electrophoretic detection system permits detection of mosaic mutants that exist in lower frequencies in the organism than does the visible system.

The degree of inclusion of cells of any one origin in a set of aggregation chimeras varies considerably (Tarkowski 1961). The possibility of losing a weaker cell with mutational damage must be considered whenever mosaic mutants are induced. Selection against certain presumably weaker phenotypes and/or strains has been demonstrated in mammalian aggregation chimeras. An example is the loss of the XO component in mosaic embryos derived from XX and XO female mice (Burgoyne and Biggers 1976). Furthermore, embryos whose paternal parents have been treated with X-rays show proliferative disadvantage in fusion chimeras (Obasaju et al. 1989). Presumably, the cells with such a disadvantage would have a significantly lower chance of contributing to the somatic and/or germ line of the individual.

Finally, mosaicism should be considered in the analysis of de novo mutations in man. Both germinal and somatic mosaics have been reported in man. "Segmental" neurofibromatosis, in which patches of cells are diseased in the affected individual, is only one example (Riccardi and Lewis 1988). Many presumed single gene de novo mutations may have their origin in parental mosaicism (Darras and Francke 1987; Byers et al. 1988). Such germ-line mosaicism potentially has its origin either in a developmentally early germ-cell mutation in the parent or in the late germ cells of the previous generation. In either case, there are important reasons for considering mosaicism when assessing familial recurrence risks and evaluating populations at risk for exposure to mutational agents.

ACKNOWLEDGMENTS

This work was supported in part by National Institute of Environmental Health Sciences contract NO-1-ES-55078. We thank Dr. F. de Serres, Dr. L.B. Russell, and Ms. L. Overton for critical reading of this manuscript. We thank Cathee Winkie for typing the manuscript.

REFERENCES

Burgoyne, P.S. and J.D. Biggers. 1976. The consequences of X-dosage deficiency in the germ line: Impaired development *in vitro* of preimplantation embryos from XO mice. *Dev. Biol.* **51:** 109.

Byers, P.H., P. Tsipouras, J.F. Bonadio, B.J. Starman, and R.C. Schwartz. 1988. Perinatal lethal osteogenesis imperfecta (type II): A biochemically heterogeneous disorder usually due to new mutations in the genes for type I collagen. *Am. J. Hum. Genet.* **42:** 237.

Cobb, R.R., J.G. Burkhart, J.S. Dubins, L.B. Barnett, and S.E. Lewis. 1990. Biochemical and molecular analysis of spontaneous and induced mutations at the mouse *Mod-1* locus. *Mutat. Res.* **234:** 1.

Darras, B.T. and U. Francke. 1987. Male germline mosaicism for DMD deletion mutation: Implications for genetic counseling. *Am. J. Hum. Genet.* **41:** A95.

Ehling, U.H. and A. Neuhäuser-Klaus. 1988. Induction of specific-locus mutations in female mice by 1-ethyl-1-nitrosourea and procarbazine. *Mutat. Res.* **202:** 139.

Falconer, D.S., I.K. Gauld, and R.C. Roberts. 1978. Growth control in chimaeras. In *Genetic mosaics and chimeras in mammals* (ed. L.B. Russell), vol. 12, p. 39. Plenum Press, New York.

Favor, J., A. Neuhäuser-Klaus, and U. Ehling. 1990. The frequency of dominant cataract, recessive specific locus mutations and mutation mosaics in F_1 mice derived from postspermatogonial germ cells treated with ethylnitrosourea. *Mutat. Res.* (in press).

Golin, J., D.J. Grunwald, F. Singer, C. Walker, and G. Streisinger. 1982. Ethylnitrosourea induced germline mutations in zebrafish *Brachydaniorerio*. *Genetics* **100:** S27.

Hitotsumachi, S., D.A. Carpenter, and W.L. Russell. 1985. Dose-repetition increases the mutagenic effectiveness of *N*-ethyl-*N*-nitrosourea in mouse spermatogonia. *Proc. Natl. Acad. Sci.* **82:** 6619.

Johnson, F.M. and S.E. Lewis. 1981. Electrophoretically detected germinal mutations induced by ethylnitrosourea in the mouse. *Proc. Natl. Acad. Sci.* **78:** 3138.

―――. 1982. The human genetic risk of airborne genotoxics: An approach based on electrophoretic techniques applied to mice. In *Genotoxic effects of airborne agents* (ed. R. Tice and K. Schaido), p. 596. Plenum Press, New York.

Johnson, F.M., G.T. Roberts, R.K. Sharma, F. Chasalow, R. Zweidinger, A. Morgan, R.W. Hendren, and S.E. Lewis. 1981. The detection of mutants in mice by electrophoresis: Results of a model induction experiment with procarbazine. *Genetics* **97:** 113.

Lee, W.R., C.J. Kirby, and C.W. Debney. 1967. The relation of germ line mosaicism to somatic mosaicism in *Drosophila*. *Genetics* **55:** 619.

Lee, W.R., G.A. Sega, and J.B. Bishop. 1969. Chemically-induced mutations observed as mosaics in *Drosophila melanogaster*. *Mutat. Res.* **9:** 323.

Lewis, S.E. and F.M. Johnson. 1983. Dominant and recessive effects of electrophoretically detected specific locus mutations. *Workshop on utilization of mammalian specific locus studies in hazard evaluation and estimation of genetic risk* (ed. F.J. de Serres and W. Sheridan), p. 267. Plenum Press, New York.

———. 1986. The nature of spontaneous and induced electrophoretically detected mutations in the mouse. In *Genetic toxicology of environmental chemicals*, part B: Genetic effects and applied mutagenesis, p. 359. A.R. Liss, New York.

Lewis, S.E., F.M. Johnson, L.C. Skow, D. Popp, L.B. Barnett, and R.A. Popp. 1985. A mutation in the β-globin gene detected in the progeny of a female mouse treated with ethylnitrosourea. *Proc. Natl. Acad. Sci.* **82:** 5829.

Obasaju, M.F., L.M. Wiley, D.J. Oudiz, O. Raabe, and J.W. Overstreet. 1989. A chimera embryo assay reveals a decrease in embryonic cellular proliferation induced by sperm from x-irradiated male mice. *Radiat. Res.* **118:** 246.

Riccardi, V.M. and R.A. Lewis. 1988. Penetrance of von Recklinghausen neurofibromatosis: A distinction between predecessors and descendants. *Am. J. Hum. Genet.* **42:** 284.

Russell, L.B. 1964. Genetic and functional mosaicism in the mouse. In *Role of chromosomes in development* (ed. M. Locke), p. 153. Academic Press, New York.

———. 1979. Analysis of the albino-locus region of the mouse. II. Mosaic mutants. *Genetics* **91:** 141.

Russell, L.B., J.W. Bangham, K.F. Stelzner, and P.R. Hunsicker. 1988. High frequency of mosaic mutants produced by N-ethyl-N-nitrosourea exposure of mouse zygotes. *Proc. Natl. Acad. Sci.* **85:** 9167.

Russell, W.L., P.R. Hunsicker, G.D. Raymer, M.H. Steele, K.F. Stelzner, and H.M. Thompson. 1982. Dose-response curve for ethylnitrosourea-induced specific-locus mutations in mouse spermatogonia. *Proc. Natl. Acad. Sci.* **79:** 3589.

Russell, W.L., E.M. Kelley, P.R. Hunsicker, J.W. Bangham, S.C. Maddux, and E.L. Phipps. 1979. Specific-locus test shows ethylnitrosourea to be the most potent mutagen in the mouse. *Proc. Natl. Acad. Sci.* **76:** 5818.

Tarkowski, A.K. 1961. Mouse chimaeras developed from fused eggs. *Nature* **190:** 857.

COMMENTS

Mohrenweiser: Susan, in the mutations that have been characterized, can anybody tell which strand the lesion is in?

Lewis: No. Two of the three that have been looked at have only been done by extrapolation—Ray Popp did this in his lab—via protein C sequencing, and the DNA sequence was deduced from that.

Liane Russell: The two mosaics that you gave that don't transmit—I believe one was IDH and one was...

Lewis: PGD

Liane Russell: Yes, but the PGD you couldn't really tell because you couldn't distinguish the parental contributions?

Lewis: I can't. But the mutant pattern, which was a slow allele, was in all the other tissues.

Liane Russell: I see. And another that was also a very low transmitter was the MOD-1, which was 3 out of some...

Lewis: Two out of 33.

Liane Russell: Now, how long after exposure were these conceived, the low transmitters or the no transmitters?

Lewis: All the ones from male treatment were from the third week of mating, so they are from spermatids. The non-transmitting *Idh-1* mutant from treated females was conceived a week after treatment.

Liane Russell: So all these low transmitters came from that?

Lewis: The hemoglobin, which was a little better transmitter, came from the second or third day after mating after females were treated. I included both male and female treateds.

Liane Russell: That's really interesting, because I guess Jack [Favor] finds a majority of his mosaics to be transmitting, right?

Favor: Yes. For the specific-locus mutants, we identified 11 with variegated phenotype, and we confirmed 9 of those by transmission.

Lewis: There is maybe one extra thing here. Because of the electrophoretic system—and I could see all three alleles actually there—I may pick up some little tiny events that might not be picked up for coat color or cataract, and thus may be getting these non-transmissions. Of course, as you pointed out, you may basically have had mosaics where you detect them as whole-body mutants because some of the rest of it is lost in extraembryonic membrane.

Mohrenweiser: In the ones where you have low transmission, is the litter size normal?

Lewis: Yes.

Analysis of Electrophoretically Detected Mutations Induced in Mouse Germ Cells by Ethylnitrosourea

JOSEPHINE PETERS,[1] JANET JONES,[1] SIMON T. BALL,[1]
AND JOHN B. CLEGG[2]
[1]MRC Radiobiology Unit
Chilton, Didcot
Oxon OX11 ORD, United Kingdom
[2]MRC Molecular Haematology Unit
John Radcliffe Hospital
Headington, Oxford OX3 9DU, United Kingdom

OVERVIEW

Germinal mutations in the mouse affecting enzymes and proteins are useful because they may provide models of human genetic disease and because they are relatively easily analyzed at the molecular level. The molecular characterization of mutations is essential for elucidating the spectrum of DNA changes induced in mammalian germ cells.

Some analysis of N-ethyl-N-nitrosourea (ENU)-induced mutations supports earlier findings that ENU causes small intragenic changes in mouse germ cells. A single amino acid substitution α127 Lys→Asn occurred in a mutant α hemoglobin, and the previously described amino acid substitution β145 Tyr→Cys in a β hemoglobin (Peters et al. 1985) has been found to be due to an A:T→G:C transition. Taken together with the results of other investigators (Popp et al. 1983; Lewis et al. 1985), the data so far indicate that ENU causes single amino acid substitutions attributable to single substitutions in DNA involving the A:T base pair.

INTRODUCTION

In recent years, unidimensional electrophoresis of enzymes and proteins has been used to detect germinal mutations in the mouse (Malling and Valcovic 1977; Soares 1979; Johnson and Lewis 1981, 1983; Pretsch and Charles 1984; Lewis and Johnson 1986; Peters et al. 1986). The method detects mutations that either lead to loss of enzyme activity or protein product or that result in a change of the net charge of the enzyme or protein; it is capable of detecting single amino acid substitutions.

One useful feature of this method is that all the gene products studied are relatively well understood biochemically, and many are well understood molecularly as well, allowing the investigation of the nature of induced and spontaneous mutational events. Furthermore, most of the genes studied have unequivocal homologs in man, and so it has become possible to compare mutation frequencies at homologous loci in mouse and man. In addition, enzyme deficiency has been demonstrated in about one third (250/750) of genetic diseases with a certain recessive mode of inheritance (McKusick 1988), and, thus, induced mutations leading to altered function of homologous enzymes and proteins in the mouse may provide models of human genetic disease.

Peters et al. (1986) reported 16 mutations induced in mouse spermatogonia by doses of 250 mg/kg or 200 mg/kg ENU and detected by electrophoresis. There were four at *Hbb* (hemoglobin β-chain), four at *Pgm-1* (phosphoglucomutase-1), five at *Gpi-1* (glucose phosphate isomerase-1), and one each at *Hba* (hemoglobin α-chain), *Pgm-2* (phosphoglucomutase-2), and *Trf* (transferrin). One of the mutants at *Hbb*, designated Hbb^{d4}, resulted in a structurally altered hemoglobin that was associated with polycythemia, increased oxygen affinity, and decreased heme-heme interactions. Amino acid analysis showed that a single substitution β145 Tyr→Cys had occurred (Peters et al. 1985). The mutant is homologous with Hb Rainier in man, in terms of amino acid substitution, and has similar physiological consequences. Single amino acid substitutions leading to altered hemoglobins in the mouse and resulting from ENU-induced mutants have been reported by other workers (Popp et al. 1983; Lewis et al. 1985).

With the long-term aim of elucidating the spectrum of DNA changes induced by mutagens in mammalian germ cells, we have carried out a direct analysis of DNA from wild-type Hbb^d and mutant Hbb^{d4} haplotypes using the polymerase chain reaction in conjunction with the amplification refractory mutation system (ARMS) (Newton et al. 1989). We also report some preliminary analysis of the mutants at *Gpi-1*, and that the mutation at *Hba*, α-hemoglobin, leads to a single amino acid substitution.

The procedures used for ENU mutagenesis have been described previously (Peters et al. 1986). Briefly, males treated with ENU were mated to T (tester) stock females, and 0.1 ml blood was taken from F_1 offspring for electrophoretic testing. The methods for analysis of mutants of *Gpi-1* are reported by Peters and Ball (1990); for amino acid analysis of globins by Peters et al. (1986); and for DNA analysis by J. Jones and J. Peters (in prep).

RESULTS

Analysis of Five Mutants of *Gpi-1*

Peters et al. (1986) found five null mutants of *Gpi-1* induced by ENU in mouse spermatogonia. Two arose in the same father and exerted similar phenotypic effects and were therefore considered to be the result of a single mutational event. Thus, only four independent mutations had been found, and of these one, $Gpi-1^{a-m1H}$, arose in the 101/H parental chromosome and the other three, $Gpi-1^{b-m1H}$, $Gpi-1^{b-m2H}$, and $Gpi-1^{b-m3H}$, arose in the C3H/HeH parental chromosome. For all four mutant alleles, the homozygous mutant genotype appeared to be lethal because intercrosses did not yield any homozygous mutant offspring (Peters and Ball 1990).

Heterozygotes for all four mutants had similar GPI-1 phenotypes in blood, in that each resembled the phenotype of the untreated parent in terms of electrophoretic mobility, but had diminished staining activity. Heterozygotes for three of the mutant alleles, $Gpi\text{-}1^{a\text{-}m1H}$, $Gpi\text{-}1^{b\text{-}m1H}$, and $Gpi\text{-}1^{b\text{-}m2H}$, had similar GPI-1 phenotypes to those found in blood for all other tissues examined, namely liver, thymus, spleen, brain, heart, and testis, and for conceptuses at all ages tested so far, from 8.5 to 16.5 days p.c. Thus, these three alleles appeared to be complete nulls. However, when freshly prepared extracts of testis or conceptuses heterozygous for $Gpi\text{-}1^{b\text{-}m3H}$ were tested, a new major enzyme was seen (Fig. 1). The interpretation was that this new isozyme was a heterodimer formed from polypeptides determined by a normal allele, $Gpi\text{-}1^a$ or $Gpi\text{-}1^b$, and the mutant allele $Gpi\text{-}1^{b\text{-}m3H}$. The heterodimeric band had similar staining intensity to homodimers attributable to either of the normal alleles, although on theoretical grounds, it is expected that heterodimers will occur twice as frequently and thus have twice the staining intensity as homodimeric bands. Our current interpretation is that the mutant GPI-BM3 polypeptide is unstable but can form heterodimers of diminished stability with a wild-type polypeptide; and that these relatively unstable heterodimers can be seen fleetingly in tissues with a rapid protein turnover. By extension, the mutant homodimer is probably even more unstable than the heterodimer, and is therefore never seen. Thus, the mutagenic lesion induced by ENU leading to the $Gpi\text{-}1^{b\text{-}m3H}$ allele must be small, because polypeptide appears to be synthesized.

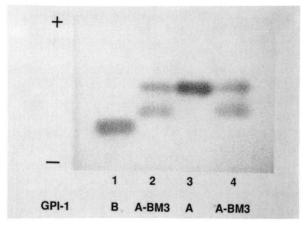

Figure 1
GPI-1 in testis extracts after cellulose acetate electrophoresis. (Lane 1) C3H/HeH, $Gpi\text{-}1^b/Gpi\text{-}1^b$; (lane 3) 101/H $Gpi\text{-}1^a/Gpi\text{-}1^a$; (lanes 2 and 4) mice from the $Gpi\text{-}1^{b\text{-}m3H}$ stock, of genotype $Gpi\text{-}1^a/Gpi\text{-}1^{b\text{-}m3H}$.

Analysis of Hba ^{c2}, a Mutant Affecting the Hemoglobin α-chain

The mutation was found during screening for induced mutations in (T × C3H/HeH)F$_1$ progeny of tester (T) stock females and C3H/HeH males treated with 250 mg/kg ENU. Figure 2 is a photograph of part of an isoelectric focusing gel showing various hemoglobin phenotypes attributable to different alleles at *Hba*. The T stock has the *Hba*^{f} haplotype that determines a single polypeptide, chain 1, and a single major band is seen after isoelectric focusing (lane 1), whereas C3H/HeH has the *Hba*^{c} haplotype that determines two distinct polypeptides, chains 1 and 4, and two major bands are seen after focusing (lane 2). In the absence of mutation, (T × C3H/HeH)F$_1$ are expected to have the HBA-FC phenotype that looks like a mixture of the parental phenotypes (lane 3). During the course of the experiment, one (T × C3H/HeH)F$_1$ was found with the HBA-FC2 phenotype (lane 5) in which the band containing chain 4 was missing and had been replaced by a band of greater anodal mobility (C2). Subsequent electrophoretic analysis of the hemoglobin from both parents showed that they had the phenotypes expected, HBA-F for the T stock mother and HBA-C for the C3H/HeH father. All other progeny of the cross were HBA-FC. Thus, the mutation appeared to have been induced in the C3H/HeH male. On backcrossing HBA-FC2 to C3H/HeH, two types of progeny were found: those

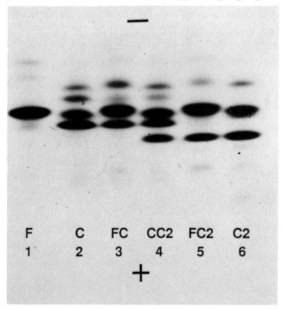

Figure 2
HBA in hemolysates after isoelectric focusing. (Lane *1*) Tester (T) stock, *Hba*^{f}/*Hba*^{f}; (lane 2) C3H/HeH, *Hba*^{c}/*Hba*^{c}; (lanes *3–6*) offspring from a mutation experiment and from mutant stocks.

with the HBA-FC2 phenotype and those with a new phenotype, HBA-CC2 (channel 4), in which three major bands were present. When mice of phenotype HBA-FC2 were intercrossed, three types of offspring were found: HBA-F, HBA-FC2, and HBA-C2 (lane 6). Thus it appears that a mutation arose in germ cells of the C3H/HeH mouse affecting the mobility of a hemoglobin band containing chain 4.

The results of amino acid analysis indicated that a single amino acid substitution had occurred and that lysine at α127 had been substituted by asparagine. At the DNA level, the simplest change that will account for this is substitution of the codon AAA (for lysine) by either AAT or AAC, both of which specify asparagine. Thus, either a transversion A:T→T:A or a transition A:T→G:C must have occurred.

Analysis of *Hbb*^{d4} Polycythemia, a Mutant Affecting the Hemoglobin β Chain

The finding of a single amino acid substitution β145Tyr→Cys (Peters et al. 1985) in the *Hbb*^{d4} mutant led to the proposal that ENU had induced a transition A:T→G:C in the tyrosine codon (TAC→TGC). We have used (J. Jones and J. Peters, in prep.) the polymerase chain reaction in conjunction with ARMS (Newton et al. 1989) to test for the exact substitution in the mutant, *Hbb*^{d4} haplotype.

Design of the Primers

The basis of the ARMS method is that the success of the polymerase chain reaction is directly related to the degree of matching between the 3′ ends of the primers and the DNA to be amplified. The probability of DNA amplification becomes less the greater are the number of mismatches at the 3′ ends, particularly if these mismatches are purine/purine (A/A, A/G, G/G) or pyrimidine/pyrimidine (T/T, C/T, C/C) (Newton et al. 1989); these findings were taken into account in the design of the primers.

The amplification primer (primer A, Fig. 3) was a 23-mer, 5′ to exon 1 of the *Hbb-b1* gene. Two reverse primers were used, both 32-mers, one of which, primer T (Fig. 3), was specific for the wild-type haplotype *Hbb*^{d} or *Hbb*^{s}, and one, primer C (Fig. 3), that was specific for the *Hbb*^{d4} haplotype. Both primers had 3′ ends that align with the DNA codon specifying the amino acid at β145. Thus, primer T has a 3′ sequence AT̲G, complementary to the tyrosine codon TAC found in wild-type DNA; and primer C has a 3′ sequence AC̲G complementary to a putative TGC codon for cysteine in *Hbb*^{d4}. To ensure that the primers were specific for either wild-type or mutant DNA, a purine/purine (C/T) mismatch was introduced into each primer four bases from the 3′ end (Fig. 3). Thus, there was a single mismatch between the primer and the DNA it was designed to amplify, but two mismatches between the primer and the alternative DNA.

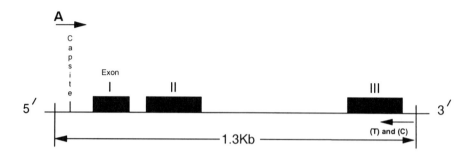

Primer A
CCTCACATTTGCTTCTGACATAG

Primer T
A T̲ G T T G A T T T G G G G G A A A G G A G G A G G A C G G A C
Amino acid   Tyr

Primer C
A C̲ G T T G A T T T G G G G G A A A G G A G G A G G A C G G A C
Amino acid   Cys

Figure 3
Structure of the *Hbb-b1* gene. (Primer A) The amplification primer; (Primer T) the reverse primer specific for wild-type DNA; and (Primer C) the reverse primer specific for mutant DNA. Bases that are underlined indicate sites of possible mismatch.

Allele-specific Amplification of Wild-type *Hbb* [d4] and Mutant *Hbb* [d4] DNA

As shown in Figure 3, successful amplification would result in the formation of a 1.3-kb DNA fragment. A 1.3-kb fragment was derived from the amplification of wild-type DNA only when reverse primer T, specific for wild-type DNA, in conjunction with amplification primer A, was used (Fig. 4). There was no evidence of amplification of wild-type DNA when primer C, the *Hbb*[d4]-specific primer, was used. Conversely, a 1.3-kb DNA fragment from *Hbb*[d4] was amplified only if primer C was used in conjunction with primer A, but not if primer T was used (Fig. 4). Thus, the reverse primers were specific for either wild-type or mutant DNA. Thus, in the *Hbb*[d4] mutation, the tyrosine codon TAC has been replaced with the cysteine codon TGC, and so a transition A:T→G:C has occurred.

DISCUSSION

ENU was considered to induce small intragenic changes in mouse germ cells because several of the mutations recovered in the recessive visible specific locus

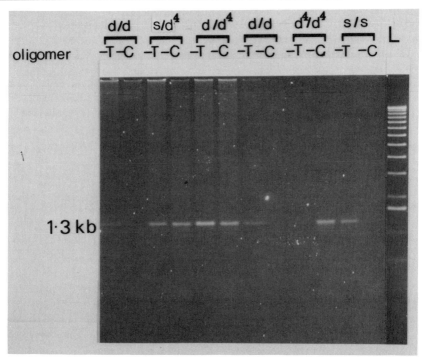

Figure 4
Agarose gel showing amplified DNA. For convenience in labeling, the alleles are indicated by their superscripts alone, and not by the complete gene symbol. Wild-type DNA (d/d, d/s) is amplified with primer T but not primer C; mutant DNA (d^4/d^4) is amplified with primer C but not primer T. DNA from heterozygotes s/d^4 and d/d^4 is amplified with both primers T and C. L indicates the lane loaded with 1-kb ladder, the marker DNA.

test appeared to be intermediate alleles (Russell et al. 1979; Russell 1986). The finding of mutations leading to altered electrophoretic mobility of enzymes and proteins (Johnson and Lewis 1981, 1983; Pretsch and Charles 1984; Lewis and Johnson 1986; Peters et al. 1986) supports this view, because mutations resulting in altered mobility must be sufficiently small to enable functional enzyme or protein to be synthesized. In germ cells, ENU has been found to induce point mutations (Schöneich and Braun 1986), single amino acid substitutions (Popp et al. 1983; Lewis et al. 1985), and now single base substitutions in the DNA (J. Jones and J. Peters, in prep.).

Four mutations affecting globin polypeptides induced in mouse germ cells by ENU have been analyzed for amino acid substitutions (Table 1). Two of the mutant alleles determine altered β-globin polypeptides, and two determine altered α-globins. Three of the mutations, Hbb^{d4}, Hbb^{c2}, and Hba^{g2}, were induced in sperma-

Table 1
Analysis of Mutations Induced by ENU in Mouse Germ Cells

Haplotype	Polypeptide	Amino acid substitution	DNA substitution	Reference
Hbb^{d4}	β globin	β145 Tyr→Cys	TAC→TGC	Peters et al. (1985); J. Jones and J. Peters (in prep.)
Hbb^{s2}	β globin	β60 Val→Glu	GTG→GAG (GAA)	Lewis et al. (1985)
Hba^{c2}	α globin	α127 Lys→Asn	AAA→AAT or AAC	This paper
Hba^{g2}	α globin	α89 His→Leu	CAC→CTC (CCT, CTA, CTG)	Popp et al. (1983)

togonia and the fourth, Hbb^{s2}, was induced in oocytes. For each of the mutations, a single amino acid substitution has occurred, a single substitution in the DNA sequence is postulated, and, in each example, the A:T base pair appears to have mutated. Thus, for both Hbb^{s2} and Hba^{g2}, the simplest possibly mutational mechanism would be a transversion A:T→T:A; and either a transversion A:T→T:A or a transition A:T→G:C would account for the α127 Lys→Asn substitution in Hba^{c2}. From DNA analysis of the Hbb^{d4} mutation, it has now been shown that ENU has induced a transition A:T→G:C (J. Jones and J. Peters, in prep.). In summary, of the four mutants analyzed, one is a transition A:T→G:C; two, and maybe three, are probably transversions A:T→T:A.

In studies of bacteria, *Drosophila*, and human cells, the most abundantly occurring lesion produced by ENU is the transition G:C→A:T (Lee et al. 1987; Richardson et al. 1987; Eckert et al. 1988; Zielenska et al. 1988; Pastink et al. 1989). However, whereas ENU induces transitions G:C→A:T and A:T→G:C almost exclusively in bacteria (Richardson et al. 1987; Zielenska et al. 1988), a significant proportion of transversions, particularly A:T→T:A transversions, occurs in *Drosophila* and human cells (Lee et al. 1987; Eckert et al. 1988; Pastink et al. 1989). In human cells, the transversion A:T→T:A seems to be especially important, because 20% of all the ENU-induced mutations were of this type (Eckert et al. 1988). Therefore, it is of some interest that from the limited data on ENU-induced mutations in mouse germ cells, it appears that the A:T→T:A transversion may also be an important mutational mechanism in the mouse. However, the full spectrum of ENU-induced mutational damage in mouse germ cells will become clear only when more mutants have been analyzed.

Whereas G:C→A:T and A:T→G:C transitions most probably result from unrepaired O^6-alkylguanine and O^4-alkyladenine adducts, the mechanism(s) leading to the formation of transversions is less clear.

The ARMS for mutational analysis used in some of the work reported here is relatively simple but requires knowledge of the likely DNA sequence change, from knowledge of the amino acid change. When this information is lacking, sequencing the mutant may be the only way of establishing the exact change induced by the mutagen.

REFERENCES

Eckert, K.A., C.A. Ingle, D.K. Klinedinst, and N.R. Drinkwater. 1988. Molecular analysis of mutations induced in human cells by N-ethyl-N-nitrosourea. *Mol. Carcinog.* **1**: 50.

Johnson, F.M. and S.E. Lewis. 1981. Electrophoretically detected germinal mutations induced in the mouse by ethylnitrosourea. *Proc. Natl. Acad. Sci.* **78**: 3138.

———. 1983. The detection of ENU-induced mutants in mice by electrophoresis and the problem of evaluating the mutation rate increase. In *Utilization of mammalian specific locus studies in hazard evaluation and estimation of genetic risk* (ed. F. J. de Serres and W. Sheridan), p. 95. Plenum Press, New York.

Lee, C.S., D. Curtis, M. McCarron, C. Love, M. Gray, W. Bender, and A. Chovnick. 1987. Mutations affecting expression of the *rosy* locus in *Drosophila melanogaster*. *Genetics* **116**: 55.

Lewis, S.E. and F.M. Johnson. 1986. The nature of spontaneous and induced electrophoretically detected mutations in the mouse. *Prog. Clin. Biol. Res.* **209B**: 359.

Lewis, S.E., F.M. Johnson, L.C. Skow, D. Popp, L.B. Barnett, and R.A. Ropp. 1985. A mutation in the β-globin gene detected in the progeny of a female mouse treated with ethylnitrosourea. *Proc. Natl. Acad. Sci.* **82**: 5829.

Malling, H.V. and L.R. Valcovic. 1977. A biochemical specific locus mutation system in mice. *Arch. Toxicol.* **38**: 45.

McKusick, V.A. 1988. *Mendelian inheritance in man*, eighth edition. Johns Hopkins University Press, Baltimore.

Newton, C.R., A. Graham, L.E. Heptinstall, S.J. Powell, C. Summers, N. Kalsheker, J.C. Smith, and A.F. Markham. 1989. Analysis of any point mutation in DNA. The amplification refractory mutation system. *Nucleic Acids Res.* **17**: 2503.

Pastink, A., C. Vreeken, M.J.M. Nivard, L.L. Searles, and E.W. Vogel. 1989. Sequence analysis of N-ethyl-N-nitrosourea-induced *Vermilion* mutations in *Drosophila melanogaster*. *Genetics* **123**: 123.

Peters, J. and S.T. Ball. 1990. Analysis of mutations of glucose phosphate isomerase-1, *Gpi-1*, in the mouse. *Prog. Clin. Biol. Res.* (in press).

Peters, J., S.T. Ball, and S.J. Andrews. 1986. The detection of gene mutations and their analysis. *Prog. Clin. Biol. Res.* **209B**: 367.

Peters, J., S.J. Andrews, J.F. Loutit, and J.B. Clegg. 1985. A mouse β-globin mutant that is an exact model of hemoglobin Rainier in man. *Genetics* **110**: 709.

Popp, R.A., E.G. Bailiff, L.C. Skow, F.M. Johnson, and S.E. Lewis. 1983. Analysis of a mouse α-globin gene mutation induced by ethylnitrosourea. *Genetics* **105**: 157.

Pretsch, W. and D.J. Charles. 1984. Detection of dominant enzyme mutants in mice: Model studies for mutations in man. *IARC Sci. Publ.* **59**: 361.

Richardson, K.K., F.C. Richardson, R.M. Crosby, J.A. Swenberg, and T.R. Skopek. 1987. DNA base changes and alkylation following *in vivo* exposure of *Escherichia coli* to *N*-methyl-*N*-nitrosourea and *N*-ethyl-*N*-nitrosourea. *Proc. Natl. Acad. Sci.* **84**: 344.

Russell, L.B. 1986. Information from specific-locus mutants on the nature of induced and spontaneous mutations in the mouse. In *Genetic toxicology of environmental chemicals. Part B: Genetic effects and applied mutagenesis* (ed. C. Ramel et al.), p. 437. A.R. Liss, New York.

Russell, W.L., E.M. Kelly, P.R. Hunsicker, J.W. Bangham, S.C. Maddux, and E.L. Phipps. 1979. Specific-locus test shows ethylnitrosourea to be the most potent mutagen in the mouse. *Proc. Natl. Acad. Sci.* **76**: 5818.

Schöneich, J. and R. Braun. 1986. Mutagenesis in germinal tissues of the mouse and the influence of DNA repair. *Hereditas* **104**: 166.

Soares, E.R. 1979. TEM-induced gene mutations at enzyme loci in the mouse. *Environ. Mutagen.* **1**: 19.

Zielenska, M., D. Beranek, and J.B. Guttenplan. 1988. Different mutational profiles induced by *N*-nitroso-*N*-ethylurea: Effects of dose and error-prone DNA repair and correlations with DNA adducts. *Environ. Molec. Mutagen.* **11**: 473.

COMMENTS

Lewis: I am very interested in your missing homodimer phenomenon. You have a heterodimer proper and then you are missing the homodimer from the mutant. We have one. It's at IDH-1 and it was of spontaneous origin.

Peters: We have another spontaneous Gpi like that as well where there is a heterodimer and a missing homodimer. You find that situation in the red cell, and in the red cell you get very little protein turnover—well, there's no new protein synthesis—and you see just that picture. That is a stability mutant.

Lewis: That's what I suspected, although we haven't done rigorous studies.

Smithies: It might not be missing. It may be there, but just not enzymatically active.

Peters: Yes. I suppose the heterodimer would work. You would still expect the heterodimer to work.

Smithies: Yes. You can't assume that the protein isn't there; you can only assume that it has no activity.

Lewis: Right, but I'll tell you one thing, there is a big streak associated with this. You're right, it has to be done, but my suspicion is that it's unstable. The homozygous for that missing heterodimer mutant has got a streak going up the gel.

Peters: If we keep the extracts for more than a few hours, we can't find the band again. That's why I would expect that one to be unstable.

Smithies: Yes.

Mohrenweiser: Do you get conceptions from the homozygotes?

Peters: No. That's lethal.

Mohrenweiser: Is there at least an embryo formed?

Peters: Yes.

Mohrenweiser: So that you could pluck out DNA?

Peters: Yes. We know that for one of them you can find embryos up to about 8 days. We were hoping that some of the others might last a bit longer, because it's much easier to pick out conceptuses when they're older than 8 days. In fact, the conceptuses that lack Gpi are dying by 8 days, and just the trophoblasts last a few days longer than that.

Lewis: All the Gpi mutants are null mutants. True nulls are lethal.

Peters: Well, we would actually expect them to be lethal by about 6 days because the Gpi is switched on at 3 1/2 days and the maternal Gpi is meant to last 5 1/2–6 days. In fact, we were quite surprised that the conceptuses that are null-null homozygotes, true real null-null homozygotes, can actually last until 8 1/2–9 days. It's quite surprising in that there are still trophoblasts there. This is the work of John West.

Investigating Inborn Errors of Phenylalanine Metabolism by Efficient Mutagenesis of the Mouse Germ Line

J. DAVID MCDONALD, ALEXANDRA SHEDLOVSKY, AND WILLIAM F. DOVE
McArdle Laboratory for Cancer Research
University of Wisconsin
Madison, Wisconsin 53706

OVERVIEW

It has long been desirable to produce mutant laboratory animal strains to serve as models for human diseases. This desire is now a reality in the laboratory mouse through the use of highly efficient germ-line mutagens like ethylnitrosourea (ENU). A project designed to produce mice defective in phenylalanine (PHE) catabolism has yielded four such strains. The cardinal symptom of impaired PHE metabolism is elevated blood PHE levels, or hyperphenylalaninemia (HPH). The HPH-1 strain is deficient in GTP-cyclohydrolase (GTP-CH); the *hph-1* mutation has been mapped to mouse chromosome 14. Exposure to PHE not only produces a severe and chronic HPH, but also causes interesting effects on the breeding performance of female mice. The recessive mutations in the HPH-2 and HPH-4 strains, although not well characterized, appear to be allelic. The HPH-5 strain is deficient in PHE hydroxylase (PAH) activity, and the *hph-5* mutation has been mapped to chromosome 10 at or near the PAH locus. The ready isolation of these strains demonstrates that, given effective screens, any desired mutation can be produced with the high-efficiency mouse germ-line mutagenesis afforded by ENU. This research provides some useful mouse models for specific human diseases.

INTRODUCTION

Using the specific locus test, ENU was found to produce the highest rate of mutation among a number of mutagenic agents tested (Russell et al. 1979). The mutation frequency was later shown to be approximately doubled by fractionating the ENU dose over several exposures (Hitotsumachi et al. 1985). These discoveries have permitted production of mouse mutants to investigate murine genetics, development, metabolism, and biology (Bode 1984; Justice and Bode 1988; Shedlovsky et al. 1988). The success of these efforts has set the stage for ENU mutagenesis to provide mouse models for a number of human diseases.

Phenylketonuria (PKU), one of several inborn errors of PHE catabolism (the hyperphenylalaninemias), was one of the first metabolic disorders demonstrated to have a heritable component (Folling 1934). Considerable research has been carried out to characterize the biochemical mediators of PHE metabolism in mammalian systems and to define the particular deficiencies associated with each HPH syndrome (for review, see Scriver et al. 1988, 1989).

PHE catabolism occurs predominantly in the liver. The initial and rate-limiting step is hydroxylation to form tyrosine. This reaction, catalyzed by PAH (Fisher et al. 1972), requires one equivalent of tetrahydrobiopterin (BH_4) per equivalent tyrosine produced (Kaufman and Levenberg 1959). BH_4 is oxidized during the course of the PAH reaction, is regenerated by dihydropteridine reductase (DHPR) (Kaufman 1971), and is synthesized, through a number of intermediates, from GTP. The initial and rate-limiting step in the synthesis of BH_4 is catalyzed by the enzyme GTP-CH (Milstein and Kaufman 1986). Of the human hyperphenylalaninemias, extreme deficiency in PAH activity (PKU) is by far the most common biochemical deficiency (Kaufman 1958). DHPR and GTP-CH deficiencies account for approximately 3% of human hyperphenylalaninemias.

METHODS

Mice were treated with ENU according to the method of Shedlovsky et al. (1986). They were then mated in a three-generation breeding protocol to produce potential homozygous mutant animals to screen for the desirable phenotype of either spontaneously elevated serum PHE levels in neonates (Bode et al. 1988) or the inability of weanlings to effectively clear a PHE challenge (McDonald et al. 1990) administered by intraperitoneal injection of aqueous PHE solution to deliver a 1 mg PHE/g body weight exposure. Blood levels of PHE were determined 3.5 hours after challenge using the Guthrie assay (Guthrie and Susi 1963). Subsequent litters from the parents of Guthrie-positive animals were tested to establish heritability of the HPH phenotype.

RESULTS

From 349 offspring of ENU-treated animals, four independent HPH mutant strains have been isolated; each mutation is fully penetrant and inherited as an autosomal recessive.

HPH-1, isolated on the basis of its neonatal HPH, was the first such mutation isolated. The mutant is defective in PHE clearance unless BH_4 is administered (McDonald and Bode 1988). By enzyme assay, HPH-1 mutants were shown to be deficient in hepatic GTP-CH activity but not in PAH and DHPR activities (McDonald et al. 1988).

The *hph-1* mutation was induced in a (C57BL/6J x CBA/CaJ) F_1 and was mapped (Bode et al. 1988) in an interspecific cross between *Mus musculus domesticus* and *Mus spretus*/Pas. Interspecific backcross offspring were typed for their HPH status and also for other previously mapped polymorphic genetic loci. The complete concordance in the segregation of *Np-1* and *hph-1* in 50 such backcross animals localizes *hph-1* to the proximal region of chromosome 14. In an independent mapping experiment, the *hph-1* mutation was shown to be linked to *Es-10*, another chromosome-14 locus.

Ad libitum exposure to 25 mg PHE/ml in their drinking water results in death of *hph-1* homozygotes but not of heterozygotes or wild-type animals. Ad libitum exposure of mutant females to 5 mg PHE/ml interferes with fertility: No offspring survive the first few hours, and some pregnancies are aborted before term. At 2.5 mg PHE/ml, no aborted pregnancies have been noted, and some offspring have survived to adulthood.

The remaining HPH mutants have all been isolated on the basis of inability to efficiently clear a PHE challenge. HPH-2 and HPH-4 are not well characterized. Complementation tests and further genetic crosses indicate that the two mutations may be allelic (data not shown). The biochemical lesions in these mutants have not yet been identified. In vitro enzyme assays reveal normal levels of hepatic PAH and DHPR activities in homozygous mutants, and their ability to clear a PHE challenge is not augmented by BH_4 supplementation (data not shown).

In vitro assay reveals a deficiency of hepatic PAH in HPH-5 mutants. In contrast, hepatic DHPR activities are normal, and BH_4 supplementation does not enhance PHE clearance (Table 1). Liver homogenates were prepared for PAH and q-DHPR assays according to the method of Zannoni (1976). Total protein levels were determined by the biuret method (Gornall et al. 1949), and the PAH activities were determined according to the method of Parniak and Kaufman (1981). q-DHPR activity in liver homogenates was determined according to the method of Arai et al.

Table 1
Enzyme Activities and PHE Clearance

Genotype	Liver PAH activity[a]	Liver q-DHPR activity[b]	PHE clearance index[c]	
			$+BH_4$	$-BH_4$
+/+	62 ± 6	6.1 ± 0.4	14 ± 3	13 ± 5
[hph-5/hph-5] and [+/+][d]	42 ± 1	n.d.	n.a.	n.a
hph-5/+	19 ± 2	n.d.	n.d.	n.d.
hph-5/hph-5	5 ± 5	5.4 ± 0.3	87 ± 4	85 ± 6

n.d. indicates not done. n.a. indicates not applicable.
[a]$\Delta O.D._{340nm}$ min^{-1} mg total protein^{-1} x 10^{-3}.
[b]$\Delta O.D._{550nm}$ min^{-1} μg total protein^{-1} x 10^{-3}.
[c][Serum PHE at 2 hr after injection/serum PHE at 1 hr after injection] x 10^2.
[d]50:50 mixture.

Table 2
Mapping *hph-5* to Chromosome 10

	Chromosome 10 locus tested	No. of progeny examined	No. of recombinants	
			expected	observed
Inbred line cross	*Sl*	147	73.5	6
Interspecific cross	*Pah*	29	14.5	0

(1982). Serum PHE clearance was followed by spectrofluorometric PHE determination (McCaman and Robins 1962). Animals were injected intraperitoneally at 1 mg PHE/g body weight with 25 mg/ml aqueous PHE. Serum was obtained by centrifugation of whole blood. The effect of BH_4 supplementation was determined as described previously by McDonald and Bode (1988).

An interspecific mapping experiment places the *hph-5* mutation at or near the *Pah* locus on mouse chromosome 10. Furthermore, the mutation has been mapped to chromosome 10 by an interstrain cross that demonstrated linkage to the *Sl* locus (Table 2). (*hph-5* +/+ *Sl*)F_1 mice were obtained by mating 129/Sv-*Sl*/+ females to a male homozygous for *hph-5* and not carrying *Sl*. Progeny exhibiting the Sl coat color were backcrossed to *hph-5* homozygotes. The resulting offspring were scored for both HPH and Sl. (*hph-5*/+*Pah^M*/*Pah^S*)F_1 females were produced by mating homozygous mutant *Mus musculus domesticus* females to *Mus spretus*/Pas males. *M. m. domesticus* and *M. spretus* carry an *Eco*RI restriction site polymorphism detected by a mouse *Pah* cDNA. We designate the *M. spretus* allele as *Pah^S* and the *M. m. domesticus* allele as *Pah^M*. F_1 females were backcrossed to homozygous mutant *M. m. domesticus* males. The progeny from this cross were typed for HPH and *Pah*. All 12 animals scored as non-HPH were *Pah^M*/*Pah^S*, and all 17 scored as HPH were homozygous for the *Pah^M* allele.

hph-5 homozygotes are viable and fertile when exposed to 25 mg PHE/ml ad libitum. Effects on neonatal development are currently under intensive study.

DISCUSSION

Although a number of other researchers have used ENU with success to provide mouse mutants to address different biological problems, this research substantiates that ENU is a very effective tool for producing specific desired mutant strains of biomedical importance.

By considering the catabolism of PHE in mammalian systems, one would predict that there are at least 7 loci mutable to an HPH phenotype and, thus, given a per locus frequency of 1 in 1500, one would expect about 1 independent mutation for every 215 gametes examined. Our observation of 3 distinct HPH mutants isolated from 347 gametes screened is more frequent than expected and may reflect an underestimate of the number of *hph* loci based on strictly biochemical considerations. However, the apparent repeat isolation (*hph-2* and *hph-4*) in our small sample suggests that the number of loci is indeed small. These considerations have not taken into account the possibility of hypermutable loci. Regardless of the explanation for this phenomenon, it is clear that ENU has been a very effective tool in isolating mutations that result in HPH. By taking these findings together with the results of other experimenters in the area of mouse ENU mutagenesis, a strong argument results that ENU can be used to isolate mutations in virtually any mouse gene.

HPH-1 mutants are characterized biochemically by a deficiency in hepatic GTP-CH activity. This feature indicates that these mutants may represent a model for atypical PKU, the human HPH syndrome characterized by this same biochemical deficiency. In humans, a deficiency in GTP-CH causes an even more serious disease syndrome than classic PKU, due to the pathology that accrues not only from chronic HPH, but also from catecholamine neurotransmitter deprivation. Although there is no apparent pathology in homozygous mutants when they are maintained on a standard laboratory diet, dramatic effects are observed when these mice are exposed to PHE in their drinking water. At 25 mg PHE/ml ad libitum, homozygous mutants are preferentially killed, while heterozygotes and wild-type animals survive and thrive. This observation suggests that *hph-1* homozygotes may provide a good model for the pathology that characterizes GTP-CH deficiency in humans. When the level of PHE exposure is reduced, homozygous mutants survive, but females have impaired fertility. Perhaps this effect can provide some insights into the problem of maternal PKU. Further reducing the ad libitum exposure level permits the survival of some offspring. We are currently working to define a dose at which 50% of developing fetuses survive to term (i.e., the ED_{50}). These mice will be examined for evidence of a characteristic pathology.

hph-1 has been genetically mapped to the proximal portion of mouse chromosome 14. This finding may have very important implications regarding the location of the human GTP-CH structural gene. *Np-1* is one member of a gene pair (along with *Ifng*) that appears to delineate a conserved region between mouse and man. Therefore, the location of the murine *hph-1* mutation may furnish an important clue concerning the location of the unmapped human gene.

The HPH-5 strain exhibits a hepatic deficiency of PAH activity. This is the same biochemical phenotype as the most common human HPH syndrome, classical PKU. It is possible that, for this reason, this mutant strain will serve as a biomedical model for at least some aspects of this important human disorder. We are currently investigating whether this mutant exhibits some of the central nervous system pathology characteristic of classic PKU in humans. Furthermore, we are exploring the effects of elevated PHE on the developing fetus in homozygous mutant mothers and on neonates. Finally, we are attempting to employ the mutant as a recipient in somatic gene therapy experiments.

The success of this pilot project opens the door on some exciting future applications of high-efficiency mutagenesis experiments designed to produce mouse disease models. Now that it has been demonstrated that a focused search for mutants can yield success, more wide-ranging searches can be undertaken. Potential homozygotes can be screened for many inborn errors of metabolism simultaneously by determining the organic-acid levels in their urine. There are literally hundreds of important disease phenotypes that can be detected in this manner. One would predict that individual mutant strains would appear at a high frequency.

In the more immediate future, we are undertaking to induce and isolate new alleles at the *hph-1* and *hph-5* loci in order to better understand the range of phenotypes that can be produced in the mouse by mutation at these loci. In particular, we are interested in studying the phenotypes of null alleles at these loci. Because we have identified conditions in which both HPH-1 and HPH-5 strains are fertile, we can detect new alleles at the loci by a specific locus screen much more efficiently than we were able to detect the initial mutations.

REFERENCES

Arai, N., K. Narisawa, H. Hayakawa, and K. Tada. 1982. Hyperphenylalaninemia due to dihydropteridine reductase deficiency: Diagnosis by enzyme assay on dried blood spots. *Pediatrics* **70**: 426.

Bode, V.C. 1984. Ethylnitrosourea mutagenesis and the isolation of mutant alleles for specific genes located in the *T* region of mouse chromosome 17. *Genetics* **108**: 457.

Bode, V.C., J.D. McDonald, J.-L. Guenet, and D. Simon. 1988. A mouse mutant with hereditary hyperphenylalaninemia induced by ethylnitrosourea mutagenesis. *Genetics* **118**: 299.

Fisher, D.B., R. Kirkwood, and S. Kaufman. 1972. Rat liver phenylalanine hydroxylase, an iron enzyme. *J. Biol. Chem.* **247**: 5161.

Folling, A. 1934. Uber Ausscheidung von Phenylbrenztraubensaure in den Harn als Stoflwechselanomalis in Verbindung mit Imbezillitat. *Hoppe-Seyler's Z. Physiol. Chem.* **227**: 169.

Gornall, A.G., C.J. Bardawill, and M.M. David. 1949. Determination of serum proteins by means of the biuret reagent. *J. Biol. Chem.* **177**: 751.

Guthrie, R. and A. Susi. 1963. A simple phenylalanine method for detecting phenylketonuria in large populations. *Pediatrics* **32**: 328.

Hitotsumachi, S., D.A. Carpenter, and W.L. Russell. 1985. Dose-repetition increases the mutagenic efficiency of *N*-ethyl-*N*-nitrosourea in mouse spermatogonia. *Proc. Natl. Acad. Sci.* **82**: 6619.

Justice, M.J. and V.C. Bode. 1988. Three ENU-induced alleles of the murine quaking locus are recessive lethal mutations. *Genet. Res.* **51**: 95.

Kaufman, S. 1958. Phenylalanine hydroxylase cofactor in phenylketonuria. *Science* **128**: 1506.

―――. 1971. The phenylalanine hydroxylating system of mammalian liver. *Adv. Enzymol.* **35**: 245.

Kaufman, S. and B. Levenberg. 1959. Further studies on the phenylalanine hydroxylation cofactor. *J. Biol. Chem.* **234:** 2683.

McCaman, M.W. and E. Robins. 1962. Fluorometric method for the determination of phenylalanine in serum. *J. Lab. Clin. Med.* **55:** 885.

McDonald, J.D. and V.C. Bode. 1988. Hyperphenylalaninemia in the *hph-1* mouse mutant. *Pediat. Res.* **23:** 63.

McDonald, J.D., V.C. Bode, W.F. Dove, and A. Shedlovsky. 1990. Pah^{hph-5}: A mouse mutant deficient in phenylalanine hydroxylase. *Proc. Natl. Acad. Sci.* (in press).

McDonald, J.D., R.G.H. Cotton, I. Jennings, F.D. Ledley, S.L.C. Woo, and V.C. Bode. 1988. Biochemical defect in the *hph-1* mouse mutant is a deficiency in GTP-cyclohydrolase activity. *J. Neurochem.* **50:** 655.

Milstein, S. and S. Kaufman. 1986. The biosynthesis of tetrahydrobiopterin in rat brain. In *Pteridines and folic acid derivatives* (ed. B.A. Cooper and V.M. Whitehead), p. 169. De Gruyter, Berlin.

Parniak, M.A. and S. Kaufman. 1981. Rat liver phenylalanine hydroxylase. *J. Biol. Chem.* **256:** 6876.

Russell, W., E. Kelly, P. Hunsicker, J. Bangham, S. Maddux, and E. Phipps. 1979. Specific locus test shows ethylnitrosourea to be the most potent mutagen in the mouse. *Proc. Natl. Acad. Sci.* **76:** 5818.

Scriver, C.R., S. Kaufman, and S.L.C. Woo. 1988. Mendelian hyperphenylalaninemia. *Annu. Rev. Genet.* **22:** 301.

------. 1989. The hyperphenylalaninemias. In *The metabolic basis for inherited disease* (ed. C.R. Scriver et al.), p. 495. McGraw-Hill, New York.

Shedlovsky, A., T.R. King, and W.F. Dove. 1988. Saturation germline mutagenesis of the murine *T* region including a lethal allele at the quaking locus. *Proc. Natl. Acad. Sci.* **85:** 180.

Shedlovsky, A., J.-L. Guenet, L.L. Johnson, and W.F. Dove. 1986. Induction of recessive lethal mutations in *T/t-H-2* region of the mouse genome by a point mutagen. *Genet. Res.* **47:** 135.

Zannoni, V.G. 1976. Liver phenylalanine hydroxylase assay. *Biochem. Med.* **16:** 251.

COMMENTS

Arnheim: The *hprt*-negative mice passed their psychological test with flying colors, as I understand it. Have you subjected any to behavioral tests?

McDonald: No, we haven't. Probably the most common question I get about these mice is, "are they retarded?" because that's one of the kinds of pathology. I don't know because it's not really my field. We're starting some collaborations with people who are accustomed to doing maze learning and things like that, but we don't call upon our mice to do too many difficult things. If they can eat and reproduce, then they're okay for us.

Hecht: With the *hph*-1 mutants, what phenotypic problems do the animals have as far as lacking that cofactor?

McDonald: That's a good question. The cofactor machinery, the reductase and the synthetic pathway, are also involved in synthesizing all of the catecholamine neurotransmitters. Serotonin, dopa, and all of those require tetrahydrobiopterin. Possibly that's why it's such a severe phenotype. We don't know. We know that we can kill them by exposing them to phenylalanine; they die relatively rapidly, too. As far as characteristic pathology, we haven't looked at that enough to be able to state that this is modeling the human syndrome.

Hecht: The fertility problems could be through a different pathway, too.

McDonald: Yes.

Peters: I was going to ask you about the mutation frequency.

McDonald: Actually, we are getting these mutants at about twice the rate that we would predict. There are 7 loci mutable to this phenotype, and so, one would expect (since these are single ENU exposures, the 1 in 1500 number is the number that is operant here) to get them about 1 every 200 or so. We have gotten 4 now in about 350, so we're getting them at about twice the rate. I don't know what to make of that. Maybe we've just been lucky.

Liane Russell: What mutagenesis regimen are you using?

McDonald: Single exposure, and then we wait. These are poststerile period animals, so we've never really found it necessary to give a multiple exposure.

Bishop: What dose?

McDonald: The first animal was an F_1 that could tolerate a higher exposure. The dose was 250 mg/kg. All of the subsequent animals were an inbred line called BTBR, and the dose was 200 mg/kg.

Lewis: As you know, I'm interested in your maternal experiments. Have you ever mated a +/+ male to one of these homozygous females?

McDonald: That has been done once. I was pondering whether I should show that because we only have one female on that trial, and the litter lived. So, this would be counterindicative, although it is very preliminary, of a true maternal PKU model, because true maternal PKU is independent of the fetal genotype; only the maternal genotype is important. This one result would tend to indicate that fetal genotype may be important too.

Lewis: You've got to do more of them, and, even better, mate a heterozygous male to it and see if there's any difference as to who comes out.

McDonald: Right. Yes.

Favor: How difficult are the enzyme assays to carry out? If it's not very difficult you could just screen for enzyme activity reduction rather than use the amino acid levels as your phenotype.

McDonald: Right. In heterozygotes?

Favor: Screen for new mutations of whatever in heterozygotes.

McDonald: It's much easier to screen for this elevation of blood phenylalanine. There is a very easy assay called the Guthrie assay with which you can screen literally 100 animals at a time. In every one of these mutants that we have characterized biochemically, once you know what you are looking for you can see a heterozygous phenotype; but it's a low-probability thing to go looking for that heterozygous phenotype because your false-positive rate comes up and you spend all your time running down false-positives.

Favor: With enzyme activity?

McDonald: Correct.

Favor: It depends. Pretsch identifies fairly well enzyme activity mutations. The other thing is if you have unlimited mouse space.

McDonald: We've never needed unlimited mouse space for this project.

Davisson: With respect to the fetal genotype versus the mother's, how long have you followed that litter that survived?

McDonald: We still have them. They're still alive.

Davisson: We've been working with a new allele of toxic milk, and that's a maternal effect on the progeny, and there is an effect of the fetal genotype, but they eventually all die before we can look at them.

William Russell: You might in your experiments use a regimen like 4 x 100, which seems to be much more effective.

McDonald: Right. We're also interested in looking at other mutagens, because now that we can use the specific-locus assay directly, we can afford to give a little on the efficiency of mutagenesis and maybe get a more desirable type of mutational lesion, because, as you saw, in each one of these that we characterized there was some residual activity. As opposed to what Jo [Peters] and Susan [Lewis] have gotten, we have not been very good at getting complete knockout, complete null mutations. We've been getting residuals.

William Russell: Using the specific-locus approach, it would be nice if you could use a closely linked marker to separate your new mutants from the old ones.

McDonald: The ideal thing would be a RFLP at the locus. As you know, between inbred lines those are sometimes hard to find.

Bishop: You mentioned that the females have a maternal toxicity that occurs in humans with mothers for PKU, and you said they had fetal malformations. What is that fetal malformation?

McDonald: It's similar, as far as I understand it, to the malformations one sees in classic PKU, so it's severe mental retardation, microencephaly, and some internal problems as well.

Bishop: The models developed, the four mutants that you have, they are different genetically?

McDonald: Correct.

Bishop: How do all those relate back to the maternals, and have you looked at the fetuses that are coming off of those? You don't have any comparative animal model, do you? Or are they turning out to be the same model?

McDonald: We have looked for maternal effects—I hesitate to call them maternal effects because they're not strictly maternal effects. That's why I called them "pregnancy outcome." We don't see it in *hph*-5, but we do see some kind of an effect on pregnancy outcome in 2 and 4, but it doesn't look like it's phenylalanine-dependent. On certain feeds, they don't breed; on others, they do. It's almost like this is permissive and this is restrictive. We have done feed shift experiments and we can even see it happening.

Bishop: Which one is the PKU that's 96%?

McDonald: That is *hph*-5. We are still looking, but we have not yet seen any of these effects on fertility in *hph*-5. That's still a developing story. It's a relatively new mutant.

Fritz: When you're looking for your ED_{50} for phenylalanine, have you ever tried, instead of just giving the ad libitum phenylalanine in the water, having two bottles, one pure water and one phenylalanine, to see if the animals can discriminate?

McDonald: Yes. Actually, the phenylalanine water is quite bitter. So, I would predict that, given a choice, they would really go to the pure water. Not only would the phenylalanine water make them sick, but I think a wild-type mouse would avoid it if possible. We have thought of going to the NutraSweet for that very reason.

Bridges: I was a little puzzled when you said that your mutation rate seemed to be twice what you would expect. I can't see how you can know what to expect unless you know how many AT base pairs are at risk. How else can you know what to expect?

McDonald: This is based on numbers that have come to us from the specific locus screen. Those are based on a number of targets, so it is an average per locus frequency. But that is a "hit." On that basis, you're saying that if a gene is spread out, it's a larger target.

Bridges: We are dealing with a mutagen that appears to be specific for AT, and so in any gene what is crucial is the number of AT base pairs at sites that would give the phenotype you can detect. That is going to vary widely. From cellular systems we know this happens, so a factor of two is nothing, actually.

McDonald: Okay. Maybe we've picked the right system then.

Lewis: There is a fair amount of per locus data both in the visible test and the electrophoretic test, and for dominant cataracts. This is totally aside from gene size. We seem to have a new member of the "AT Club," don't we? Anyway, six out of six is pretty good. The point is, we don't even know the gene size, the composition, or whatever. What he is just looking at is what has been gotten in other systems. This is really interesting because it's relevant disease loci that do seem to be at least as susceptible, as far as a ball park figure goes, as the so-called "tester" type genes.

Mohrenweiser: Have you looked at the kinetic properties of the 2 and 4 mutants?

McDonald: Yes.

Mohrenweiser: So it's not only the activity that is normal, but also the kinetics; in other words, you don't have a K_m variate?

McDonald: Correct. They're not thermally instable either. We have looked at a lot of things like that. They perform quite normally in vitro.

Mohrenweiser: Have you considered looking at the rest of the Guthrie tests? Once you have the mouse, the blood, etc., there is a whole range of Guthrie tests you can do.

McDonald: Yes. This was kind of a pilot project to show that you could say, "I'm interested in this disease phenotype" and produce it. Now that we have shown that, we are opening up to a less focus. We have always kept our focus in this project—less focus now. There is some nice instrumentation for testing organic gases in the urine in which you are essentially screening for hundreds of important disease loci in a single sample. We are pursuing that with John Wolf also at the Weissman Institute. It's a metabolic way.

Favor: Are you saying that *hph*-2 and -4 are at a locus other than what you had expected to start with?

McDonald: It's at a different locus from either 1 or 5, so it's an independent locus.

Favor: It's not one of the enzyme loci that you had expected to start with?

McDonald: Only *Pah* had been mapped when we started this, so I really can't answer that.

Favor: No. I was saying you had expected seven enzyme loci that would result in HPH.

McDonald: Right.

Favor: These two seem to be transport mutations?

McDonald: We're proceeding on that assumption because all of the enzyme results say that they're normal activities.

Favor: So your expectation based on the number of structural enzyme loci has been increasing?

McDonald: No, because when I said seven, I was including a transport mutant.

Liane Russell: Could you tell us again why you think there are seven?

McDonald: Well, there are six enzymes that we know are involved, and then there is a common carrier for all aromatic amino acids to the hepatocyte where this has to happen.

Liane Russell: So you have essentially found the transport one, and two out of the other six, you think?

McDonald: We think, yes.

Factors Affecting the Nature of Induced Mutations

LIANE B. RUSSELL, WILLIAM L. RUSSELL, EUGENE M. RINCHIK, AND
PATRICIA R. HUNSICKER
Biology Division, Oak Ridge National Laboratory
Oak Ridge, Tennessee 37831-8077

OVERVIEW

The recent considerable expansion of specific-locus-mutation data has made possible an examination of the effects of germ-cell stage on both quantity of mutation yield and nature of mutations. For chemicals mutagenic in post-stem-cell stages, three patterns have been identified according to the stages in which they elicit maximum response: (1) early spermatozoa and late spermatids; (2) early spermatids; and (3) differentiating spermatogonia. The majority of chemicals tested fall into pattern 1. Chemicals that are also mutagenic in stem-cell spermatogonia do not preferentially belong to any one of these three categories.

For only one chemical (chlorambucil [CHL]) has an entire set of mutations been analyzed molecularly. However, the results of genetic and molecular analyses of genomic regions surrounding six of the specific-locus markers allow us to conclude that any mutation that causes lethality of homozygotes (in the case of d, prenatal lethality, specifically) must involve one or more loci in addition to the marked one. Such mutations have been classified as "large lesions" (LL), the remainder as "other lesions" (OL). Analysis of the data shows that, regardless of the nature of the chemical (pattern-1, -2, or -3), (1) LLs constitute a very low proportion of the mutations induced in either stem-cell or differentiating spermatogonia, and (2) LLs constitute a high proportion of mutations induced in postmeiotic stages. Chemicals that are active in both pre- and postmeiotic stages produce LL or OL mutations, depending on cell stage.

The distribution of mutations among the seven marked loci was also examined. For mutations induced in stem-cell spermatogonia, this spectrum was similar for all groups of chemicals. The distribution for differentiating spermatogonia generally resembled this spectrum. Since only a very small fraction of spermatogonial mutations are LLs, this spectrum may be indicative of the relative sizes of the target loci. On the other hand, the spectrum for postmeiotically induced mutations departs in several ways from that for the other groups.

On the basis of both lesion size and distribution of mutations among loci, we conclude that germ-cell stage, rather than identity of the chemical, is the major determinant of the nature of chemically induced mutations, and that the distinction is between pre- and postmeiotic rather than between stem-cell and post-stem-cell stages.

INTRODUCTION

Not too many years ago, the classification of agents as mutagens or nonmutagens constituted the chief focus of interest in several areas of mutagenesis research. It has long been clear, however, that where germ cells are concerned, identification of a mutagen is only the first step in the assessment of genetic risk. Exposure to the mutagen involves many variables, among which the *type* of germ cell exposed has probably the strongest influence on both the quantitative and qualitative outcomes.

The fact that the number of potential mutagens is almost infinite provides a strong impetus to search for commonalities or relationships in patterns of effects. Are there, for example, correlations between certain arrays of phenotypic effects detectable in offspring of mutagenized individuals and certain structural types of genetic lesions? Is the nature of genetic lesions determined primarily by the mutagen, or by the germ-cell type that is exposed to the mutagen? In this paper, we examine the latter question.

It would be ideal if data were available to compare all germ-cell stages in both sexes. Unfortunately, there is little information on induction of specific-locus mutations in females. Only four chemicals have been tested (Cattanach 1982; Ehling 1984; Ehling and Neuhäuser-Klaus 1988c), and only two of them were reported positive. This paucity of data is all the more regrettable in the light of recent preliminary evidence that, for certain classes of mutagens, the sexes may differ greatly with regard to induction of dominant lethals (Katoh et al. 1990). For males, the past 6 years have seen a considerable expansion in the specific-locus mutagenesis data for spermatogenic cell stages. By 1983, 16 chemicals had given conclusive test results in stem-cell spermatogonia, but only 6 in post-stem-cell stages; by 1989, the respective numbers were 30 and 14 (L.B. Russell 1990).

EFFECT OF GERM-CELL STAGE ON QUANTITY OF MUTATION YIELD

In the specific-locus test (W.L. Russell 1951), mice of a strain homozygous for seven recessive markers are mated with exposed (or control) mice that carry the wild-type genes at the corresponding loci. Recessive mutations at, or involving, any of the marked loci are recovered in the first-generation offspring and are generally propagated in stocks that are analyzable genetically or molecularly. An examination of specific-locus-test results for 14 chemicals that have been conclusively tested in *both* spermatogonial stem cells and post-stem-cell stages shows none that was positive in the former but not the latter (Table 1). On the other hand, six chemicals that were negative in stem cells were positive in summed post-stem-cell stages; of the four that were positive in both sets of stages, two (triethylenemelamine [TEM] and methylnitrosourea [MNU]) yielded lower mutation rates in stem cells than they did in the most responsive post-stem-cell stage(s) (Cattanach 1966, 1967; W.L.

Table 1
Specific-locus Test Outcomes for Chemicals That Have Given Conclusive Results in Both Stem-cell Spermatogonia and Post-stem-cell Stages

Chemical[a]	Stem-cell spermatogonia	Relation[b]	Post-stem cells
ENU	+	≈	+
MNU	+	<	+
PRC	+	≈	+
TEM	+	<	+
AA	−		+
CHL	−		+
MLP	−		+
CPP	−		+
DES	−		+
EMS	−		+
6MP	−		−
ADR	−		−
PLA	−		−
UR	−		−

[a] ENU, ethylnitrosourea (W.L. Russell et al. 1979; Ehling and Neuhäuser-Klaus 1984; W.L. Russell, unpubl.); MNU, methylnitrosourea (W.L. Russell and Hunsicker 1983; W.L. Russell, unpubl.); PRC, procarbazine hydrochloride (Ehling and Neuhäuser 1979; Kratochvilova et al. 1988); TEM, triethylenemelamine (Cattanach 1966, 1967); AA, acrylamide monomer (L.B. Russell, unpubl.); CHL, chlorambucil (L.B. Russell et al. 1989); MLP, melphalan (L.B. Russell, unpubl.); CPP, cyclophosphamide (L.B. Russell et al. 1981; Ehling and Neuhäuser-Klaus 1988a); DES, diethyl sulfate (Ehling and Neuhäuser-Klaus 1988b); EMS, ethyl methane sulfonate (L.B. Russell et al. 1981; 6MP, 6-mercaptopurine (L.B. Russell and Hunsicker 1987); ADR, adriamycin (L.B. Russell, unpubl.); PLA, platinol (L.B. Russell, unpubl.); UR, urethane (L.B. Russell et al. 1987).

[b] Comparison of mutation rate in stem-cell spermatogonia with that in most responsive post-stem-cell stage. (<) Rate lower in stem-cell spermatogonia.

Russell and Hunsicker 1983; W.L. Russell, unpubl.). Only with two chemicals does the stem-cell yield of mutations match that from either the most responsive post-stem-cell stage (ethylnitrosourea [ENU], W.L. Russell et al. 1979; W.L. Russell, unpubl.) or from post-stem-cell stages overall (procarbazine hydrochloride [PRC], Ehling and Neuhäuser 1979; Kratochvilova et al. 1988).

Stem-cell spermatogonia are located in the basal compartment of the seminiferous tubule, closest to blood and lymph circulation. The mutagenesis results showing a generally much higher response of post-stem-cell than stem-cell stages cannot, therefore, be explained in terms of mutagen transport, unless one assumes that the action of the chemicals tested to date is intensified by passage through the

"Sertoli-cell barrier" (L.D. Russell, this volume) to reach postspermatogonial stages. It has been suggested that stem-cell spermatogonia have effective repair systems, and this has been demonstrated for ENU, a chemical that is mutagenic in these cells (W.L. Russell 1986).

The procedure of sequential matings, which was first used in mouse radiation mutagenesis and has been a recommended protocol for specific-locus studies with chemicals (L.B. Russell et al. 1981), yields results indicative of effects in successive germ-cell stages. Each of the 11 chemicals that have given positive specific-locus-test results in summed post-stem-cell stages has been found to elicit differential response of different substages.

Within the post-stem-cell period, three distinct patterns of maximum response can be identified (Fig. 1). Most of the chemicals (7 of the 11) elicit their maximum

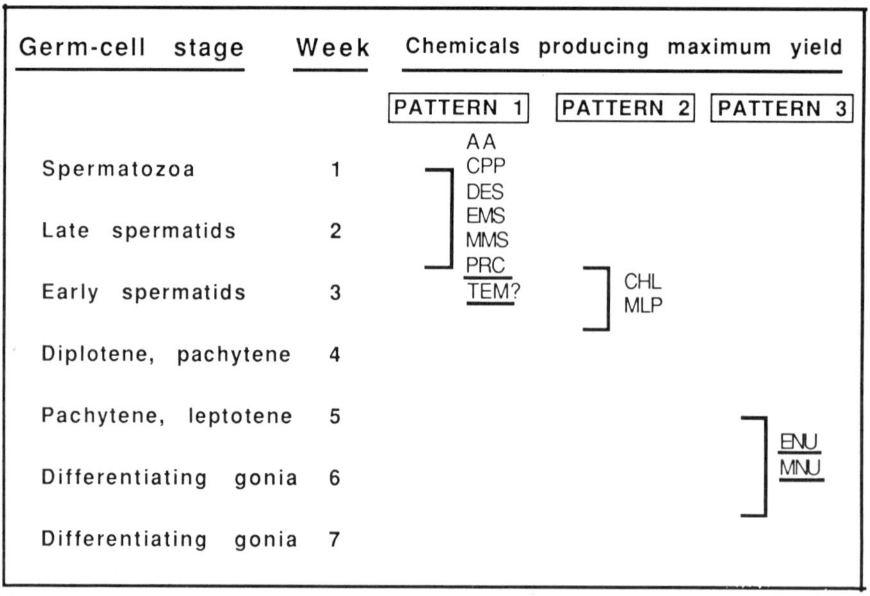

Germ-cell stage	Week	Chemicals producing maximum yield		
		PATTERN 1	PATTERN 2	PATTERN 3
Spermatozoa	1	AA CPP DES EMS MMS PRC TEM?	CHL MLP	ENU MNU
Late spermatids	2			
Early spermatids	3			
Diplotene, pachytene	4			
Pachytene, leptotene	5			
Differentiating gonia	6			
Differentiating gonia	7			

Figure 1
Germ-cell-stage sensitivity patterns for chemicals that produce a positive response in post-stem-cell stages of spermatogenesis. Chemicals are listed opposite the post-stem-cell stage(s) that yield maximum frequencies of specific-locus mutations. The assignment of TEM to pattern 1 is questionable. Those post-stem-cell-positive chemicals that also elicit a positive response in stem-cell spermatogonia are underlined. (Note: MMS, methyl methanesulfonate, has not been conclusively tested in stem-cell spermatogonia. Stem-cell-positive chemicals that have not been tested in post-stem-cell stages are not shown.) Weeks shown represent successive 7-day intervals between exposure and conception. For abbreviations and references, see footnote a to Table 1.

response in early spermatozoa and late spermatids (pattern 1). *Early* spermatids yield by far the highest rate of mutations following exposure to CHL (L.B. Russell et al. 1989) (pattern 2), and work in progress (L.B. Russell, unpubl.) indicates that the related compound, melphalan (MLP), will follow the same pattern. Finally, MNU and ENU have their greatest effect in differentiating spermatogonia (and possibly preleptotene spermatocytes?), i.e., in premeiotic, though still post-stem-cell stages (pattern 3). It should be noted that no chemicals have been discovered that produce maximum mutation yields from exposed pachytene or leptotene spermatocytes.

Both of the pattern-3 compounds also elicit positive response from spermatogonial stem cells, although the response for MNU is weak. Among the 9 *post*meiotically active chemicals (patterns 1 and 2), two (PRC and TEM) elicit a positive response in stem cells. Thus, post-stem-cell-mutagenic chemicals that are also positive in stem-cell spermatogonia do not preferentially belong to any one of the three pattern groups.

NATURE OF SPECIFIC-LOCUS MUTATIONS

Recent and continuing advances in the genetic and molecular characterization of the genomic regions surrounding the loci used as markers in the specific-locus test are providing ever-increasing opportunities for examining the factors that might affect the *nature* of mammalian germ-line mutations. Extensive genetic analyses have been carried out for large sets of radiation-induced mutations involving three of the seven loci (*c, d,* and *se*) (L.B. Russell 1971; L.B. Russell et al. 1982). Complementation studies utilizing combinations of overlapping deletions, each of which is homozygous lethal, yielded information on the phenotypes associated with complete ablation of these loci. Animals in which either *se* or *c* is ablated are completely viable—indicating that the genes at these two loci are not vital ones—and resemble *se/se* and *c/c*, respectively. The null phenotype in the case of *d* is juvenile lethal and dilute opisthotonic (like d^{op}/d^{op}). Subsequent molecular studies confirmed that the mutations assumed to be deletions on the basis of the genetic evidence were indeed deleted for DNA sequences (Rinchik et al. 1986; L.B. Russell and Rinchik 1987; L.B. Russell 1989). It can therefore be assumed that any *prenatally lethal* mutation affecting *d*, or any nonviable mutation affecting *se* or *c*, must involve loci additional to the marked ones.

Studies currently in progress with *b*-locus mutations (E.M. Rinchik, unpubl.), *a*-locus mutations (L.B. Russell; R.P. Woychik, both unpubl.), and *p*-locus mutations (L.B. Russell and C.S. Montgomery; D.K. Johnson and E.M. Rinchik, both unpubl.) make it highly probable that, as in the case of *se* and *c*, the null phenotype is viable in each case. That is, any *lethal* mutation involving *b, a,* or *p* can also be concluded to involve loci additional to the marked ones. Since complementing overlapping deletions are characterized by the null phenotype for each of the markers, any homozygous viable mutation having a phenotype other than null (e.g., a c-locus mutation producing some pigment) is probably an intragenic one (L.B. Russell and Rinchik 1987). Thus, for at least six of the seven specific-locus markers, the viability/lethality and phenotype of homozygotes can provide clues on whether large or small genomic lesions have been produced by the mutagen. In the case of s, the phenotype associated with complete ablation of the locus is not yet known; the possibility that a homozygous-lethal phenotype might be produced from intragenic as well as multigenic lesions can, therefore, not be discounted.

As molecular analysis of the markers and the regions surrounding them progresses, it will be possible to refine the characterization of mutations beyond that derivable on the basis of homozygous phenotype. A recent example is the molecular study of mutations induced by CHL (L.B. Russell et al. 1989). Southern-blot analyses with probes at, or closely linked to, c, b, and d were performed for ten of the mutations (Rinchik et al. 1990). All eight of the mutations (two c, three b, and three d) that had arisen in post-stem-cell stages of spermatogenesis were found to be deleted for DNA sequences, whereas two d-locus mutations that had arisen in spermatogonial stem cells were not deleted for either of two d-region probes.

The study provided information beyond the mere diagnosis of the deletional nature of the CHL post-stem-cell mutations. For example, in the case of the c-locus mutations, all sequences normally recognized by a tyrosinase cDNA clone were found to be completely deleted. Since the coding sequences for the mouse tyrosinase gene extend over a 70-kb length of genomic DNA, the CHL-induced c deletions are now known to be at least 70 kb long. That they are, in fact, probably considerably longer was demonstrated by the finding that they are also deleted for p23.3, a probe flanking the *Emv-23* integration site, which is ≤0.5 cM distal to c. The d-locus deletions turned out to be a heterogeneous group. All three were deleted for sequences recognized by probe p94.1, less than 1 cM proximal to d; but only two of them were also deleted for sequences recognized by p0.3, a probe derived from genomic DNA 3' to the integration of *Emv-3*, the provirus thought to be responsible for the original d mutation.

FACTORS AFFECTING THE NATURE OF SPECIFIC-LOCUS MUTATIONS

Analysis of mutations induced by ionizing radiations has indicated that, for a given type of radiation, there is a strong effect of germ-cell stage on the nature of mutations; and for a given germ-cell stage, there is an effect (although not as large) of type of radiation (L.B. Russell and Rinchik 1987). Are similar effects of cell type and/or mutagen also demonstrable for chemicals? If so, will one or the other of these factors be found to predominate? To answer these questions, we have examined available information on the nature of mutations induced by different chemicals in different germ-cell stages.

To date, CHL is the only chemical mutagen for which a whole set of mutations has been analyzed with molecular probes. For all other data sets, information on the phenotype of homozygotes can, however, be used to assess the nature of the mutations. For the subsequent analysis we have classified mutations as large lesions if they are homozygous lethal (this makes, for the time being, the assumption that genes at the *s* locus are, like those at the other markers, non-vital genes). Lethals not included in the large-lesion classification are *dilute-opisthotonic* juvenile lethals (d^{op} [L.B. Russell 1971]), which represent the null state at the *d* locus, without giving evidence of involvement of neighboring loci (Jenkins et al. 1989). Mutations not classified as large lesions have been designated simply as other lesions. Although we would have preferred to separate the viable nulls from the viable non-nulls, this distinction has sometimes not been made in the literature being reviewed. Viable nulls may also be deletions extending beyond the marker locus (as proved for *c* mutations by E.M. Rinchik et al., in prep.), although generally smaller than those associated with the lethals; non-nulls have a high probability of being intragenic lesions (see above).

Tables 2, 3, and 4 summarize information on large-lesion (LL) and other-lesion (OL) mutations at each genetic marker for chemicals that have given positive results, respectively, in spermatogonial stem cells, differentiating spermatogonia, and postmeiotic stages. For the sum of all loci, Figure 2 shows the relative proportions of LL and OL mutations induced in these various germ-cell stages; information for ENU and MNU (the pattern-3 chemicals, see Fig. 1) is shown separately from that for all other chemicals.

It is clear that mutations induced in stem cells include very few large lesions, regardless of whether the mutagen is a pattern-3 or a pattern-1 chemical (3.5 and 4.8%, respectively; or, 0.4 and 1.7% if *s* is omitted [see above]). Large lesions are, likewise, very rare (2.1%; or, 0 if *s* is omitted) among mutations induced in differentiating spermatogonia, where data are available only for pattern-3 chemicals. In contrast, the proportion of LLs is very high among mutations induced in postmeiotic stages by pattern-1 and -2 chemicals (66.7%), and apparently also by pattern-3 chemicals (50%, although the sample, in this case, is small); the respective proportions are 65.8 and 40%, if *s* is omitted. (That pattern-3 chemicals are, in fact, capable of producing significant proportions of LL mutations in cells other than spermatogonia is confirmed by ENU results for oocytes [Ehling and Neuhäuser-Klaus 1988c].) The comparison between stem cells and postmeiotic stages can also be made for the pattern-1 chemical PRC (the TEM data are too limited for this purpose). The proportion of LLs among PRC-induced mutations is only 4.3% for stem cells, but is 10 times higher in postmeiotic stages, in parallel with the comparisons discussed above. All available data thus support the conclusion that germ-cell stage (rather than chemical) is the major determinant of the nature of chemically induced mutations. (It should be noted that both ENU and PRC are capable of producing mosaic mutants when administered to late secondary oocytes

Table 2
Distribution by Locus and Lesion Size of Mutations Induced in Spermatogonial Stem Cells

Chemical[a]	a		b		c		d		d se		se		p		s		Total		
	LL[b]	OL[b]	LL	OL	LL	OL	LL	OL	LL	OL	LL	OL	LL	OL	LL	OL	LL	OL	?[c]
MMC[1]	0	1	0	2	0	1	0	3[d]	0	0	0	0	0	4	0	1	0	12	1
TEM[2]	1	0	0	0	0	0	0	3	0	0	0	0	0	1	0	0	1	4	1
PRC[1]	0	1	0	9	0	7	0	8[d]	0	0	0	3	0	15	2	1	2	44	9
ENU[1]	0	4	0	7	0	7	0	17[d]	0	0	0	6	0	24	4	0	4	65	92
ENU[3]	0	4	0	32	0	25	0	50	0	0	0	11	0	51	4	8	4	181	30
MNU[4]	0	0	0	0	0	2	0	1	0	0	0	0	1	0	0	0	1	3	0
Total	1	10	0	50	0	42	0	82	0	0	0	20	1	95	10	10	12	308	134
Total mutations[e]																			
ENU, MNU	15		58		53		89		0		21		122		22			380	
Other	4		13		10		14		0		3		23		7			74	

[a] Abbreviations: MMC, mitomycin-C; for all others, see footnote a to Table 1. References: [1]Ehling and Neuhäuser-Klaus (1984); [2]Cattanach (1966); [3]W.L. Russell (1979), W.L. Russell (1982), W.L. Russell (unpubl.); [4]W.L. Russell and Hunsicker (1983), W.L. Russell (unpubl.).
[b] (LL) Large genomic lesions; (OL) other lesions (see text).
[c] No information available on viability of homozygotes.
[d] It is assumed that dilute mutations described as "lethal" in these publications were d^{op}, which are classified as OL (see text). The numbers of such mutations for MMC, PRC, and ENU (Ref. 1) were, respectively, 3, 6, and 13, the remainder being described as "viable." Of the 50 tested dilute mutations observed in the ENU experiments of WLR (Ref. 3), 37 were d^{op}, 9 d^x, and 4 d; none was prenatally lethal. All dilutes observed in TEM and MNU experiments were d^{op}.
[e] Including those for which no information is available on viability of homozygotes.

Table 3
Distribution by Locus and Lesion Size of Mutations Induced in Differentiating Spermatogonia

Chemical[a]	\multicolumn{2}{c}{a}		b		c		d		d se		se		p		s		Total[d]	
	LL[c]	OL[c]	LL	OL	LL	OL	LL	OL	LL	OL	LL	OL	LL	OL	LL	OL	LL	OL
ENU[b]	0	0	0	2	0	2	0	1	0	0	0	0	0	0	0	0	0	6
MNU[b]	0	4	0	10	0	5	0	5	0	0	0	2	0	9	1	1	1	36
									Total mutations									
ENU, MNU	4		12		7		6		0		2		10		2		43	

[a] (ENU) Ethylnitrosourea; (MNU) methylnitrosourea.
[b] Reference: W.L. Russell, unpubl.
[c] (LL) Large genomic lesions; (OL) other lesions (see text).
[d] All mutations were tested for viability of homozygotes.

Table 4
Distribution by Locus and Lesion Size of Mutations Induced in Postmeiotic Stages

Chemical[a]	a		b		c		d^c		$d\,se$		se		p		s		Total		
	LL[b]	OL[b]	LL	OL	LL	OL	LL	OL	LL	OL	LL	OL	LL	OL	LL	OL	LL	OL	?[d]
MMS[1]	0	0	3	0	0	0	0	0	1	0	1	0	4	1	1	3	10	4	3
EMS[1]	0	0	0	1	0	0	0	0	0	0	0	0	0	0	0	0	0	1	15
CPP[2]	0	0	0	0	1	1	0	0	0	0	1	1	1	0	0	0	3	2	3
TEM[3]	1	0	0	0	0	0	0	2	0	0	0	0	0	0	0	0	1	2	1
DES[4]	0	0	0	0	1	0	0	0	0	0	0	0	3	1	1	0	5	1	2
PRC[5]	0	1	0	1	0	0	0	0	0	0	0	0	0	1	2	0	2	3	0
AA[6]	0	0	1	1	0	0	0	0	1	0	0	0	1	0	0	1	3	2	0
CHL[7]	0	0	2	0	2	0	2	0	0	1	0	1	0	1	3	0	10	2	2
ENU[8]	0	0	0	0	0	0	0	1	0	0	0	0	0	0	1	0	1	2	3
MNU[8]	0	0	0	0	0	0	0	0	0	0	0	0	2	1	0	0	2	1	1
Total	1	1	6	3	4	1	2	3	2	1	2	2	11	6	10	4	37	20	30

Total mutations[e]

	a	b	c	d^c	$d\,se$	se	p	s	Total
ENU, MNU	0	1	0	3	0	0	4	2	10
Other[f]	2	20	6	7	4	4	20	24	87

Since most publications have classified tested mutations only as either stem-cell or post-stem-cell-induced, the data for chemicals other than ENU and MNU may include a small fraction induced in differentiating spermatogonia.

[a] Abbreviations: MMS, methyl methanesulfonate; for all others, see footnote a to Table 1. References: [1]Ehling and Neuhäuser-Klaus (1984); [2]Ehling and Neuhäuser-Klaus (1988a); [3]Cattanach (1966, 1967); [4]Ehling and Neuhäuser-Klaus (1988b); [5]Ehling and Neuhäuser (1979); [6]Kratochvílová et al. (1988); [7]L.B. Russell (unpubl.); [8]L.B. Russell et al. (1989); other lesions (see text).
[b] (LL) Large genomic lesions; (OL) other lesions (see text).
[c] Wherever d^{op} mutations were described (1, 2, and 4 in TEM, ENU, and MNU experiments, respectively), they are classified as OL (see text).
[d] No information available on viability of homozygotes.
[e] Including those for which no information is available on viability of homozygotes.
[f] Includes the following MLP-generated mutations (not yet tested for viability of homozygotes: 2, 2, 2, and 4, involving b, d, $d\,se$, and s, respectively.

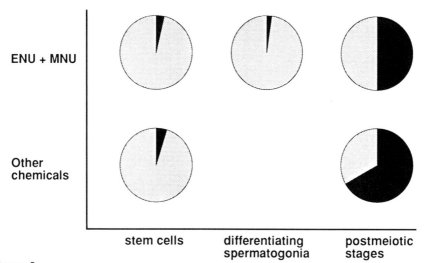

Figure 2
Relative proportions of large lesions, LL, and other lesions, OL (see text), among specific-locus mutations induced in various germ-cell stages by ENU and MNU and by other chemicals. Based on data for all loci in Tables 2, 3, and 4; LL proportions are represented by dark sectors.

[L.B. Russell et al. 1988], primary oocytes [Ehling and Neuhäuser-Klaus 1988c], or meiotic stages of spermatogenesis [J. Favor, pers. comm.], and that, in at least the first of these cases, OL-type mutations, primarily, are induced.)

Another indicator of the quality of mutations is the distribution (spectrum) among the marked loci. The information from Table 1 is plotted in Figure 3; that from Tables 2 and 3 in Figure 4. For stem-cell spermatogonia (Fig. 3), there is a remarkable similarity between the spectra produced by pattern-3 and other chemicals; both are characterized by relatively low frequencies of a-, se-, and s-locus mutations, and absence of $d\ se$'s. Figure 4 contrasts the distribution of mutations induced in differentiating spermatogonia (by pattern-3 chemicals) with that induced in postmeiotic stages (by pattern-1 and -2 chemicals). The former is not dissimilar to the spectrum for stem-cell mutations shown in Figure 3. The postmeiotic pattern, on the other hand, departs from this spectrum in a number of ways: the occurrence of $d\ se$'s, a considerable increase in mutations at s, and an apparent reduction at some other loci.

If, in fact, most mutations induced in stem-cell spermatogonia are small lesions, the spectrum shown in Figure 3, similar for the various chemicals, may be roughly indicative of the relative sizes of the target loci. The increase in the frequency of s-locus mutations observed when postmeiotic stages are mutagenized (Fig. 4), i.e.,

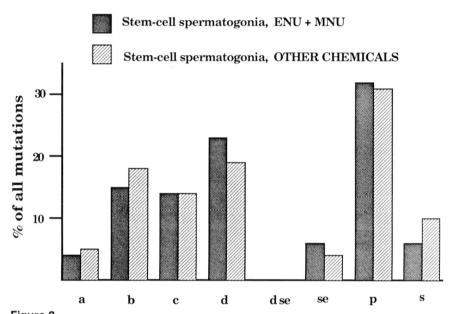

Figure 3
Distribution among the marked loci of mutations induced in stem-cell spermatogonia; based on data in Table 2.

when relatively large lesions are being induced, may indicate that the loss of genomic material in the vicinity of s is more readily tolerated in heterozygotes than is the loss of material in the vicinity of the other loci; the corollary would be that deletions involving s are, on average, larger than those involving other loci. A test of this suggestion must await the discovery of probes for the s region. Genetic and/or molecular analysis of this region may also help to answer the question why a higher proportion of mutations at s than at other loci is lethal; indeed, even with experimental conditions that produce very low frequencies of large genomic lesions, half of all s-locus mutations are lethal, suggesting that certain intra-locus, as well as multilocus, lesions might produce this phenotype.

CONCLUSIONS

The findings that the germ-cell stage of induction (rather than the identity of the chemical) is the major determinant of the *nature* of chemically induced mutations, and that the distinction appears to be between pre- and postmeiotic rather than between stem-cell and post-stem-cell stages, suggest several possible mechanistic

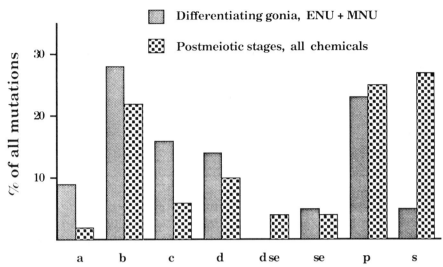

Figure 4
Distribution among the marked loci of mutations induced in differentiating spermatogonia (dark shading; based on data in Table 3) and in postmeiotic stages (checkered shading; based on data for all chemicals in Table 4).

determinants. Among these are chromatin configuration, diploidy versus haploidy, and the existence of different repair opportunities in mitotic versus nonmitotic cells. Although selection against LL mutations induced in premeiotic stages may play some role in their virtual absence from such stages, it should be noted that LL mutations induced in *post*meiotic stages, when propagated in breeding stocks, are in most cases *not* selected against during spermatogenesis. It is also noteworthy that very low LL frequencies can be found in instances where chemicals cause no measurable elimination of premeiotic cells and where selection is thus not a factor (e.g., in the case of MNU exposure of differentiating spermatogonia).

In addition to their implications for mechanisms of mutagenic action, the findings may also provide some shortcuts to the assessment of genetic risk from chemicals. A major element in risk assessment is the nature of the organismic effects produced by the induced mutations. If magnitude or type of phenotypic effects can be shown to be different, on average, for LL and OL arrays of mutations, it will be possible to make predictions of risk on the basis of germ-cell stage(s) that would be exposed, rather than having to measure phenotypic effects separately for each chemical. Finally, the findings can be put to use in optimizing conditions for the use of chemicals in the production of *desired types* of mutations that can serve as tools for analyses of genome structure and function.

ACKNOWLEDGMENTS

This research was jointly sponsored by the Office of Health and Environmental Research, U.S. Department of Energy, under contract DE-AC05-84OR21400 with the Martin Marietta Energy Systems, Inc., and by the National Institute of Environmental Health Sciences under IAG No. 22Y01-ES-10067. The submitted manuscript has been authored by a contractor of the U.S. Government under contract No. DE-AC05-84OR21400. Accordingly, the U.S. Government retains a nonexclusive, royalty-free license to publish or reproduce the published form of this contribution, or allow others to do so, for U.S. Government purposes.

REFERENCES

Cattanach, B.M. 1966. Chemically induced mutations in mice. *Mutat. Res.* **3**: 346.

———. 1967. Induction of paternal sex-chromosome losses and deletions and of autosomal gene mutations by the treatment of mouse post-meiotic germ cells with triethylenemelamine. *Mutat. Res.* **4**: 73.

———. 1982. Induction of specific-locus mutations in female mice by triethylenemelamine (TEM). *Mutat. Res.* **104**: 173.

Ehling, U.H. 1984. Methods to estimate the genetic risk. In *Mutations in man* (ed. G. Obe), p. 292. Springer-Verlag, Berlin.

Ehling, U.H. and A. Neuhäuser. 1979. Procarbazine-induced specific-locus mutations in male mice. *Mutat. Res.* **59**: 245.

Ehling, U.H. and A. Neuhäuser-Klaus. 1984. Dose-effect relationships in germ-cell mutations in mice. In *Problems of threshold in chemical mutagenesis* (ed. Y. Tazima et al.), p. 15. Environmental Mutagenesis Society of Japan, Mishima.

———. 1988a. Induction of specific-locus and dominant-lethal mutations by cyclophosphamide and combined cyclophosphamide-radiation treatment in male mice. *Mutat. Res.* **199**: 21.

———. 1988b. Induction of specific-locus and dominant-lethal mutations in male mice by diethyl sulfate (DES). *Mutat. Res.* **199**: 191.

———. 1988c. Induction of specific-locus mutations in female mice by 1-ethyl-1-nitrosourea and procarbazine. *Mutat. Res.* **202**: 139.

Jenkins, N.A., M.C. Strobel, P.K. Seperack, D.M. Kingsley, K.J. Moore, J.A. Mercer, L.B. Russell, and N.G. Copeland. 1989. A retroviral insertion in the dilute (d) locus provides molecular access to this region of mouse chromosome 9. *Prog. Nucleic Acid Res. Mol. Biol.* **36**: 207.

Katoh, M., K.T. Cain, L.A. Hughes, L.B. Foxworth, J.B. Bishop, and W.M. Generoso. 1990. Female-specific dominant lethal effects in mice. *Mutat. Res.* (in press).

Kratochvilova, J., J. Favor, and A. Neuhäuser-Klaus. 1988. Dominant cataract and recessive specific-locus mutations detected in offspring of procarbazine-treated male mice. *Mutat. Res.* **198**: 295.

Rinchik, E.M., L.B. Russell, N.G. Copeland, and N.A. Jenkins. 1986. Molecular genetic analysis of the dilute-short ear *(d-se)* region of the mouse. *Genetics* **112**: 321.

Rinchik, E.M., J.W. Bangham, P.R. Hunsicker, N.L.A. Cacheiro, B.S. Kwon, I.J. Jackson, and L.B. Russell. 1990. Genetic and molecular analysis of chlorambucil-induced germline mutations in the mouse. *Proc. Natl. Acad. Sci.* **87:** 1416.

Russell, L.B. 1971. Definition of functional units in a small chromosomal segment of the mouse and its use in interpreting the nature of radiation-induced mutations. *Mutat. Res.* **11:** 107.

―――. 1989. Functional and structural analyses of mouse genomic regions screened by the morphological specific-locus test. *Mutat. Res.* **212:** 23.

―――. 1990. Patterns of mutational sensitivity to chemicals in poststem-cell stages of mouse spermatogenesis. In *Mutation and the environment, part C: Somatic and heritable mutation, adduction, and epidemiology* (ed. M.L. Mendelsohn and R.J. Albertini). Alan Liss, New York. (In press.)

Russell, L.B. and P.R. Hunsicker. 1987. Study of the base analog 6-mercaptopurine in the mouse specific-locus test. *Mutat. Res.* **176:** 47.

Russell, L.B. and E.M. Rinchik. 1987. Genetic and molecular characterization of genomic regions surrounding specific loci of the mouse. *Banbury Rep.* **28:** 109.

Russell, L.B., C.S. Montgomery, and G.D. Raymer. 1982. Analysis of the albino-locus region of the mouse: IV. Characterization of 34 deficiencies. *Genetics* **100:** 427.

Russell, L.B., J.W. Bangham, K.F. Stelzner, and P.R. Hunsicker. 1988. High frequency of mosaic mutants produced by N-ethyl-N-nitrosourea exposure of mouse zygotes. *Proc. Natl. Acad. Sci.* **85:** 9167.

Russell, L.B., P.R. Hunsicker, E.F. Oakberg, C.C. Cummings, and R.L. Schmoyer. 1987. Tests for urethane induction of germ-cell mutations and germ-cell killing in the mouse. *Mutat. Res.* **188:** 335.

Russell, L.B., P.B. Selby, E. von Halle, W. Sheridan, and L. Valcovic. 1981. The mouse specific-locus test with agents other than radiation: Interpretation of data and recommendations for future work. *Mutat. Res.* **86:** 329.

Russell, L.B., P.R. Hunsicker, N.L.A. Cacheiro, J.W. Bangham, W.L. Russell, and M.D. Shelby. 1989. Chlorambucil effectively induces deletion mutations in mouse germ cells. *Proc. Natl. Acad. Sci.* **86:** 3704.

Russell, W.L. 1951. X-ray-induced mutations in mice. *Cold Spring Harbor Symp. Quant. Biol.* **16:** 317.

―――. 1982. Factors affecting mutagenicity of ethylnitrosourea in the mouse specific-locus test and their bearing on risk estimation. In *Environmental mutagens and carcinogens* (ed. T. Sugimura et al.), p. 59. University of Tokyo Press, Japan, and Alan R. Liss, New York.

―――. 1986. Positive genetic hazard predictions from short-term tests have proved false for results in mammalian spermatogonia with all environmental chemicals so far tested. In *Genetic toxicology of environmental chemicals* (ed. C. Ramel et al.), p. 67. Alan R. Liss, New York.

Russell, W.L. and P.R. Hunsicker. 1983. Extreme sensitivity of one particular germ-cell stage in male mice to induction of specific-locus mutations by methylnitrosourea. *Environ. Mutagen.* **5:** 498.

Russell, W.L., E.M. Kelly, P.R. Hunsicker, J.W. Bangham, S.C. Maddux, and E.L. Phipps. 1979. Specific-locus test shows ethylnitrosourea to be the most potent mutagen in the mouse. *Proc. Natl. Acad. Sci.* **76:** 5918.

COMMENTS

Lewis: With X rays, is *s* not one of the more mutable loci in spermatogonia, as well as in postgonial stages?

Liane Russell: Yes. It's the highest. However, the *overall* frequency of LL mutations induced in stem-cell spermatogonia is much higher after radiation than after chemical treatments—24% after X- or gamma-ray exposures and 35% after neutron exposures, as opposed to 4% or 5% after exposure to chemicals. So it fits the pattern we have discussed: the higher the frequency of LLs, the higher the proportion of *s* among the mutations.

Generoso: The large-lesion mutations induced in the postmeiotic stages with ENU do not seem to be in agreement with the mosaicism.

Liane Russell: The postmeiotically induced ENU mutations I discussed were detected in whole-body mutants. In addition, some mosaic mutants are found, both at Oak Ridge and at Neuherberg, after exposure of male postmeiotic stages. Mosaics are also inducible by ENU (as well as by procarbazine) in primary oocytes; and we find very high frequencies of mosaics after ENU treatment of zygotes (that is, from exposure at the end of the second meiotic division of the oocyte). Our mosaics are largely, if not entirely, due to *small* mutational lesions. And I believe that's true of many of your mosaics too, isn't it, Jack?

Favor: Yes. A lot of them are intermediates. So they are point mutations most likely.

Liane Russell: I think ENU can do two kinds of things. First, it can make a small lesion in a single DNA strand, and this will be detected as a mosaic, except when it occurs in mitotic cells, such as spermatogonia, where DNA synthesis and cell division soon cause both strands of the descendent cell to carry the same lesion. Second, under conditions existing in postmeiotic cells, it can also cause large (and presumably two-stranded) mutational lesions.

Handel: From one of your earlier slides it appears that prophase meiotic spermatocytes are relatively resistant. With the chemicals examined, no one of them had its "hottest" or most effective activity in those stages.

Liane Russell: You mean the pachytene and leptotene spermatocytes?

Handel: Yes, they appear to be relatively resistant, that is, for mutations assayed by the specific-locus test. Some people here have induced mutations in a number of different ways and have induced a number of different kinds of mutations—specifically, maybe, dominant lethals or heritable translocations. Is that a common finding?

Generoso: I don't really know. You see, when you come down to the meiotic and postmeiotic stages and you see large deficiency-related mutations, then you think of chromosomal-aberration-related events. In many of these experiments, the chemicals killed those stages and you really cannot generate very many offspring from those. So I doubt if the stage has been tested sufficiently. I know that there are chemicals that will break chromosomes at those stages.

Handel: I think what we are asking is a fairly complex question: Are these stages more sensitive to cell killing, and is there anything different about the chromatin at these stages affecting the kinds of mutations that we can induce? You're saying certainly we can induce translocations?

Generoso: Yes. We have induced translocations in pachytene with TEM. At this point, the question is whether, at these stages, the cell killing and the mutagenic events have the same etiology; I really don't know.

Handel: I think most of the cell-killing data that I'm aware of suggest cell killing in spermatogonial stages, or that spermatogonia are more susceptible.

Adler: Differentiating spermatogonia are more susceptible.

William Russell: That depends very much on the chemical. ENU is, at least in the differentiating spermatogonia, far more cytotoxic than MNU. This is strange, because with regard to whole-animal toxicity, ENU is much *less* damaging.

Liane Russell: There was practically no MNU cytotoxicity in differentiating spermatogonia. You may remember the slide that showed mutations in serial matings after MNU exposure. The number of offspring was practically the same in each successive week of mating.

Bridges: It seems to me, looking at your early slides, if I had a group of chemicals and I wanted to induce mutations in germ cells, then the stage in which I am least likely to induce anything is the stem-cell gonia compared to all the other stages that come after that. However, that's the one stage that is not protected by the Sertoli cell barrier.

Liane Russell: Exactly.

Bridges: Now, I wonder if you would like to comment on that, which is not a widely known fact except among mouse geneticists.

Liane Russell: I should have mentioned that; because people often shrug off negative germ-cell mutagenesis results—particularly in cases where short-term-test results are positive—by concluding that the active metabolite has not reached the mouse germ-cell target. And yet, those cells *least* likely to "see" the chemical, the postgonial stages, are, as Bryn points out, much more often mutagenized than are the stem-cell spermatogonia, which are *not* protected by the Sertoli-cell barrier.

Handel: But the differentiating spermatogonia are in that category also, of not being protected by the barrier.

Moses: So what?

Liane Russell: So, if anything, we might guess that they (like the stem cells) would be more sensitive to mutation induction than the ones that are on the other side of the Sertoli barrier.

Moses: No, on the contrary. The barrier may or may not be a factor. The damage that is done could very well be lost. In other words, the primary spermatocyte behind the barrier may be as badly affected, or even more so, in terms of mutations, except those are never picked up because the toxicity is so great that the damaged cells are lost.

Liane Russell: However, we have several examples (from different chemicals) of cells behind the barrier (meiotic and postmeiotic stages) in which there is little or no cell killing and yet a high mutation rate; and other examples, too, of high mutation rates *despite* cell killing.

Bridges: There must be data on cytotoxic response of the stem cells from dominant lethal studies. Are they that much more sensitive than the later stages?

Liane Russell: No.

Bridges: So the data don't seem to support that argument. The data would suggest that the damage is actually repaired rather than leading to the death of that stem cell.

Liane Russell: In fact, most of the chemical mutagenesis studies are done at dose levels that do not kill stem cells. One generally works at the highest dose that one can use without killing stem cells. So, I don't think we are on the *down* part of humped dose-effect curves (like the ones that Bruce Cattanach has shown). We are not beyond the hump for most of the chemicals that have been studied.

William Russell: If the only chemical we looked at was ENU, which is highly mutagenic in stem-cell spermatogonia and only barely mutagenic in the postmeiotic stages, the argument would clearly seem to be in favor of the barrier. But ENU is a freak chemical in a way; many more chemicals are more mutagenic in meiotic and postmeiotic stages than in the stem-cell gonia.

Nonmutational Genetic Effects on Early Development

Effect of Imprinting on Analysis of Genetic Mutations

JULIE DELOIA AND DAVOR SOLTER
The Wistar Institute of Anatomy and Biology
Philadelphia, Pennsylvania 19104

INTRODUCTION

Mammalian development begins with the fusion of the egg and sperm and mixing of two seemingly equivalent genetic components from each parent. As a consequence of fertilization, oocyte meiosis resumes, diploidy is restored, and cleavage ensues. Although parthenogenetic activation can mimic these early events and produce grossly normal embryos, parthenogenotes cannot complete development. Failure of parthenogenotes is due to a functional uniqueness of the male and female genomes, the phenomenon known as genetic imprinting.

The notion of genetic imprinting is supported by results obtained from pronuclear transfer experiments (McGrath and Solter 1984a; Surani et al. 1984). Pronuclei of each parental origin can be distinguished because of their size and position within the zygote. Therefore, enucleated zygotes can be reconstituted in any combination: gynogenones with two female-derived pronuclei; androgenones with two male-derived pronuclei; or normal manipulated embryos with a pronucleus from each parent. Neither gynogenones nor androgenones can complete development. Gynogenones die shortly after implantation as well-formed embryos with diminutive extraembryonic membranes. In contrast, androgenones have fairly normal extraembryonic membranes but lack the embryo proper and die earlier than gynogenones, seldom surviving implantation. Normal manipulated embryos survive to birth. Aggregation chimeras between androgenetic and gynogenetic 4-cell embryos can survive up to 15 days but still do not complete development, demonstrating a cell-autonomous defect (Surani et al. 1987). In these chimeric embryos, cells derived from gynogenones are found primarily in the embryo proper, and cells from androgenones are found primarily in the extraembryonic membranes.

Establishment and Maintenance of Genomic Imprinting

The mechanism by which genomic imprinting arises remains enigmatic. Any postulated mechanism must account for the following observations: The genomic imprint must be capable of surviving multiple rounds of DNA replication and cell divisions yet retain the capacity for erasure during gametogenesis; the imprint must mark a chromosome or chromosome region for identification by *trans*-acting factors (in mice, only the paternally derived X chromosome is inactivated in the extraembryonic lineage of the conceptus); a genetic imprint must be able to affect the level of gene expression at the imprinted allele. Variation in the level of DNA methylation has been hypothesized to account for imprinting in mammals. The general level of methylation found in diplotene oocytes is much lower than that found in spermatocytes, spermatids, and sperm, and methylation levels of the same sequences in blastocysts are intermediate, suggesting maintenance of prefertilization methylation levels following several cell divisions (Sanford et al. 1987). In addition to endogenous genes, integrated transgenes have been used to follow methylation changes when the transgene is inherited from a male or female. Several transgenic lines show parent-of-origin-dependent methylation differences sustained throughout development; however, the majority of these differentially methylated transgenes are never transcribed (for review, see Surani et al. 1988; McGowan et al. 1989; Sapienza et al. 1989).

Effect of Imprinting on Gene Expression

The salient characteristic of an imprinted gene is differential expression dependent on parental origin. Mutations at the more active allele would appear dominant, whereas mutations at the less active allele would be inconsequential. The murine mutation T^{hp}, a small deletion including the Tme (T maternal effect) locus on chromosome 17, is lethal only when maternally inherited. That this effect is nuclear (*cis*) and not cytoplasmic (*trans*) was demonstrated by pronuclear transfer experiments (McGrath and Solter 1984b). The simplest explanation for this inheritance pattern is that only the maternally derived Tme gene is expressed. A mutation in a paternally derived Tme allele would go undetected, since this allele is not expressed. In cases of genetic imprinting, the effect of genotype on phenotype can be misleading.

A theoretical pedigree based on the inheritance of an imprinted allele that is active only when paternally inherited, the opposite of the Tme gene just discussed, is depicted in Figure 1. Mutations at this locus are transmitted by both parents, but only paternally inherited mutations affect phenotype. Were mutations at this allele only phenotypically detected, then one would incorrectly assume a dominant mutation arising de novo in member IIe. Without the availability of molecular probes to detect mutations at this allele, the entire left side of the pedigree would be lost for data analysis. Only if IIIc or IVc sired affected progeny would the explanation of an autosomal dominant defect come into question. This type of pedigree where a dominant phenotype occurs only with uniparental inheritance has been referred to as dominance modification.

The existence of dominance modification should be taken into account when designing protocols for generating mutations in mice. Generally, a mutagen is administered to a single male, which is then mated to several females. The progeny from this mating are screened for phenotypic changes. If the targeted gene is

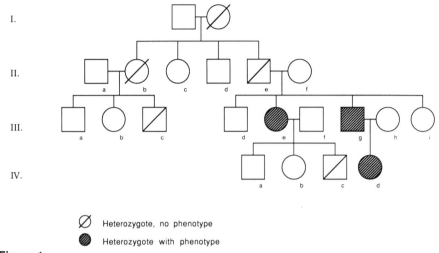

Figure 1
Hypothetical pedigree for an imprinted allele. Open symbols represent individuals who do not carry the imprinted locus; a line through the symbol denotes individuals carrying the imprinted allele with a normal phenotype; hatched symbols denote individuals that carry the imprinted allele and have an affected phenotype.

imprinted, i.e., the paternally derived allele is normally transcriptionally inactive, then the mutation frequency of this gene would be artificially low. In fact, a mutation in an imprinted gene expressed only when maternally inherited would be missed if females were not also administered the mutagen and progeny were not tested. Alternatively, in cases where a paternally inherited allele is transcriptionally more active than its maternally derived homolog, mutations would be detected more frequently, giving the appearance of a hypermutable site in the genome.

Several human syndromes have been described that fall into the category of imprinting depicted in Figure 1 (Reik 1989). Two of these syndromes have been localized to the same region of chromosome 15 and are thought to be allelic. Prader-Willi syndrome (PWS) is characterized by mental retardation, hypotonia, hyperphagia, and severe obesity. Cytological analysis revealed a deletion of 15q11-q13 exclusively on the paternal chromosome in 60% of cases (Ledbetter et al. 1987). In the remaining 40% of patients with no obvious deletions, molecular DNA markers were used to demonstrate the presence of two copies of a maternally inherited chromosome 15 (maternal heterodisomy or uniparental inheritance) (Nicholls et al. 1989). Although a specific gene mutation in the maternally derived genes cannot be ruled out, it seems likely that PWS in these individuals resulted from the absence of a paternally inherited chromosome 15.

Patients with Angelman syndrome (AS) carry the same cytological deletion as that found in Prader-Willi patients; however, the deletion always occurs on the maternally derived chromosome. The phenotype present in AS differs from that found in PWS. AS is characterized by seizures, inappropriate laughter, ataxic gait, and jerky arm movements. Mental retardation is the one feature in common between PWS and AS. Because the deletion involved in both AS and PWS is sizeable, it is not possible to rule out the involvement of different genes in the establishment of the two phenotypes. Alternatively, the two phenotypes could result from a variation in expression of a single gene resulting from imprinting with subsequent uniparental inheritance.

Approaches to Identifying Imprinted Genes

Study of imprinting has been hindered by lack of a cloned imprinted gene. Several avenues to obtain such a gene are open. In mice, loci that show clear imprinting are surprisingly rare; indeed, Tme is the only example of such a locus, and this region of chromosome 17 has not yet been cloned. Analyses of progeny derived from matings of mice bearing Robertsonian and other types of translocations have revealed at least eight regions of noncomplementarity in the mouse genome. That is, uniparental disomy of these regions results in embryonic lethalities or dramatic changes in body size and shape (for review, see Cattanach 1986). Molecular cloning of these regions is possible but not entirely feasible, since the genetic material included in the different translocations is large, and no specific imprinted gene within any of these regions has been identified. Our laboratory is currently investigating imprinting at the molecular level in a line of transgenic mice (DeLoia et al. 1989). In this line, paternal inheritance of the transgene is associated with skull and digit deformity of variable severity. No deformity has ever been found when the transgene is inherited from the mother. Transgene expression does not occur until adulthood, eliminating the possibility that a transgene product is interfering with normal development. It appears that this transgene has integrated into an endogenous gene expressed when paternally inherited whose gene product is necessary for proper skull and digit formation. Using the transgene as a molecular entry into this region of the mouse genome, we should be able to isolate a gene or genes that show differential expression dependent on parental inheritance.

ACKNOWLEDGMENTS

This work was supported in part by U.S. Public Health grants HD-17720, HD-21355, and HD-23291 from the National Institute of Child Health and Human Development. J.D. was supported by grant CA-09171 from the National Cancer Institute.

REFERENCES

Cattanach, B.M. 1986. Parental origin effects in mice. *J. Embryol. Exp. Morphol.* (suppl.) **97:** 137.

DeLoia, J., M. Bucan, J. Price, and D. Solter. 1989. Use of a transgenic mouse line to study genomic imprinting. In *Abstracts from meeting on Mouse Molecular Genetics*, p. 132. E.M.B.L., Heidelberg.

Ledbetter, D.H., F. Greenberg, V.A. Holm, and S.B. Cassidy. 1987. Conference report: Second annual Prader-Willi syndrome scientific conference. *Am. J. Med. Genet.* **28:** 779.

McGowan, R., R. Campbell, A. Peterson, and C. Sapienza. 1989. Cellular mosaicism in the methylation and expression of hemizygous loci in the mouse. *Genes Dev.* **3:** 1669.

McGrath, J. and D. Solter. 1984a. Completion of mouse embryogenesis requires both the maternal and paternal genomes. *Cell* **37:** 179.

———. 1984b. Maternal T^{hp} in the mouse is a nuclear, not cytoplasmic, defect. *Nature* **308:** 550.

Nicholls, R.D., J.H.M. Knoll, M.G. Butler, S. Karem, and M. Lalande. 1989. Genetic imprinting suggested by maternal heterodisomy in non-deletion Prader-Willi syndrome. *Nature* **342:** 281.

Reik, W. 1989. Genomic imprinting and genetic disorders in man. *Trends Genet.* **5:** 331.

Sanford, J.P., H.J. Clark, V.M. Chapman, and J. Rossant. 1987. Differences in DNA methylation during oogenesis and spermatogenesis and their persistence during early embryogenesis in the mouse. *Genes Dev.* **1:** 1039.

Sapienza, C., J. Paquette, T.H. Tran, and A. Peterson. 1989. Epigenetic and genetic factors affect transgene methylation imprinting. *Development* **107:** 165.

Surani, M.A.H., S.C. Barton, and M.L. Norris. 1984. Development of reconstituted mouse eggs suggests imprinting of the genome during gametogenesis. *Nature* **308:** 548.

———. 1987. Influence of parental chromosomes on spatial specificity in androgenetic-parthenogenetic chimaeras in the mouse. *Nature* **326:** 395.

Surani, M.A.H., W. Reik, and N.D. Allen. 1988. Transgenes as molecular probes for imprinting. *Trends Genet.* **4:** 59.

COMMENTS

Lyon: Regarding this business of the retinoblastoma and whether it is a maternal or paternal deletion: Does that have to be imprinting, or could it just be a differential mutation rate?

Solter: There is no way to distinguish between paternal imprinting and, let's say, increased mutation rate in sperm. Both are going to produce the same result—loss of heterozygosity and duplication of the paternal chromosome.

Handel: Do we have any evidence that the imprint or the sexual switch is reversed in somatic cells?

Solter: No.

Handel: We really know very little about this phenomenon, what it means in molecular and biochemical terms. We all tend to think of imprinting of active versus inactive genes, and we construct imprinting maps. Perhaps, what those maps mean is not that there are regions of the genome that are imprinted and other regions that are not; they may just mean that there are regions where we can detect the differences. Maybe we put blinders on ourselves when we think of this only in terms of active and inactive genes.

Solter: I cannot think of how one would even find it, but imprinting might make the imprinted gene more active than the nonimprinted gene. Unless we have endogenous imprinted genes available that we can then follow through multiple generations, and see what happens to them, and how this correlates with expression, there is practically nothing we can do.

Wiley: Are all genes imprinted?

Solter: It is possible, but there is no evidence for or against it.

Smithies: There is some evidence that methylation can be one of the means of imprinting. Would you like to comment on that, or other mechanisms?

Solter: It is so far the only molecular or biochemical mechanism we have. An imprinted, inactive allele is usually hypermethylated in comparison with its nonimprinted allele. It is very likely that methylation is the final step of the imprinting phenomenon. The gene gets imprinted, eventually the imprinted gene is inactivated, and then finally the gene gets methylated.

Wyrobek: Is it possible that some of these deletions that preferentially come through the male or female have associated with them some haploid-specific genes, and that the gametes aren't formed only in either the mother or the father, and that may be why they aren't seen in the progeny?

Solter: That is possible in sperm, but, certainly, there is no way we can have haploid gene expression during oogenesis, since oocytes are never haploid.

Arnheim: Was your transgene on chromosome 5 within a region that has already been mapped? Was it imprinted?

Solter: Yes. The transgene inserted on chromosome 5, between *Pgm-1* and *c-kit*. Chromosome 5 is not imprinted according to data derived from translocation studies. These studies detect those imprinted genes which cause lethality or very severe morphological changes. It is thus conceivable that relatively mild malformations with incomplete penetrance would be undetected.

What Is the Radiosensitive Target of Mammalian Gametes and Embryos at Low Doses of Radiation?

LYNN M. WILEY
Division of Reproductive Biology and Medicine
Department of Obstetrics and Gynecology
School of Medicine
University of California
Davis, California 95616

OVERVIEW

The chromosomes have been regarded as the radiosensitive target of ionizing radiation in mammalian gametes. However, recent evidence from the embryo aggregation chimera assay is inconsistent with a nuclear target for mouse male germ cells irradiated with 0.01 Gy to 0.05 Gy. Six weeks after irradiation, sperm from these irradiated males confer a proliferative disadvantage to their progeny embryos if these embryos are paired with control embryos to form aggregation chimeras. These sperm would have been type B spermatogonia at the time of irradiation. However, the DNA repair that occurs in premeiotic stages of spermatogenesis would be expected to eliminate mutations from these doses. In addition, the expected mutation frequencies from these doses are considered too low for sperm from as few as five to ten irradiated males to confer a consistent significant effect on progeny embryos. In this chapter, I discuss the chimera assay and results it has yielded from preimplantation embryos irradiated in vitro and male germ cells irradiated in vivo. The chapter closes with the speculation that a radiosensitive target for very low doses of irradiation to the male germ cell is the plasma membrane, whose response to the irradiation can be demonstrated as an embryonic cell proliferation disadvantage.

INTRODUCTION

Why Use Mouse Embryo Aggregation Chimeras as an Assay System?

A chimera is an individual composed of cells originating from more than one fertilized egg. The proportion of the different cell types comprising the chimera is determined by two factors: (1) the initial starting proportions of the different cell types and (2) the relative cell proliferation rates of the different cell types.

The rationale for using mouse embryo aggregation chimeras to detect exposure to ionizing radiation rests on two studies published more than 10 years ago. In the first study, published by Horner and McLaren (1974), 2-cell embryos were cultured to the blastocyst stage in the chronic presence of 5 nCi/ml tritiated thymidine ([^{3}H]TdR) (Fig. 1). These blastocysts attained the same number of cells per embryo as did control blastocysts, but upon transfer to foster mothers, none of the blastocysts cultured in this concentration of [^{3}H]TdR yielded live-born young. This result showed that irradiation during preimplantation development that produced no apparent effect on embryonic cell proliferation (the conventional parameter for monitoring embryo viability) could later result in postimplantation embryonic death.

In the second study, published by Kelly and Rossant (1976), 8-cell blastomeres were incubated in 0.25 μCi/ml [^{3}H]TdR for 2 hours and then either self-aggregated to form homologous chimeras or aggregated with control blastomeres to form heterologous chimeras (Fig. 1). After an incubation period to allow two to three cell cycles to occur, the chimeras had attained the late morula/early blastocyst stage and were processed for autoradiography and for obtaining embryo cell numbers. The number of progeny cells contributed per 8-cell blastomere was the same for homologous chimeras composed of just [^{3}H]TdR-exposed blastomeres or just control blastomeres. However, in heterologous chimeras composed of thymidine-exposed and control blastomeres, the thymidine-exposed blastomeres contributed significantly fewer progeny cells to the chimera than did control blastomeres.

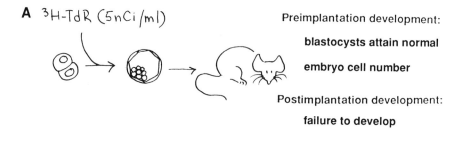

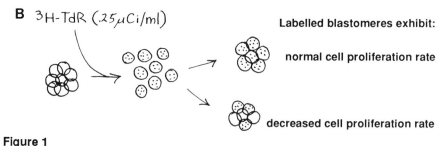

Figure 1
(A) Schematic of experimental results reported by Horner and McLaren (1974). (B) Schematic of experimental results reported by Kelly and Rossant (1976).

In other words, cell proliferation by the blastomeres irradiated by tritium was unaffected in the homologous chimeras, but significantly depressed in heterologous chimeras in which irradiated blastomeres were challenged by direct cell-cell contact with unirradiated blastomeres. As stated by these authors, "This competitive situation provides a very stringent test of their viability and vigor" (Kelly and Rossant 1976). Within the context of the Horner and McLaren (1974) study, however, these results obtained with the chimeras suggest that irradiation producing no apparent effect on embryo viability during preimplantation development could confer upon the irradiated embryo a proliferative disadvantage when challenged by direct cell-cell contact with an unirradiated embryo in an aggregation chimera.

The Chimera Assay: Technical Aspects

Chimera assays are initiated with 4-cell embryos (Fig. 2) because they are the most radiosensitive of the mouse preimplantation stages to wave-type radiation (X-rays; Goldstein et al. 1975). Each chimera consists of two embryos, one of which is prelabeled with the viable dye fluorescein isothiocyanate (FITC; Overstreet 1973) so that its progeny cells can be distinguished from those of its partner embryo. This dye has no apparent effect on preimplantation embryo viability or on subsequent postimplantation embryo viability following transfer to foster mothers (Ziomek 1982). Zonae pellucidae are removed by a brief treatment with acidified phosphate-buffered saline, treated with phytohemagglutinin to promote embryo adhesion, and then aggregated into pairs to form three types of chimeras; (1) *homologous control* chimeras, containing one FITC-labeled control embryo and one non-FITC-labeled

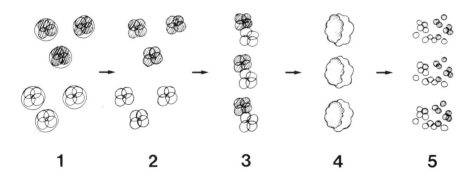

Figure 2
Four-cell embryos are labeled with FITC as desired (*1*), treated with acidic phosphate-buffered saline to remove their zonae pellucidae (*2*), aggregated into pairs using phytohemagglutinin (PHA) (*3*), and cultured 40–45 hr (chimera incubation period) (*4*). Chimeras are then partially dissociated using calcium-free medium to obtain proliferation ratios (*5*).

control embryo; (2) *homologous experimental* chimeras, containing one FITC-labeled experimental embryo and one non-FITC-labeled experimental embryo; and (3) *heterologous* chimeras, containing one FITC labeled control embryo and one non-FITC-labeled experimental embryo.

The chimeras are cultured for 40–45 hours to allow for 2–3 cell cycles, partially dissociated into cell clusters with calcium-free medium, and then viewed under phase-contrast optics for obtaining total chimera cell number and under epifluorescence illumination to obtain the number of non-FITC-labeled cells. These two numbers are expressed as a *proliferation ratio* (unlabeled cell number:total cell number). A ratio of 0.5 indicates that both partner embryos contributed equal numbers of progeny cells to the chimera, which is the observation for both types of homologous chimeras. A ratio less than 0.5 for heterologous chimeras indicates that the unlabeled experimental embryo contributed fewer progeny cells than did the FITC-labeled control embryo.

Coefficients of variation are typically 10–15% using five chimeras of each type per dose group for in vitro exposures and per animal for in vivo exposures. We allow from 5 to 15 animals per dose group for in vivo exposures with the animal as the statistical unit.

RESULTS

In Vitro Irradiation of 4-Cell Embryos

In this experiment (Obasaju et al. 1988), the principal question was whether we would observe a decrease in proliferation ratios in aggregation chimeras constructed from intact embryos irradiated with X-irradiation instead of 8-cell blastomeres irradiated with [^{3}H]TdR as in the study by Kelly and Rossant (1976). Four-cell embryos received an acute dose of X-irradiation in vitro from a Maximar 250-kVp therapeutic X-ray generator operated at a dose rate of 0.32 Gy/minute. The dose range was 2.00 Gy–0.05 Gy, with 2.00 Gy approaching the ED_{50} for 4-cell embryos (2.32 Gy; Goldstein et al. 1975) and 0.05 Gy falling below an estimated dose delivered in the Kelly and Rossant study (0.19 Gy; T. Straume, pers. comm.). Irradiated embryos were assayed by the chimera assay as described above for a cell proliferation disadvantage.

For heterologous chimeras, proliferation ratios decreased along with increasing radiation dose (Table 1). No decreases in proliferation ratios were observed when irradiated embryos were merely cocultured with control embryos (Table 2). This result showed that the decreasing ratios observed in the chimera assay required direct cell-cell contact between irradiated and control embryos. In a separate experiment with an irradiation dose of just 0.01 Gy, there was still a significant decrease in the proliferation ratio for heterologous chimeras (data not shown; M.F. Obasaju and L. Wiley, unpubl.).

Table 1
Cellular Contribution per Embryo in Aggregation Chimeras between Irradiated and Control Embryos

Chimera (no.)	No. of experiments	No. cells contributed per embryo (mean ± S.D.)		Proliferation ratio
		irradiated	control	
0 Gy←→0 Gy (32)	5	17.4 ± 5.5	17.9 ± 5.6	0.49 ± 0.03
0 Gy←→0.05 Gy (30)	5	12.9 ± 6.4	16.5 ± 6.4	0.43 ± 0.01
0 Gy←→0.25 Gy (23)	4	13.9 ± 6.0	19.8 ± 7.2	0.41 ± 0.08
0 Gy←→0.50 Gy (21)	4	14.2 ± 2.9	21.5 ± 4.9	0.40 ± 0.04
0 Gy←→1.00 Gy (23)	4	15.5 ± 3.2	27.7 ± 8.0	0.37 ± 0.06
0 Gy←→2.00 Gy (17)	3	10.3 ± 2.8	21.6 ± 5.6	0.33 ± 0.06

Proliferation ratios were compared by ANOVA and found to be significantly different from each other. Control embryos were labeled with FITC. Adapted from Obasaju et al. (1988).

In Vivo Irradiation of the Male Germ Cell

In this experiment (Obasaju et al. 1989), the principal question was whether the decrease in proliferation ratios observed above with 4-cell embryos irradiated in vitro could be transmitted by the male germ cell irradiated in vivo. Male mice received an acute dose of X-rays from a Maximar 250-kVp therapeutic X-ray generator operated at a dose rate of 0.19 Gy/min. The dose range was 1.80 Gy–0.05 Gy (air doses), which was similar to that used for irradiating 4-cell embryos in vitro. Beginning 1 week after irradiation, the males were bred once weekly for a 9-week test breeding period, and their progeny embryos were tested in the chimera assay for a cell proliferation disadvantage in terms of proliferation ratios.

Beginning with week 4 of the test breeding period, there was a dose-dependent decrease in proliferation ratios with increasing dose (Table 3). In addition, there was a temporal pattern to the interval during which the greatest decreases in proliferation ratios occurred (Table 3). This pattern coincided with the known pattern of radiosensitivities of the different stages of spermatogenesis, with week 6 of the test breeding period exhibiting the greatest radiosensitivity.

This experiment was repeated using a dose range of 0.01 Gy–0.005 Gy to identify the chimera assay's detection limit. A dose of 0.01 Gy still yielded a significant decrease in proliferation ratios for week 6 of the test breeding period (Table 4). No

Table 2
Effect of Coculturing Irradiated and Control Embryos on Mean Embryo Cell Number at 95 Hours Post-hCG

Embryos labeled/unlabeled (no.)	Cell number per embryo (mean ± S.D.)		Proliferation ratio
	labeled	unlabeled	
0 Gy/0 Gy (12)	26.8 ± 8.9	31.4 ± 12.4	0.54 ± 0.1
0 Gy/0.05 Gy (14)	35.4 ± 7.5	39.0 ± 10.1	0.54 ± 0.07
0.05 Gy/0.05 Gy (13)	32.2 ± 8.9	34.8 ± 15.9	0.52 ± 0.12

Adapted from Obasaju et al. (1988).

decreases in proliferation ratios were produced by 0.005 Gy over the 9-week test breeding period, suggesting that the detection limit of the assay under these experimental conditions for detecting exogenous irradiation to the male germ cell lies between 0.01 Gy and 0.005 Gy.

DISCUSSION

Ionizing radiation to mouse germ cells produces genetic alterations ranging from point mutations to gross chromosomal anomalies (Carrano et al. 1978; BEIR 1980).

Table 3
Proliferation Ratios of Heterologous Chimeras from Chimera Assays of Sperm from X-irradiated Males

Week after mating	Proliferation ratios (Gy ←→ Gy)[a]			
	0 ←→ 0	0 ←→ 0.05	0 ←→ 0.29	0 ←→ 1.73
4	0.50	0.46	0.44[b]	0.41[b]
5	0.50	0.48	0.43[b]	0.38[c]
6	0.49	0.42[b]	0.40[b]	0.29[c]
7	0.50	0.41[b]	0.40[b]	0.31[c]
8	0.50	0.48	0.49	0.40[b]
9	0.50	0.50	0.49	0.45

Adapted from Obasaju et al. (1989).
[a] The doses are values corrected for tissue attenuation.
[b] Significantly different from control ratio ($p < 0.05$).
[c] Significantly different from control ratio ($p < 0.01$).

Table 4
Proliferation Ratios of Heterologous Chimeras from Assays of Sperm from Males Irradiated with 0.005 Gy and 0.01 Gy of Gamma Rays

Week after mating	Proliferation ratios[a] (mean ± S.D.) (Gy ←→ Gy) (no. of animals)		
	0 ←→ 0	0 ←→ 0.005	0 ←→ 0.01
4	0.50 ± 0.03 (14)	0.48 ± 0.05 (6)	0.50 ± 0.03 (5)
5	0.52 ± 0.03 (12)	0.50 ± 0.06 (8)	0.51 ± 0.03 (11)
6	0.50 ± 0.04 (17)	0.50 ± 0.04 (6)	0.46 ± 0.05[b] (11)
7	0.50 ± 0.04 (17)	0.52 ± 0.04 (7)	0.46 ± 0.10 (9)

Data from P. Warner et al. (in prep.). Irradiation source was ^{137}Cs.
[a] Data pooled from two experiments.
[b] Significantly different from control ratio ($p < 0.05$).

Genetic alterations including point mutations and aneuploidies tend to become underrepresented in aggregation chimeras (Mintz 1964; Spiegelman 1978; Epstein et al. 1982). It is possible, then, that genetic alteration might underlie the cell proliferation disadvantage exhibited by the irradiated 4-cell embryos and by embryos produced from irradiated male germ cells in the chimera assays. However, as stated in the Overview, the mutation frequency at these low doses would probably be too low to produce a consistent, significant biological effect that was detectable by the small numbers of embryos and males irradiated in these experiments. Furthermore, DNA repair would probably have eliminated any radiation-induced alterations sustained by either the embryos or by the male germ cells. Nonetheless, it could be postulated that doses of radiation as low as 0.01 Gy and 0.05 Gy might produce mutations whose quantity and/or quality render them undetectable by the DNA repair system(s) in these cell types.

The effect was long-lived—6 weeks—on cell proliferation produced by sperm from males irradiated with 0.01 Gy or 0.05 Gy, and this is consistent with a genetic origin of the embryonic cell proliferation disadvantage. However, it is important to remember that the zygote receives additional organelles other than the nucleus from the fertilizing sperm, notably, mitochondria and a significant amount of plasma membrane that becomes incorporated into the plasma membrane of the fertilized oocyte. Since paternal mitochondria become degraded during early cleavage and their numbers are insignificant (Szollosi 1965) relative to the number of maternal mitochondria (~92,000 per oocyte; Piko and Matsumoto 1976), they are unlikely to exert effects on embryo viability and/or development.

Paternal plasma membrane antigens persist on the surface of the embryo as late as the blastocyst stage, however, and are abundant enough for specific antibodies to bind them and mediate complement-dependent lysis (Menge and Fleming 1978). In addition, cell-surface components on the mature spermatozoan, including epitopes unique to the male germ cell (Millette and Bellvé 1977) and sulfogalactoglycerolipid (Kornblatt 1979) are inserted into the developing male germ cell during the early spermatocyte stage of development, which immediately follows (or overlaps?) the type B spermatogonial stage that exhibits the greatest sensitivity to radiation in the chimera assay. Might it not be possible, then, that these low doses of radiation were altering some aspect(s) of the synthesis and/or assembly of such plasma membrane components that occur in type B spermatogonia, and that these alterations conferred a plasma-membrane-associated cell proliferation disadvantage in progeny embryos in the chimera assay?

Invoking the plasma membrane (or synthesis and/or assembly of its components) as a radiosensitive target has been proposed before (Koteles 1986; Feinendegen et al. 1988). Proposing the plasma membrane as a target would be consistent with the observed requirement by a decrease in proliferation ratios for cell-cell contact between irradiated and control blastomeres (Kelly and Rossant 1976) and embryos (Obasaju et al. 1988). In the broader context of germ-cell biology, however, consideration of the plasma membrane as a target draws attention to issues regarding the epigenetic aspects of germ-cell development and what influences such aspects might have on the developing embryo. Experiments are currently under way to explore these issues as well as to determine whether the chimera assay is reflecting a nuclear or extranuclear target in the male germ cell and early embryo.

ACKNOWLEDGMENTS

The author acknowledges the expert technical assistance of Marie Suffia and Kevin Walsh in these experiments. In addition, she acknowledges Michael Femi Obasaju, Deborah Oudiz, and Petra Warner for both their hands-on participation in the actual experiments and the many stimulating discussions regarding conceptual development involved in this work. Finally, she acknowledges the critical comments and valuable suggestions of Andrew Wyrobek, Tore Straume, and James W. Overstreet during the course of this work. Research work was supported by National Foundation grant 15-115 to L.M.W.

REFERENCES

Biological Effects of Ionizing Radiation Committee (BEIR). 1980. Genetic effects. In *The effects on populations of exposure to low levels of ionizing radiation*, p. 71. National Academy Press, National Research Council, Washington, D.C.

Carrano, A.V., L.H. Thompson, P.A. Lindl, and J.L. Minkler. 1978. Sister chromatid exchange as an indicator of mutagenesis. *Nature* **271**: 551.

Epstein, C.J., S.A. Smith, T. Zamora, J.A. Sawicki, T.R. Magnuson, and D.R. Cox. 1982. Production of viable adult trisomy 17 <—> diploid mouse chimeras. *Proc. Natl. Acad. Sci.* **79**: 4376.

Feinendegen, L.E., V.P. Bond, J. Booz, and H. Muhlensiepen. 1988. Biochemical and cellular mechanisms of low-dose effects. *Int. J. Radiat. Biol.* **53**: 23.

Goldstein, L.S., A.I. Spindle, and R.A. Pedersen. 1975. X-ray sensitivity of the preimplantation mouse embryo in vitro. *Radiat. Res.* **62**: 276.

Horner, D. and A. McLaren. 1974. The effect of low concentrations of [^{3}H]-thymidine on pre- and postimplantation mouse development. *Biol. Reprod.* **11**: 553.

Kelly, S.J. and J. Rossant. 1976. The effect of short-term labelling in [^{3}H]thymidine on the viability of mouse blastomeres: Alone and in combination with unlabelled blastomeres. *J. Embryol. Exp. Morphol.* **35**: 95.

Kornblatt, M.J. 1979. Synthesis and turnover of sulfogalactoglycerolipid, a membrane lipid, during spermatogenesis. *Can. J. Biochem.* **57**: 255.

Koteles, G.J. 1986. The plasma membrane as radiosensitive target. *Acta Biochim. Biophys. Hung.* **21**: 81.

Menge, A.C. and C.H. Fleming. 1978. Detection of sperm antigens on mouse ova and early embryos. *Dev. Biol.* **63**: 111.

Millette, C.F. and A.R. Bellvé. 1977. Temporal expression of membrane antigens during mouse spermatogenesis. *J. Cell Biol.* **74**: 86.

Mintz, B. 1964. Formation of genetically mosaic mouse embryos and early development of "lethal (^{12}t/^{12}t) <—> normal" mosaics. *J. Exp. Zool.* **61**: 273.

Obasaju, M.F., L.M. Wiley, D.J. Oudiz, O. Raabe, and J.W. Overstreet. 1989. A chimera embryo assay reveals a decrease in embryonic cellular proliferation induced by sperm from X-irradiated male mice. *Radiat. Res.* **118**: 246.

Obasaju, M.F., L.M. Wiley, D.J. Oudiz, L. Miller, S.J. Samuels, R.J. Chang, and J.W. Overstreet. 1988. An assay using embryo aggregation chimeras for the detection of non-lethal changes in X-irradiated mouse preimplantation embryos. *Radiat. Res.* **113**: 289.

Overstreet, J.W. 1973. The labelling of living rabbit ova with fluorescent dyes. *J. Reprod. Fertil.* **32**: 291.

Piko, L. and L. Matsumoto. 1976. Number of mitochondria and some properties of mitochondrial DNA in the mouse egg. *Dev. Biol.* **49**: 1.

Spiegelman, M. 1978. Fine structure of cells in embryos chimeric for mutant genes at the T/t locus. In *Mosaics and chimaeras in mammals* (ed. L.B. Russell), p. 59. Plenum Press, New York.

Szollosi, D.G. 1965. The fate of sperm middle-piece mitochondria in the rat egg. *J. Exp. Zool.* **159**: 367.

Ziomek, C.A. 1982. The use of fluorescein isothiocyanate (FITC) as a short-term lineage marker in the peri-implantation mouse embryo. *Wilhelm Roux's Arch. Dev. Biol.* **191**: 37.

COMMENTS

Hecht: We developed an antibody to an α-tubulin that appears to be testis specific. In collaboration with Gerald Schatten, we have been following this α-tubulin epitope after fertilization. We find the sperm axoneme survives up to the blastocyst stage in the mouse. So, very clearly, not only is the nucleus getting in, and perhaps some of the membranes, but the entire axoneme seems to survive.

Wiley: I'm trying to figure out how that would work here. It segregated to a single cell?

Hecht: Yes, which appears to be randomly segregating.

Wiley: So this tubulin is not reutilized in other microtubular structures?

Hecht: No. Well, that was the intent of the experiment. The antibody only recognizes the sperm α-tubulin; it doesn't recognize any of the egg α-tubulins. Seemingly, the α-tubulin at least stays together as the axoneme up to the blastocyst stage, and then it could be reutilized, but it simply disappears at that point.

Wiley: There was an abstract that appeared in the *Journal of Cell Biology* in 1986 that was done by J. Bennett. It came out of Roger Pedersen's lab, where they followed to see if there was any correlation between which blastomere after the first cleavage inherited the axoneme and how many progeny cells did those blastomeres then contribute toward the inner cell mass or trophectoderm, and if there was any differential inheritance as far as the germ layers or the principal stem cell lines were concerned. In that study, the blastomere that inherited the axoneme from the sperm contributed more progeny blastomeres to the inner cell mass than the trophectoderm. I'm going to have to look at Schatten's study to see how he got a difference, just a random segregation. In the earlier work, it seems that that blastomere contributes more progeny to the inner cell mass. It may be that that blastomere itself is random. Its progeny have a preferential distribution toward the inner cell mass. Since this development is clonal, and we analyze these before they begin to cavitate, I'm not sure how that would affect cell proliferation rates in these cells.

Generoso: In this system, what's the lowest ratio that you have observed so far?

Wiley: 0.2.

Generoso: You never have seen a case where you brought it down to the ground?

Wiley: No, we didn't want to. The whole reason that we chose the initial dose range of 2 Gy to 0.1 was to try to approximate the ED_{50} that Goldstein had found for four-cell embryos in vitro. We didn't want to kill the cells. The whole idea was to try to detect radiation exposure that would not normally otherwise be manifested as far as development was concerned. So we didn't want to have any overt sign of damage.

Generoso: The reason I asked that, if you are looking for something that's missing or something that's destroyed, maybe you need to—

Wiley: Kill them. Maybe we should.

Favor: To try to separate effects that might be nuclear versus nonnuclear, have you ever tried waiting a generation between male treatment and constructing the chimeras to look for proliferative differences?

Wiley: No.

Favor: So that would confine it to a nuclear effect.

Wiley: Well, we had planned to do that.

Fritz: Didn't you show that by week 8 the effect was gone?

Wiley: Yes.

Fritz: I don't quite understand what you mean about prolonging the experiment.

Wiley: You would take those embryos that were produced 6 weeks postirradiation, allow them to develop into adults, breed the males, and test their progeny 4-cell embryos in the assay.

Fritz: From the calculation roughly of 6 weeks, between 6 and 7 weeks maximum time, it looks like the cell that was affected by the irradiation was either a B type spermatogonium or one of the A types. Do you wish to postulate that a plasma membrane component made then is still on the sperm surface?

Wiley: I'm just throwing that out on table, being brave.

Fritz: You can't rule it out. I think it's not very plausible, because most of the plasma membrane components on the sperm surface are made subsequently.

Wiley: Most, but not all.

Lonnie Russell: Most of the autoantigenic ones, not necessarily the run of the mill ones.

Fritz: It's impossible to say quantitatively what's going on, because there have been many that disappear and there are some that appear. What Lonnie just said is absolutely correct. All of the antigenic determinants come later, all the ones that you could get immunologically.

Wiley: Yes. I don't know. I'm just putting it out on the table.

Selby: I'm curious about the dose-response curve. It seems a little surprising that you can get a significant effect with as little as 1 rad. When you go all the way up to 2 Gy, you're not getting it down any more than, say, to 0.2. What do you think is going on to explain that?

Wiley: I am a developmental biologist, and I had no training in dosimetry or in toxicology, so I am putting this up. I'm hoping that you can explain this to me.

Smithies: There's one thing that isn't diluted in a way in this system, and that's the two strands of DNA that are in any chromosome. Just for the moment, let's assume there is no sister chromatid exchange or anything. Some of the chromosomes in every sperm are still unreplicated—now, not all of them in any sperm, but on the average, segregating all of them, and imagine each time you get two daughters—there still is something that was there at the time that it was in the spermatogonia, and even if there is some damage on the DNA, it still could be a complete piece of damage. I'm trying to convey the feeling that you have to have something that isn't diluted too much to be able to go so far. There are pieces of DNA which, in a sense, haven't been diluted, at least on some of the chromosomes.

Wiley: The thing I can't figure out after learning that repair occurs later than week 6 is why wouldn't any kind of DNA lesion be repaired, especially a small one? I don't see why it's not repaired.

Adler: You're sampling 25 sperm, or in your other experiment it's 50. The sample size is so low that you would never dream of picking up the genetics.

Wiley: That's another point, yes, but it's so long-lived.

Adler: Not at 1 rad.

Wiley: That's why I'm putting out this extragenetic, extranuclear hypothesis, because it's hard for me to reconcile this with the genetic effect, too.

Developmental Anomalies: Mutational Consequence of Mouse Zygote Exposure

WALDERICO M. GENEROSO,[1] JOE C. RUTLEDGE,[2] AND JOHN ARONSON[3]
[1]Biology Division, Oak Ridge National Laboratory
Oak Ridge, Tennessee 37831-8077
[2]Department of Laboratory Medicine, University of Washington
Seattle, Washington 98105
[3]The Wistar Institute of Anatomy and Biology
Philadelphia, Pennsylvania 19104-4268

OVERVIEW

One of the objectives of mutagenesis research in mice is to enrich our knowledge of basic mammalian biology. The practical goal is to apply this knowledge to the problems of human health. The research described here exemplifies this philosophy. The observation that certain mutagens induced high incidences of fetal anomalies and death following exposure during the zygote stage is a new phenomenon in mutagenesis and experimental embryopathy. The mechanism for the induction of zygote-derived anomalies appears to be genetic, but it is not of the conventional type. These zygote-derived anomalies resemble the large class of stillbirths and sporadic defects in humans that are of unknown etiology. The zygote research in mice presents an opportunity for studying the molecular and cellular pathogenesis of this class of defects.

INTRODUCTION

Congenital anomalies have varied etiology. It is generally believed that most are caused either by genetic factors in parental germ cells or as a result of disruptions in normal cellular processes during development. The former refers to genetic alterations in prefertilization germ cells, the latter to conditions during the period of major organogenesis. A new concept for the origin of developmental defects in mice was reported recently (Generoso et al. 1987). It is based on mutagenesis and portrays the zygote as an important stage for experimental induction of embryopathy. Many of the zygote-derived developmental defects may have some bearing on the occurrence of certain stillbirths and common sporadic defects in humans. In this paper, we summarize the varied responses to different mutagens and discuss possible mutagenic mechanisms for induction of certain developmental defects.

The stage of development between sperm entry and first cleavage division is the zygote stage. It lasts about a day in the mouse. In the experiments described in this paper, the duration of the mating period was restricted to 30 minutes, beginning at the time when the newly ovulated eggs were in the ampulla of the oviduct, thus synchronizing the time of fertilization. Exposures to mutagens were done either at 1, 6, 9, or 25 hours after the end of the 30-minute mating period. These postmating intervals correspond, respectively, to the time of sperm entry, early pronuclear stage, pronuclear DNA synthesis, and early two-cell stage (Krishna and Generoso 1977) for the stock of females used in these studies. For comparison, the response in females treated within 4 days prior to mating (preovulatory oocytes) was also determined. All females were killed for uterine analysis on the seventeenth day of gestation, and live fetuses were examined for external developmental defects.

RESULTS

The mutagens included in these reports are ethylene oxide (EtO), ethyl methanesulfonate (EMS), ethylnitrosourea (ENU), triethylenemelamine (TEM), diethyl sulfate (DES), methyl methanesulfonate (MMS), and X-rays. Data for EtO, EMS, ENU, and TEM were published earlier (Generoso et al. 1987, 1988). Data for X-rays, MMS, and DES are part of larger studies that will be published elsewhere.

The doses used for each chemical induced early embryonic death when the zygotes were exposed at 1, 6, or 9 hours after mating as indicated by increases in resorption bodies. There are, however, differences between certain mutagens in the relative responses at these three time points. For example, EtO, EMS, and DES are markedly more effective at 1 and 6 hours than at 9 hours, whereas TEM, ENU, and MMS produced similar responses at all stages. With the exception of X-rays, for which data is available only for 2.5 hours postmating, all other mutagens produced reasonably high levels of resorption bodies at 6 hours postmating (Table 1). Thus, for all mutagens except X-rays, the following characterization of late developing anomalies is based only on data obtained from conceptuses that were exposed at this postmating interval, which corresponds to early pronuclear stage.

Inducibility and Classification of Developmental Anomalies

Resorption bodies represent embryonic death near the time of implantation. For X-ray, MMS, TEM, and ENU, induced death of conceptuses was expressed exclusively at this time (Table 1). For EtO, EMS, and DES, on the other hand, there were, in addition, high incidences of mid-gestation (presence of placenta and embryonic mass that lacked eye pigment) and late-gestation death (death occurred at about day 11 of gestation or later, characterized by at least the presence of eye pigment) (Fig. 1). Among living fetuses, significant increases in the frequency of those with defects were induced by each mutagen, but it is clear that certain mutagens differed

Table 1
Death of Conceptuses and Defects among Living Fetuses

Treatment [a]	Resorption bodies (%)	Mid- and late-gestation deaths (%)	Live fetuses examined	Live fetuses with defect (%)
EtO	53	29	87	37
EMS	31	31	231	29
DES	24	37	157	24
MMS	25	2.5	451	3.8
ENU	49	2	33	15
TEM	57	2.2	112	10
X-ray[b]	58–66	1.2–2.0	326	4.6
Control	5	1.5	3436	1.3

[a] 6 hr postmating except for X-ray, which was 2.5 hr.
[b] Data for 150 R and 200 R doses were pooled.

markedly in efficacy. For example, the incidence of defective fetuses was only marginally increased in the case of X-ray and MMS and remarkably high in the case of EMS, EtO, and DES. This difference does not appear to be associated with the incidence of resorption bodies, as demonstrated by DES versus MMS, nor with the overall incidence of death of conceptuses, as demonstrated by EMS and DES on one hand versus X-ray on the other.

Detailed classification of external defects (Table 2) revealed that the relatively higher incidences of fetuses with defects in the case of EMS, EtO, and DES were attributable largely to increases in two classes of defects: hydrops (Fig. 1) and limb/tail defects. It should be noted, however, that all mutagens significantly increased these and other classes of defects. There appears to be a more even distribution of the various types of defects in the case of X-rays, ENU, TEM, and MMS.

Cellular Target for EMS Effects

Katoh et al. (1989) demonstrated that the EMS-induced mid- and late-gestation deaths and malformations are a direct effect on the zygote and not an indirect effect through the mother. The question of which component of the zygote is the important target was addressed by conducting pronuclear transfer experiments basically according to the method of McGrath and Solter (1983). Male and female pronuclei that were treated with EMS 6 hours after mating were both transferred beginning 6 hours later to treated or untreated enucleated zygotes. In addition, untreated nuclei were transferred to enucleated treated zygotes. Results shown in Table 3 indicate that the characteristic effects were produced only when both the pronuclei and the cytoplasm were exposed to EMS.

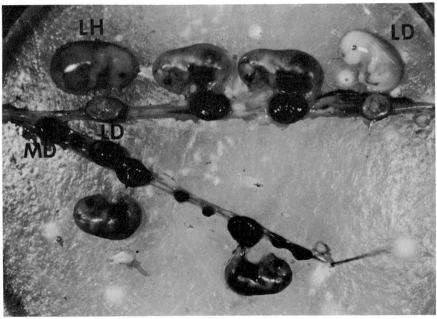

Figure 1
Characteristic anomalies induced by EtO, EMS, and DES. LH is a live hydropic fetus, LD are fetuses that died after 11 days gestation, and MD is an embryo that died at mid-gestation prior to day 11. (Reprinted, with permission, from Rutledge and Generoso 1989.)

DISCUSSION

Possible Mutagenic Mechanisms for Zygote-derived Developmental Anomalies

All mutagens induced a variety of developmental defects (Table 2). However, on the basis of the classes of defects induced, the mutagens may be grouped into two classes depending on whether or not they induce a high incidence of mid- and late-gestation deaths and a high incidence of hydropic and limb and tail defects. The high inducers are EMS, EtO, and DES; the low inducers are X-ray, MMS, TEM, and ENU. It should be remembered that the mothers received the treatment. Thus, the possibilities that the responses were caused either directly by mutagenic damage to the zygote or indirectly through maternal toxicity need to be considered. For EMS, it was confirmed that the effect is directly on the zygote and not on the mother. In a reciprocal egg transfer experiment, it was found that the distinctive anomalies were observed only in the direction in which the donor but not the recipient females were treated (Katoh et al. 1989). This result, and the fact that EMS

Table 2
Distribution of Fetal Anomalies

Defects	EtO	EMS	DES	MMS	ENU	TEM	X-ray	Control
Hydropia	13.8	9.1	6.4	0.7	0	0.9	1.2	0.03
Abdominal wall	8.0	2.2	3.2	1.1	3.0	3.6	2.5	0.1
Limb and/or tail	16.1	12.1	12.7	0.2	0	3.6	0.3	0.1
Eyes	0	7.8	2.5	2.7	9.1	4.5	1.8	0.1
Exencephaly	0	0	1.9	1.3	6.1	0	0.6	0.7
Cleft palate	4.6	1.3	0	0.2	0	0.9	0.6	0
Others	9.0	3.0	0	0	0	1.8	0	0.7
Fetuses examined	87	231	157	451	33	112	326	3436
Fetuses with defect (%)	37	29	37	3.8	15	10	4.6	1.3

Anomalies of live fetuses referred to in Table 1. Figures are percentages of live fetuses examined. Some fetuses that had more than one defect were included in each category that applies.

is a mutagen, indicate that the zygote effect is genetic in nature. It may be assumed that the anomalies induced by EtO or DES also arose as a result of direct genetic effects on the zygotes.

It is not likely that the mid- and late-gestation deaths and the hydropic and limb and tail defects characteristic of these mutagens were caused by point mutations and small deficiencies, since ENU and X-rays do not induce them. It was determined in the case of EMS and EtO that these anomalies are also not caused by structural or numerical chromosomal aberrations (Katoh et al. 1989). Therefore, although evidence suggests that the EMS, EtO, and DES effects are genetic in nature, they are not of the conventional types.

For ENU, the varied malformations appear to be associated with point mutational events. It was recently reported that ENU is highly effective in inducing intragenic specific-locus mutations during the zygotic stages (Russell et al. 1988).

Table 3
Pronuclear Transfer

Zygote composition[a]	Number of implants	Resorption bodies (%)	Mid- and late-gestation death (%)
N^+C^+	113	41	10
N^+C^-	124	41	0.8
N^-C^+	99	37	0

[a] (N) male and female pronuclei; (C) enucleated zygote; (+) EMS-treated; (–) untreated.

For X-rays, it is possible that the low incidence of malformations is attributable to certain chromosomal imbalances and small deficiencies, since chromosome aberrations were induced in the male and female genomes of exposed zygotes (Russell 1965, 1968; Russell and Montgomery 1966).

The difference between the two classes of mutagens may be associated with specific adducts formed with DNA. It is perhaps not coincidental that EMS, EtO, and DES all bind primarily to the N^7 position of guanine. However, so does MMS, which is a low inducer of anomalies. The obvious difference between these mutagens is that MMS is a methylating agent, whereas the other three are either ethylating or hydroxyethylating agents. It is therefore possible that N^7-ethyl and N^7-hydroxyethyl guanine are the important adducts responsible for the induction of the developmental anomalies that are characteristic of the class. This possibility is under further investigation.

Relevance to Problems of Human Developmental Anomalies

Many of the developmental anomalies produced in the zygote experiments, particularly those observed in the EMS, EtO, and DES studies, resemble the large class of sporadic anomalies in humans for which the etiology is unknown. Examples are cleft palate, omphalocoele, club foot, hydrops, and stillbirth (Czeizel 1985; Oakley 1986; Rutledge and Generoso 1989). It is possible that these classes of human developmental defects have, in part, similar pathogenesis to those induced experimentally in mice. Thus, one mechanism for the pathogenesis of this class of defects in humans could be a genetic one, but not of the conventional type. The experimental system involving the mouse zygotes presents an opportunity for probing into the underlying molecular and cellular mechanisms of the pathogenesis of this class of developmental defects. The vulnerability of the zygotes to mutagens demonstrated in mice raises questions regarding the susceptibility of human zygotes to drugs and environmental chemicals. The possibility that many of the spontaneously occurring stillbirths and common sporadic defects in humans are zygote-derived is intriguing.

ACKNOWLEDGMENTS

This research was sponsored jointly by the National Toxicology Program under National Institute of Environmental Health Sciences interagency agreement Y-01-ES-20085 and the Office of Health and Environmental Research, U.S. Department of Energy, under contract DE-AC05-84OR21400 with Martin Marietta Energy Systems, Inc. The submitted manuscript has been authored by a contractor of the U.S. Government. Accordingly, the U.S. Government retains a nonexclusive, royalty-free license to publish or reproduce the published form of this contribution, or allow others to do so, for U.S. Government purposes.

REFERENCES

Czeizel, A. 1985. Teratoepidemiology. In *Prevention of physical and mental congenital defects,* part B: Epidemiology, early detection and therapy, and environmental factors (ed. M. Marois), p. 7. A.R. Liss, New York.

Generoso, W.M., J.C. Rutledge, K.T. Cain, L.A. Hughes, and P.W. Braden. 1987. Exposure of female mice to ethylene oxide within hours after mating leads to fetal malformation and death. *Mutat Res.* **176:** 269.

Generoso, W.M., J.C. Rutledge, K.T. Cain, L.A. Hughes, and D.J. Downing. 1988. Mutagen-induced fetal anomalies and death following treatment of females within hours after mating. *Mutat. Res.* **199:** 175.

Katoh, M., N.L.A. Cacheiro, C.V. Cornett, K.T. Cain, J.C. Rutledge, and W.M. Generoso. 1989. Fetal anomalies produced subsequent to treatment of zygotes with ethylene oxide or ethyl methanesulfonate are not likely due to the normal genetic causes. *Mutat. Res.* **210:** 337.

Krishna, M. and W.M. Generoso. 1977. Timing of sperm penetration, pronuclear formation, pronuclear DNA synthesis, and first cleavage in naturally ovulated mouse eggs. *J. Exp. Zool.* **202:** 245.

McGrath, J. and D. Solter. 1983. Nuclear transplantation in the mouse embryo by microsurgery and cell fusion. *Science* **220:** 1300.

Oakley, G.P. 1986. Frequency of human congenital malformations. *Clin. Perinatol.* **13:** 545.

Russell, L.B. 1965. Death and chromosome damage from irradiation of preimplantation stages. *Ciba Found. Symp.* p. 17.

―――. 1968. The use of sex-chromosome anomalies for measuring radiation effects in different germ-cell stages of the mouse. In *Effects of radiation on meiotic systems,* p. 27. International Atomic Energy Agency, Vienna.

Russell, L.B. and C.S. Montgomery. 1966. Radiation-sensitivity differences within cell-division cycles during mouse cleavage. *Int. J. Radiat. Biol.* **10:** 151.

Russell, L.B., J.W. Bangham, K.F. Stelzner, and P.R. Hunsicker. 1988. High frequency of mosaic mutants produced by *N*-ethyl-*N*-nitrosourea exposure of mouse zygotes. *Proc. Natl. Acad. Sci.* **85:** 9167.

Rutledge, J.C. and W.M. Generoso. 1989. Fetal pathology produced by ethylene oxide treatment of the murine zygote. *Teratology* **39:** 563.

COMMENTS

Adler: Have you done some aneuploidy investigations on these embryos?

Generoso: This is exactly why we started the experiment. We thought that if we could disrupt the metaphase 2 stage experimentally, we could induce chromosome malsegregation. In fact, we were very excited when we saw these late deaths, because this was exactly what we were expecting for trisomies. We looked at the first cleavage, second cleavage, fourth cleavage, and mid-gestation anomalies for chromosome number and structural anomalies, and found nothing.

Favor: With ENU treatment did you get an effect?

Generoso: We have malformations. In the ENU experiment, we had only 33 fetuses, but in 15% of those, the living fetuses have malformations of one sort or another, but not of the same type as those produced by EMS, EtO, and DES.

Favor: But it showed the same time response?

Generoso: No. Mutagens like ENU and TEM have a broader spectrum, a broader stage effect. At 1 hour, 6 hours, 9 hours, 12 hours, and 25 hours, they are all affected. We are studying the spectrum of malformations that are produced by these mutagens when we treat at 1 cell or 2 cells or a later stage embryo.

Moses: Do your observations rule out the possibility that the agents are acting on certain sequences of DNA that may include promoters of key genes that are expressed—should be expressed—normally at different times during development, but that have altered expression? Is that the kind of mechanism that could be envisioned?

Generoso: Yes. There will be several of those that one can think of right now. The results that we have would not rule out, in fact, looking at the mitochondrial and nuclear DNA interruptions. The integration of transposons is being looked at as a possible mechanism. The mouse genome is replete with all kinds of transposons, and the zygote may be a unique system for the integration of transposons because the egg cytoplasm may be loaded with copies of endogenous transposons, so that at the time you treat the zygote, you may induce integration. Many different possible mechanisms could be visualized on the basis of our results.

Wyrobek: How do you interpret your requirement for an exposed cytoplasm?

Generoso: One possibility is mitochondrial DNA. According to a recent report by D.C. Wallace in *Trends in Genetics,* mitochondrial DNA mutations have been associated with neuromuscular diseases.

Wyrobek: Are you saying that you think it's multiple targets—I mean, there are some nuclear targets and some cytoplasmic targets and with any one of them you get the effect? Is that how you are thinking about this?

Generoso: It seems that both nuclei and cytoplasm have to be treated.

Hecht: The egg has about 10^5 mitochondria, and there is no mitochondrial DNA replication until about 5 days after fertilization. You have to somehow come up with a model where you have enough targets hit so that the distribution of those mitochondria will cause problems.

Generoso: Events that occur even during oocyte development have a long-lasting impact on the development of the embryo. Therefore, I would not necessarily dismiss that the damage to the maternal components is just temporary.

Manipulation of the Mouse Genome by Homologous Recombination

OLIVER SMITHIES AND BEVERLY H. KOLLER
Department of Pathology
University of North Carolina
Chapel Hill, North Carolina 27599-7525

OVERVIEW

Specific genes can now be altered in predetermined ways in the germ line of mice. This manipulative capability derives from the bringing together of two technologies. The first of these is the isolation from blastocysts of embryonic stem (ES) cells that can be cultured in vitro (Evans and Kaufman 1981; Martin 1981), yet can be returned to blastocysts for continued development in vivo to give fertile chimeras that are able to transmit the ES cell genome to their progeny (Bradley et al. 1984). The second technology is the ability to alter in a predetermined way a native chromosomal gene in living cells by homologous recombination between the chosen target gene and DNA introduced into the cell (Smithies et al. 1985). Targeted modification of chosen genes by homologous recombination (gene targeting) is now being achieved regularly in ES cells (Doetschman et al. 1987, 1988; Thomas and Capecchi 1987; Mansour et al. 1988; Joyner et al. 1989; Koller and Smithies 1989), and transfer of the modified genes to the mouse germ line has been reported in a few instances (Koller et al. 1989; Schwartzberg et al. 1989; Thompson et al. 1989). There appear to be no serious obstacles to the overall procedure's becoming routine in a relatively short time. We present here some of our experiences in two examples.

INTRODUCTION

Gene targeting requires the introduction of DNA into a recipient cell that has nucleotide sequences in common with the target gene. The common exogenous and endogenous sequences then become aligned in the cell, and homologous recombination between them can occur. Depending on the geometrical arrangement of the common sequences in the incoming DNA, the result of the recombination can be an insertion of the exogenous sequences into the target locus (an O-type event) or a replacement of target sequences by the exogenous DNA (an Ω-type event). The first example we describe involves an O-type event; the second is an Ω-type.

The frequency of obtaining recombinants in a gene-targeting experiment is usually low. Finding these recombinants, therefore, requires the use of some selection procedure that will allow recombinants to survive preferentially, or the use of a screening procedure that will allow recombinants to be identified in the presence of many nonrecombinants, or the use of some combination of selection and screening.

Direct selection has been used to achieve targeted modifications in the genes for hypoxanthine phosphoribosyltransferase, HPRT (Doetschman et al. 1987, 1988; Thomas and Capecchi 1987), and for adenine phosphoribosyltransferase, APRT (Adair et al. 1989). The first example we present here used direct selection with the HPRT locus.

Selection can also be used in a less direct way to assist in gene targeting by first enriching the culture for cells that have at least taken up the exogenous DNA into the genome of the cell. This enrichment procedure employs a directly selectable "passenger" gene placed on the same piece of DNA as the targeting sequences. Cells are selected that have incorporated the selectable passenger gene (plus the targeting sequences) at any positions in the genome where the selectable gene can be expressed. Cells that have not incorporated the DNA into their genomes in this way will be killed. The desired cells, in which the targeting sequences found and modified the target locus, will be among the few survivors, provided that the passenger gene can still be expressed after being carried into the target locus by the targeting sequences.

Our second example shows that selection with the help of a passenger gene conferring resistance to G418 can be used to assist gene targeting of the gene coding for β2-microglobulin, *B2m*, even though this gene is not overtly expressed in the target ES cell. The key feature in making passenger genes work in potentially inactive positions in the genome is probably the use of strong promoters and enhancers to drive them (see Thomas and Capecchi 1987).

Screening procedures for identifying targeted genes can range from baroque, not to say "heroic," sib-selection systems such as we used in the first targeting experiments with a native chromosomal gene, to direct colony identification procedures likely to be possible by fluorescent antibodies when targeting a gene coding for a surface antigen. Our current example uses as the means of identification a polymerase chain reaction (PCR) procedure that specifically amplifies a DNA fragment only when the correct recombination has occurred (Kim and Smithies 1988). This category of screening procedure enables recombinants to be identified in pools of cells, most of which are nonrecombinant. Smaller pools are then screened in a sib-selection protocol that eventually allows the isolation of pure recombinants. (Sib-selection methods usually kill cells during the identification, but at the same time save living siblings of the killed cells for further growth.) In our second example, we used PCR to identify pools of cells containing the desired recombinants, followed by sib-selection to isolate pure recombinants.

After the targeted modification of a chosen gene has been achieved in ES cells, the production of chimeras, followed by breeding those chimeras that are fertile, allows the eventual introduction of the targeted modification into the germ line. Our first example, with the HPRT gene, has been carried to the germ line (Koller et al. 1989), as we describe here. Our second example, with the *B2m* gene, has been carried through to the stage of chimeras (Koller and Smithies 1989), with preliminary evidence (our unpublished data) that the modified gene can be transmitted into the germ line.

RESULTS

Modification of the HPRT Gene in ES Cells

The ES cell line chosen for modification was the HPRT cell line E14TG2a (referred to below as E14), isolated by Hooper et al. (1987). The single X-linked HPRT gene in this male (XY) cell line, which was isolated from strain 129/01a mice, has suffered a deletion that removed the promoter, the first and the second exons of the HPRT gene, and an undetermined amount of 5' sequences. The correcting plasmid, designed to repair the deficiency in the manner illustrated in Figure 1, a–c, contains human and mouse sequences coding for the missing HPRT promoter and exons 1 and 2. In addition to supplying the missing sequences, the correcting plasmid contains approximately 4.5 kbp of DNA identical in sequence to exon 3 of the target HPRT gene and its surroundings. It is this identical DNA that enables the homologous recombination. The correcting plasmid (12.4 kbp in length) was cut within exon 3 to provide recombinogenic ends and was introduced into the E14 cells by electroporation. An O-type recombination occurred that inserted the plasmid sequences into the HPRT locus and restored its function. $HPRT^+$ colonies were isolated by using HAT-containing medium, which kills $HPRT^-$ cells but allows $HPRT^+$ cells to grow. Details of this work can be found in Doetschman et al. (1987). A total of 333 HAT-resistant colonies were obtained. Of the HAT-resistant colonies, 19 have now been analyzed, and 18 have the planned gene correction. Of these 18, 5 have the simplest possible change in which the 12.4 kbp correcting plasmid was inserted by a simple O-type crossover into the target locus, but not into any other extraneous sites in the ES cell genome. Two of the five colonies (98-2 and 98-12) were tested for their ability to make chimeras that could transfer the preplanned alteration to the mouse germ line.

Early Experiences with Chimeras

The outbred strain of albino mice, ICR, is very fertile, and the females have large numbers of blastocysts, particularly after superovulation. Our initial tests therefore used this strain as the source of blastocysts. Corrected ES cells (98-2) were injected into these blastocysts, and they were returned to pseudopregnant foster mothers for continued development to give chimeras. Good chimeras were obtained, as judged by their coat colors, but none transmitted the ES cell genome to their progeny. For this reason, and because the first cell line 98-2 that we tested became infected with mycoplasma, this first series of experiments was terminated.

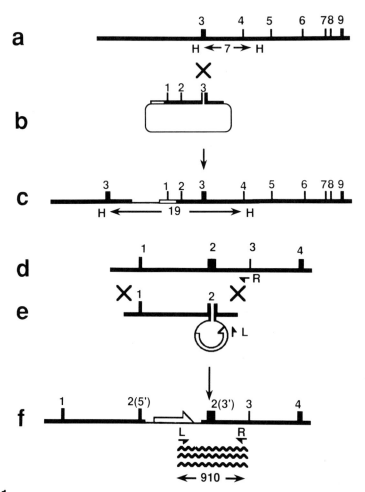

Figure 1
Targeted O-type modification of the HPRT gene (Doetschman et al. 1987) and Ω-type modification of the *B2m* gene (Koller and Smithies 1989). (*a*) The mutant HPRT gene in the ES cell line E14. Two *Hin*dIII sites, H, are separated by 7 kbp. Exons are numbered. (*b*) The O-type correcting plasmid (not to scale). (*c*) The corrected HPRT gene in the ES cell line 98-12 after targeting. The two *Hin*dIII sites are now separated by 19 kbp and have vector plasmid sequences between them. (*d*) The normal *B2m* gene in the ES cell line E14. (*e*) The Ω-type inactivating plasmid with a *Neo* gene (open arrow) in exon 2. (*f*) The inactivated *B2m* gene in ES cell lines 39b and 22a. The left and right primers (L and R), specific for the incoming DNA and the target locus, respectively, allow PCR amplification of a 910-bp fragment only after the targeting is successful.

A personal communication from Colin Stewart, European Molecular Biology Laboratory, Heidelberg (see also Schwartzberg et al. 1989), alerted us to his observation that ES cells from strain 129 mice readily enter the germ line when the recipient blastocysts are from C57BL animals. Our first experiment was therefore repeated using C57BL blastocysts and a different, noninfected but otherwise arbitrarily chosen clone (98-12), also containing the desired simple HPRT$^+$ correction. Chimeras were obtained, but they were lost at about the stage of weaning because of an infection leading to diarrhea. Their nonchimeric littermates survived.

The loss of the chimeras probably resulted from the unusual sensitivity, not known to us at the time, of strain 129 mice to a variety of infectious agents (Parker et al. 1978). The difficulty was circumvented by setting up a third series of experiments in which the foster mothers were kept in sterile microisolator cages with sterile food, water, and bedding. Cages were changed in a laminar flow hood using aseptic procedures. This third series of experiments led to the successful results described in the next section.

Transmission of the Corrected HPRT Gene through the Germ Line

In the successful series of experiments, a total of 11 surviving pups were obtained from three consecutive pregnancies. Nine of the pups (82%) were chimeras: six males and three females. Two of the chimeras died with undifferentiated tumors that were of ES cell origin, as judged by their DNA. The upper part of Figure 2 shows a photograph of four fertile male chimeras that we obtained; the central two transmitted the ES cell genome to their offspring. In a subsequent set of transfers, a heavily chimeric male was obtained who later sired a family of eight pups, all of whom received the ES cell genome. Such "jackpot" males are probably the result of the XY ES cells converting an XX blastocyst into a fertile male able to produce sperm only from his XY, and therefore ES cell-derived, germ cells.

Since male chimeras transmit an X chromosome to their female offspring and a Y chromosome to their male offspring, and since the HPRT gene is on the X chromosome, only female offspring with the agouti coat color (resulting from the dominant A^W gene carried in the 129/Ola-derived ES cells) are expected to have the altered HPRT gene. To confirm these expectations, we analyzed DNA samples from a family with male and female pups of the appropriate colors. The expected results were obtained: An agouti female pup received the modified gene; a black female did not, nor did an agouti male. Thus the complete procedure has been validated—a change in a specific gene has been made by homologous recombination, and the exact preplanned modification has been transferred to the mouse germ line. (More details of these experiments can be found in Koller et al. 1989.)

β2 Microglobulin

The 99-amino-acid protein β2 microglobulin, β2m, is an essential and virtually invariant subunit of the class I major histocompatibility antigens that occur on the surfaces of most cells in the body. β2m is coded by a single copy autosomal gene. The other subunit of class I antigens is variable in two senses: within a given mouse strain and between different strains of mice. The "within strain" variation is the consequence of the use of any of a group of linked genes (for example, *H-2K*, *H-2D*, *Qa*) to provide the variable subunit. The "between strain" variation is the consequence of the high degree of polymorphism of some of the individual variable subunit genes. The assembled class I antigens are the vehicles for presenting to the immune system self antigens, such as intracellular protein epitopes, and non-self antigens, such as viral protein epitopes, during the control of nonresponsiveness to autoantigens and of responsiveness to foreign antigens. Lack of class I antigens in a developing or mature animal is therefore likely to affect detrimentally the animal's immune response to viruses and may lead to problems of autoimmunity. Other developmental problems are possible, since the expression of some of the class I antigens appears to vary during development, and in different cell lineages. We plan to investigate these various possibilities by constructing a mouse strain lacking a functional *B2m* gene.

Inactivation of the *B2m* Gene in ES Cells

A construct of the Ω-type was made (Koller and Smithies 1989), containing about 5 kbp of sequences from the *B2m* gene as the region of homology, and also containing a G418-resistance gene (*Neo*). The G418-resistance gene, which was inserted into exon 2 of the *B2m* gene, is driven by a promoter plus enhancer that is able to function in ES cells (Thomas and Capecchi 1987). Homologous recombination between this construct and the target *B2m* locus was expected to inactivate the *B2m* gene, while concurrently converting the treated ES cell (E14) from being G418-sensitive to being G418-resistant (see Fig. 1, d–f). We used PCR to distinguish between G418-resistant colonies that are correctly targeted at the *B2m* locus and colonies that are G418-resistant due to insertion of the construct into some nonhomologous site in the genome. Two primers were synthesized, one corresponding to sequences in the incoming DNA not present in the *B2m* locus, the other corresponding to sequences in the *B2m* locus but not present in the portion of the *B2m* gene in the incoming DNA. Amplification is observed only when these two primers are together on a single piece of DNA in the manner expected after the Ω-type recombination. The construct was introduced into the E14 ES cells by electroporation. G418-resistant colonies were obtained, and small pools of them were screened by PCR to allow pools containing correctly modified cells to be identified. Retesting sibling colonies allowed identification of pure colonies of correctly modified cells.

Figure 2
(*Upper*) Four fertile male chimeras obtained with the targeted ES cell line 98-12. The center two animals transmitted the ES cell genome to their offspring (Koller et al. 1989). (*Lower*) A fertile female chimera with the "spooky" phenotype obtained with the targeted ES cell line 22a. She transmitted the ES cell genome to her offspring (B.H. Koller and O. Smithies, unpubl.).

Two experiments were carried out, and a correctly targeted ES cell line was obtained from each. Their correctness was established by Southern blotting DNA from the two cell lines, E1422a and E1439b (referred to below as 22a and 39b). Each line had the preplanned modification with a single copy of the G418-resistant gene inserted into exon 2 of the *B2m* gene, but with no additional copies at other places in the ES cell genome. The behavior of these two targeted ES cell lines was then examined with respect to chimerism and germ-line transmission.

Chimeras from the ES Cell Lines Having an Inactivated *B2m* Gene

ES cell line 39b produced 17 chimeras out of 23 live pups (75%) (Koller and Smithies 1989). However, preliminary tests have indicated that most heavily chimeric males derived from this cell line are infertile; they fail to "plug" their mates. Those that are fertile transmit the C57BL genome, but not the ES cell genome.

ES cell line 22a produced 30 chimeras out of 38 live offspring (79%) (Koller and Smithies 1989). The pups that show the 129/Ola coat color over either or both sides of their heads have abnormal eyes on the corresponding side(s). The eyes, open at birth, are opaque in older pups. Despite this abnormality, six of the "spooky"-looking animals, males and females, have transmitted the ES cell genome to their offspring, and with it the modified *B2m* locus (details to be published elsewhere). The lower part of Figure 2 shows a photograph of a spooky female chimera that transmitted the ES cell genome to some of her offspring.

DISCUSSION

The experiments we have described here demonstrate clearly that homologous recombination can be used to modify genes in ES cells, one expressed and directly selectable, one probably not expressed and only indirectly selectable. In each case, chimeras resulting from these modifications can subsequently transmit the preplanned modification through the mouse germ line. This capability of altering genes in a mammalian germ line in a predetermined manner opens the way to a virtually unlimited series of experiments whereby gene function and dysfunction can be investigated in vivo and during development. There are, however, still technical problems that need to be overcome before the procedure is routinely successful. A brief discussion of what went right and what went wrong in some of the experiments may help delineate some areas in need of further investigation.

The history of the ES cell line 98-12 (which has the corrected HPRT gene) is of interest. The original ES cell line, isolated by Evans and Kaufman (1981), was germ-line competent (Bradley et al. 1984). A descendant of this ES cell line was subcloned as E14TG2a, when a spontaneous HPRT mutant was isolated by Hooper et al. (1987), who also showed that this mutant line was germ-line competent. Doetschman et al. (1987) then isolated clones from E14TG2a that had been made HPRT$^+$ by gene targeting. One of these clones, 98-12, is still germ-line competent (Koller et al. 1989). Thus, it is clear that an ES cell line can be germ-line competent after many passages and several subclonings. The histories of the ES cell lines 39B and 22a are also of interest. Both are derived, as was 98-12, from E14TG2a, and both have been subsequently subcloned during the gene targeting. One appears to be germ-line incompetent, and the other has an abnormal phenotype, although it is germ-line competent.

We are currently attempting to find out why these several ES cell lines differ, even though all are derived from E14TG2a and their total histories are similar. Whatever the outcome of this investigation may be, we find the overall picture very encouraging. If a germ-line-proven ES cell is used in the gene targeting, and if several targeted isolates are tested, then with reasonable diligence one can expect to be able to manipulate the mouse genome in many ways by using homologous recombination.

ACKNOWLEDGMENTS

We thank Dr. Shiu Huang for some unpublished data on the germ-line transmission of the ES cell line 98-12. The color photographs were by Richard Pippin, Medical Illustrations and Photography, University of North Carolina at Chapel Hill. The work was supported by grant GM-20069 to O.S. from the National Institutes of Health. B.H.K. is a fellow of the Leukemia Society of America.

REFERENCES

Adair, G.M., R.S. Nairn, J.H. Wilson, M.M. Seidman, K.A. Brotherman, C. MacKinnon, and J.B. Scheerer. 1989. Targeted homologous recombination at the endogenous adenine phosphoribosyltransferase locus in Chinese hamster cells. *Proc. Natl. Acad. Sci.* **86:** 4574.

Bradley, A., M. Evans, M.H. Kaufman, and E. Robertson. 1984. Formation of germ-line chimaeras from embryo-derived teratocarcinoma cell lines. *Nature* **309:** 255.

Doetschman, T., N. Maeda, and O. Smithies. 1988. Targeted mutation of the *HPRT* gene in mouse embryonic stem cells. *Proc. Natl. Acad. Sci.* **85:** 8583.

Doetschman, T., R.G. Gregg, N. Maeda, M.L. Hooper, D.W. Melton, S. Thompson, and O. Smithies. 1987. Targetted correction of a mutant *HPRT* gene in mouse embryonic stem cells. *Nature* **330:** 576.

Evans, M.J. and M.H. Kaufman. 1981. Establishment in culture of pluripotential cells from mouse embryos. *Nature* **292:** 154.

Hooper, M., K. Hardy, A. Handyside, S. Hunter, and M. Monk. 1987. HPRT-deficient (Lesch-Nyhan) mouse embryos derived from germline colonization by cultured cells. *Nature* **326:** 292.

Joyner, A.L., W.C. Skarnes, and J. Rossant. 1989. Production of a mutation in mouse *En-2* gene by homologous recombination in embryonic stem cells. *Nature* **338**: 153.

Kim, H.-S. and O. Smithies. 1988. Recombinant fragment assay for gene targetting based on the polymerase chain reaction. *Nucleic Acids Res.* **16**: 8887.

Koller, B.H. and O. Smithies. 1989. Inactivating the β_2 microglobulin locus in mouse embryonic stem cells by homologous recombination. *Proc. Natl. Acad. Sci.* **86**: 8932.

Koller, B.H., L.J. Hagemann, T. Doetschman, J.R. Hagaman, S. Huang, P.J. Williams, N.L. First, N. Maeda, and O. Smithies. 1989. Germ-line transmission of a planned alteration made in a hypoxanthine phosphoribosyl transferase gene by homologous recombination in embryonic stem cells. *Proc. Natl. Acad. Sci.* **86**: 8927.

Mansour, S.L., K.R. Thomas, and M.R. Capecchi. 1988. Disruption of the proto-oncogene *int-2* in mouse embryo-derived stem cells: A general strategy for targeting mutations to non-selectable genes. *Nature* **336**: 348.

Martin, G.R. 1981. Isolation of a pluripotent cell line from early mouse embryos cultured in medium conditioned by teratocarcinoma stem cells. *Proc. Natl. Acad. Sci.* **78**: 7634.

Parker, J.C., M.D. Whiteman, and C.B. Richter. 1978. Susceptibility of inbred and outbred mouse strains to Sendai virus and prevalence of infection in laboratory rodents. *Infect. Immun.* **19**: 123.

Schwartzberg, P.L., S.P. Goff, and E.J. Robertson. 1989. Germ-line transmission of a c-*abl* mutation produced by targeted gene disruption in ES cells. *Science* **246**: 799.

Smithies, O., R.G. Gregg, S.S. Boggs, M.A. Koralewski, and R.S. Kucherlapati. 1985. Insertion of DNA sequences into the human chromosomal β globin locus via homologous recombination. *Nature* **317**: 230.

Thomas, K.R. and M.R. Capecchi. 1987. Site-directed mutagenesis by gene targeting in mouse embryo-derived stem cells. *Cell* **51**: 503.

Thompson, S., A.R. Clarke, A.M. Pow, M.L. Hooper, and D.W. Melton. 1989. Germ line transmission and expression of a corrected HPRT gene produced by gene targeting in embryonic stem cells. *Cell* **56**: 313.

COMMENTS

Woychik: Is this now or will it become a routine procedure?

Smithies: It's getting better. About five or six labs are doing germ-line transmission. Getting the targeting is relatively straightforward. The germ line is probably a matter of common sense now. I gave a few examples of some problems. Other problems, such as not letting the ES line sit in culture too long before doing the experiment, are relatively simple things that we can do something about. I used to worry about lines that have been cloned several times, but now I think the more it has been cloned and been shown to go into the germ line each time, the better it is, not the worse it is. If you can work with a line of which you have a history, then that's good. Some changes are accumulating—obviously the "spooky" changes and the infertility. On the basis of intuition, I don't think those are DNA changes. The frequency is too high. I think it's something else: a sort of imprinting, if you like. Maybe they have begun to differentiate a little. Maybe they've got methylation patterns on them that have already prevented them from accomplishing all of the changes that they need to get into the germ line. Clean them up, as it were, as it goes into the germ line and it's okay. So, I'm going to do some experiments to try to clean them up outside the germ line.

Fritz: You're actually going to investigate what the class I antigen-minus mutants look like?

Smithies: Yes, eventually, but the animals are still too young at present. We have a male that we know certainly has transmitted, and we have three animals now that we know are agouti, including some females, but they aren't old enough yet to have done the DNA on, so we don't know whether they have the gene. They have, of course, a 50/50 chance of getting it because it's an autosomal gene. So we have to breed those animals. There's no question that it's in the germ line, but we haven't got the animals, the homozygotes, yet. Heterozygotes are perfectly normal.

Favor: Are the spooky mice always spooky in the chimeras and, once you get transmissions, it's null?

Smithies: That's correct. It's conceivable that spooky is a dominant lethal and that really it doesn't come through in the next generation, you just get segregation of it. The family sizes are a little too small to be able to tell whether the litter size is normal; it probably isn't far from normal. At the moment, the indication is that spooky is not transmitted, but we can't be quite sure. There's no insertional mutation anywhere else, because there's no DNA anywhere else in the genome, other than in the target locus.

Favor: So spooky is not transmitted and it's not associated with the insertion?

Smithies: Exactly. Well, I will say it's not associated with an insertion. I know it's not transmitted, but I don't know whether that's because it gets cleaned up or because it's lethal.

Utilization of DNA Techniques in the Detection of Germ-line Mutations

Molecular Approaches to the Estimation of Germinal Gene Mutation Rates

HARVEY W. MOHRENWEISER, BRIAN A. PERRY, AND STEPHEN A. JUDD
Biomedical Sciences Division L-452
Lawrence Livermore National Laboratory
Livermore, California 94550

OVERVIEW

Estimation of the induced germinal gene mutation rate in human populations is difficult because de novo mutations, especially of functional gene loci, are rare events, and the sizes of the human populations that have been exposed to known mutagens are generally small. Thus, if statistically significant estimates of mutation rates are to be generated, it is critical that a significant body of data be obtained from each offspring included in a mutation screening study. Additionally, the assay(s) employed must be sufficiently robust to efficiently detect the spectrum of lesions that may be induced by different classes of mutagens. DNA-based techniques have the potential to overcome these problems, because it may be possible to screen the entire genome for mutational events, and the alterations in DNA structure can be analyzed directly. Two assay systems are being developed and tested for feasibility as germinal gene mutation screening strategies. One is based on denaturing gradient gel electrophoresis (DGGE) to detect nucleotide substitutions, and the second is a restriction enzyme site (RES) mapping strategy to identify DNA insertions, deletions, and rearrangements (I/D/R). Cell lines derived by clonal expansion of cells following exposure to mutagens are being screened in an initial prototype experiment. Preliminary results indicate that it should be feasible to screen for germinal gene mutations in rodent test systems and in human populations in the near future with these two techniques.

INTRODUCTION

The development of molecular biological techniques has provided the resources for detailed analysis of the alterations in gene structures that are responsible for an extensive series of human genetic diseases. The accumulated data indicate that the molecular lesions associated with genetic diseases, either "point mutations" transmitted from previous generations or lesions arising as de novo mutations, include an array of alterations in DNA structure encompassing both nucleotide substitutions and I/D/R (Mohrenweiser and Jones 1990). These alterations exist in the coding regions of genes, as well as in the introns and flanking sequences. Some of the types of lesions that have been identified and the mechanisms of subsequent manifestation as genetic diseases are outlined in Table 1. This list is not exhaustive, but it is indicative of the classes of molecular lesions that will need to be detected if a mutation screening protocol is to achieve nearly complete ascertainment of all de novo mutational events occurring in the probands being screened. A similar array of lesions are observed in the more limited number of presumably uninduced de novo germinal mutations that have been identified and characterized in humans (Mohrenweiser and Jones 1990).

Within the group of genetic diseases where the altered gene structures have been characterized, significant differences are observed in the relative frequency of different classes of molecular lesions at different loci. For example, many of the lesions at the Factor VIII gene (Antonarakis and Kazazian 1988) and most of the lesions at the αI(I)procollagen gene (Byers 1988), causing hemophilia A and osteogenesis imperfecta, respectively, are nucleotide substitutions, whereas most of the lesions of the dystrophin (Duchenne muscular dystrophy) gene (den Dunnen et al. 1987) are deletions.

Similar differences in the spectrum of mutations at different loci are observed in data from mutations in somatic cells (Thacker 1985; DeMarini et al. 1989). Spontaneous mutations at the HPRT locus are predominantly deletions and rearrangements (Vrieling et al. 1985; Thacker and Genesh 1989), whereas most mutations at the APRT locus are apparently nucleotide substitutions (Grosovsky et al. 1988). Other data from somatic cell mutagenesis studies indicate that the spectrum of mutations is a characteristic of the mutagenic agent that is used for mutation induction (Grosovsky et al. 1986; Urlaub et al. 1986; Kronenberg and Little 1989). Thus, significant differences in induced mutation rates are noted when the responses at

Table 1
Nature of Molecular Lesions Associated with Human Genetic Diseases

A. Nucleotide substitutions
1. substitutions causing amino acid substitutions
2. substitutions generating translation termination signals
3. substitutions altering RNA splice sites
4. substitutions generating cryptic splice sites
5. substitutions in transcription or translation regulatory sites

B. Insertions, deletions, and rearrangements
1. deletions or insertions involving alu repeats
2. insertions involving L1 repeats
3. deletion, duplication, or exchange of repeated segments
4. rearrangement of gene families
5. insertion or deletion of random DNA

different loci are compared (Evans et al. 1986). It is not clear if these differences in mutation rates and also the most prevalent alteration in gene structure reflect "hot spots" for induction of different types of lesions or reflect aspects of the gene related to generation of null alleles (the basis for identification of the mutation) and/or cell lethal events.

Differences in spontaneous mutation rates among loci are noted for germinal mutations, presumably reflecting a combination of the relative frequency of specific alterations in gene structure and the sensitivity of the gene to inactivation by each class of lesion (Mulvihill and Czeizel 1983). Locus-specific differences in mutation rates are also noted in mice following exposure of the parental population to mutagenic agents (Lewis and Johnson 1986; Russell and Russell 1959).

This specificity in the classes of DNA structural alterations observed at different loci and also following exposure to different classes of mutagenic agents has implications for efforts to develop methods for detecting gene mutations and subsequently estimating germinal mutation rates. First, the technique(s) must detect a range of different lesions. Second, given that the current estimate of the point mutation rate is $\sim 10^{-8}$ per nucleotide per generation (Stamatoyannopoulos and Nute 1982; Neel et al. 1988a), it will be necessary to extract a significant quantity of data from each proband, although recent identification of hypermutable loci (Jeffreys et al. 1988; Kovacs et al. 1989) may alter these calculations regarding the sample sizes or number of locus tests necessary to detect statistically significant increases in the germinal gene mutation rate. Within the context of an effort to estimate the background and induced germinal gene mutation rates, all classes of DNA structural alterations must be identified in order to have a reasonable certainty of ascertainment for each mutational event and, therefore, of generating a significant data base from each proband.

DNA-based assays have the potential to monitor the entire genome, both coding and noncoding regions, for mutational events. We are developing two strategies for detecting gene mutations. The first, a modification of the standard RES mapping or restriction fragment length polymorphism (RFLP) screening strategy, emphasizes the detection of I/D/R events and is designed to monitor a significant portion of the genome. The second method uses DGGE to detect nucleotide substitutions and focuses on more limited regions of the genome.

RESULTS

I/D/R Variation in Human Populations

The strategy employed for identification of I/D/R events has been to digest the genomic DNA with the restriction enzymes, *Eco*RI and *Bam*HI, separate the restriction fragments by electrophoresis, and transfer the fragments to membranes for subsequent probing with a series of probes. In an initial study to estimate the extent of genetic variation associated with I/D/R variants in a human population, a series of 18 probes, primarily cDNA probes, were used to screen DNA samples from 130 unrelated individuals (Mohrenweiser et al. 1989). The 18 probes screen ~40 noncontiguous DNA fragments or loci per individual because several of the probes cross-hybridize to pseudogenes as well as the functional gene or are members of gene families. These 40 loci encompass ~600 kb per individual. Approximately 3 rare I/D/R variants (19 different rare alleles occurring in 31 individuals) were identified for each 1000 loci analyzed in the survey. (An I/D/R variant is defined by variation in fragment size with >1 restriction enzyme; rare variants are alleles existing at a frequency of <1%.) Of the rare variants, 40% were detected with the pseudozeta-globin probe, a probe known to detect variation associated with differences in the number of tandem repeat elements located adjacent to the gene (Proudfoot et al. 1984). Four nucleotide substitution variant alleles, existing in 24 individuals, were identified in this survey; only one of these alleles was a rare variant. All of the variants are presumed to be inherited, although no family studies were completed.

I/D/R Mutations Detected with Functional Gene Probes

An analytical strategy similar to that outlined above to detect variation in human populations is being used to detect mutations induced in the human lymphoblastoid cell line, TK-6. Mutagenized cells were obtained by exposing a mass culture of TK-6 cells to ethylnitrosourea (ENU); individual cells were picked, without the application of any selective constraints except for the ability to divide, in order to establish new clonal cell lines (Chu et al. 1988; Hanash et al. 1988). These cell lines were then expanded to provide sufficient material for cell storage and DNA isolation. The mutation rate in these cells, based on the analysis of polypeptides by two-dimensional gel electrophoresis, is ~10^4 above background (Chu et al. 1988).

Some 31 cell lines have been screened for mutations with 8 functional gene probes that screen 18 noncontiguous fragments or loci for a total of 1116 alleles examined thus far (Table 2). One variant pattern (Fig. 1, lane 1), as compared to both control and sibling cell lines (lane 4), has been identified with the probe for the triosephosphate isomerase (TPI) gene. The identical pattern is observed in DNA from an independent DNA isolation (lane 2), although obviously the cells for the second isolation are derived from the same clone as the original sample. The altered pattern is observed when the DNA is digested with *Eco*RI (variant, lane 3; control, lane 5), but not *Bam*HI or any of 5 other restriction enzymes (data not shown). The 10.5-kb fragment previously identified as one of the 2 *Eco*RI restriction fragments associated with the functional TPI gene (Brown et al. 1985) is reduced in intensity; thus, the mutation is presumed to involve the *Eco*RI fragment that includes the 5′ portion (5′ of exon 6) of the functional gene. The RES map of the DNA from the mutant cell is consistent with either a deletion 5′ of the known regulatory region of

Table 2
Results of Screening for ENU-induced Mutations in Clonally Derived Human Lymphoblast Cell Lines with Functional Gene Probes

Locus	Number of alleles screened	Number of mutants identified
Actin	10	0
ADA	2	0
PGK	5	0
TPI	8	1
INS	2	0
DH21B	4	0
HPRT	3	0
CKMM	2	0
Total	31 cell lines x 36 alleles = 1116 tests	1

the gene or a nucleotide substitution generating a new *Eco*RI recognition site in this region. It has not been possible to obtain a more detailed description of the mutation because of the absence of mapped restriction enzyme sites in the region 5' of the gene. One other presumptive mutation at the TPI locus has been identified among 8 lymphocyte samples that were clonally derived following exposure to irradiation, but appropriate confirmation studies have not yet been completed.

Screening for Mutations at Tandem Repeat Regions

The DNA from the offspring of cells exposed to ENU was also analyzed for mutations at loci that are highly polymorphic due to variation associated with differences in the number of repeat units in a tandem array (Nakamura et al. 1987). Hypervariable loci have been observed to have high rates of both germinal (Jeffreys et al. 1988; Kovacs et al. 1989) and somatic (Armour et al. 1989) mutations. The 4 probes used in the first screening are from the set of variable number tandem repeat (VNTR) probes isolated by Nakamura et al. (1987). These 4 loci exhibit heterozygosities in human populations ranging from 0.80 to 0.95. An example of the extensive variation detected at these loci is shown in Figure 2, where DNA from 10 unrelated individuals has been digested with *Msp*I and probed with the probe pYNH24 (D2S44). At least 14 different alleles are observed; the maximum possible among 10 individuals is obviously 20 alleles. Twenty-seven cell lines have been screened for mutations with 4 hypervariable loci probes. No variation was detectable with probes for either the D2S44 or the D3S42 (pEFD64.1 probe) loci. Two variant classes (lanes 3 and 10) were detected with the probe (pEKDA2.1) for the

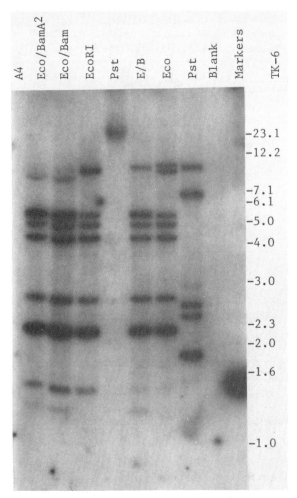

Figure 1
Analysis of a mutation at the TPI locus. (Lanes *1* and *2*) *Eco*RI/*Bam*HI-digested DNA from two preparations of DNA from the variant cell line. (Lane *3*) *Eco*RI-digested variant DNA. *(Lanes 5* and *6*) *Eco*RI/*Bam*HI- and *Eco*RI-digested control DNA. The filter has been probed with the TPI cDNA probe.

D16S83 locus in DNA digested with *Rsa*I (Fig. 3). Since both of these variants are observed in several cell lines, these are presumably spontaneous mutations that arose during expansion of the original (parental) culture and before exposure to ENU, since independent mutations would not be expected to yield identical mutant alleles. Similarly, 4 of 27 cell lines exhibited identical variant patterns when probed

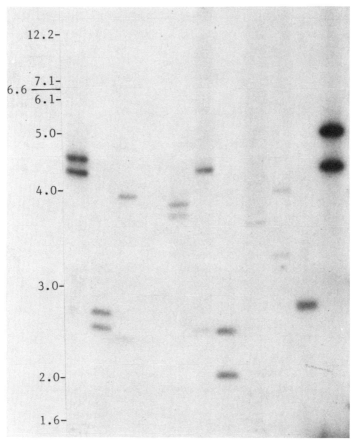

Figure 2
Variation among individuals at the D2S44 locus. DNA from ten unrelated individuals was digested with *Msp*I and probed with a probe for the D2S44 locus.

with the probe pYNZ21 (D19S*) (data not shown), again consistent with a mutation occurring in the parental cell population prior to mutagen exposure.

Screening for Nucleotide Substitution Mutations

Several methods have been developed that are capable of detecting nucleotide substitutions (Delehanty et al. 1986; OTA 1986), including DGGE and cleavage of mismatches in RNA:DNA heteroduplexes with RNase (Myers et al. 1988) or in DNA:DNA heteroduplexes with chemical reagents (Cotton and Campbell 1989).

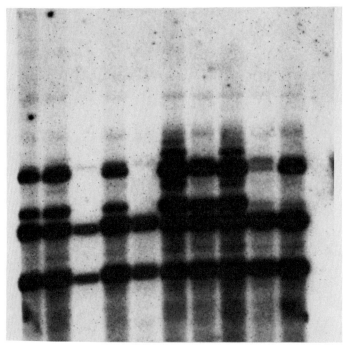

Figure 3
Variation at the D16S83 locus in a series of clonally derived human lymphoblastoid cell lines. DNA from each cell line was digested with *Rsa*I and probed as described in the Results with a probe for the D16S83 locus.

The DGGE strategy as developed by Sheffield et al. (1989) involves polymerase chain reaction (PCR) amplification of a genomic DNA segment (Saiki et al. 1988) followed by electrophoresis of the amplification products on polyacrylamide gels in the presence of an increasing concentration of denaturant (urea and formamide) (Fischer and Lerman 1983). The denaturant concentration at which homoduplex molecules begin to unwind is dependent on the nucleotide composition/sequence of the DNA molecule; heteroduplex molecules are particularly unstable relative to either of the companion homopolymers. DNA molecules differing by only a single nucleotide in several hundred bases will have different denaturation profiles (Sheffield et al. 1989; Theophilus et al. 1989). The electrophoretic mobility of a denatured molecule is significantly reduced compared to the native polymer. Examples of the use of this assay are given in Figure 4. In Figure 4A, a 123-bp region of exon 1 of the human β-globin gene (Sheffield et al. 1989) has been amplified and the products have been analyzed by DGGE. In lanes 1 and 2 are the products from the normal β-globin gene, and in lane 3 is the product of the β*S

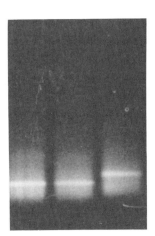 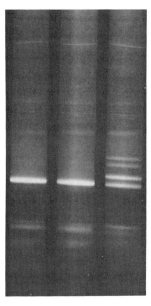

Figure 4
Analysis of variants by DGGE of amplified DNA fragments. (*Left*) Analysis of a 123-bp fragment of the β-globin gene. Lanes *1* and *2* are control DNAs, lane *3* is the β*S variant DNA. (*Right*) Analysis of a fragment of the TPI gene including exons 2–4. Lanes *1* and *2* are control DNAs; lane *3* is DNA from the TPI*N/TPI*MAN variant.

globin gene. The DNA fragments differ by a single nucleotide, T in β*S and A in β*A at position +70 of the β-globin gene.

Similar analysis has been completed of a 583-bp region including exons 2–4 of the TPI gene from an individual that is heterozygous for an isozyme electromorph at this locus (Asakawa and Mohrenweiser 1982). A single band is observed in the analysis of the exon 2–4 fragment from several unrelated individuals, indicating sequence identity of both alleles (Fig. 4B, lanes 1 and 2). The presence of two bands in the DGGE analysis of the exon 2–4 DNA fragment from the TPI*N/TPI*MAN variant individual (lane 3) is consistent with allelic variation in the DNA in the region of the TPI gene flanked by the primers chosen for PCR amplification. Only a single band is detected following DGGE analysis of amplified DNA fragments containing either exon 1 or exons 5–7 from both normal and variant individuals (data not shown). The isozyme pattern observed following electrophoresis of the cellular proteins from the variant individual is consistent with an amino acid substitution, and, therefore, a nucleotide substitution in the coding region of the gene is expected. Obviously, the specific nucleotide change in the exon 2–4 variant DNA fragment must be determined by sequencing to confirm that the heterozygosity detected by the DGGE assay is responsible for the amino acid substitution detected by isozyme analysis. This specific isozyme electromorph has only been observed in a single family during the screening of ~20,000 individuals. We are employing the PCR/DGGE strategy to screen for nucleotide substitution mutations in five regions of the DNA (~2 kb) from the same cells as screened for I/D/R mutations.

DISCUSSION

Genetic diseases constitute a significant component of the total human health cost (Czeizel and Sankaranarayanan 1984; Mohrenweiser 1987; Baird et al. 1988; Czeizel et al. 1988), and the contribution of genetic disease to the total health burden of the population has increased during the last 50 years (Costa et al. 1985). This reflects the identification of previously unrecognized genetic diseases and genetic predispositions to disease, but more importantly, the decreasing contribution of other competing factors, such as infection and malnutrition. The total contribution of de novo mutations to the disease burden is generally not well documented. Estimates have been obtained for chromosomal anomalies causing untoward pregnancy outcome and genetic disease, since most chromosomal anomalies are due to de novo mutations (Hook 1987). Thus, obtaining an estimate of the contribution of mutation to the health cost for these traits is relatively straightforward (Lyon 1985). Estimates have been made for the contribution of an increase in the gene mutation rate for dominantly inherited and sex-linked traits to the frequency of associated genetic diseases (Lyon 1985). The contribution of spontaneous de novo mutations to the health cost associated with recessively inherited gene mutations is much less well known (Searle and Edwards 1986). All of the attempts to estimate induced mutation rates and associated health costs of the increased disease burden have been limited by the absence of sufficient data (UNSCEAR 1986; Sankaranarayanan 1988). This reflects the relatively small number of individuals exposed to any "known" mutagen (e.g., Mulvihill and Byrne 1985) and/or the generally low doses of mutagen received by most individuals that subsequently reproduced following exposure to the mutagenic agent (e.g., Neel et al. 1988b). To overcome this problem of accumulating adequate data, sensitive yet robust techniques that can be used to generate a significant data base from each nuclear family must be developed.

DNA-based techniques can directly interrogate the integrity of the genetic material and thus have unique potential as a basis for viable germinal gene mutation screening strategies (Delehanty et al. 1986: OTA 1986; Mohrenweiser and Branscomb 1989). The two strategies being employed in the prototype experiment described in the Results section, RES mapping to detect I/D/R mutations and DGGE of amplified DNA segments to detect nucleotide substitutions, are both capable of detecting mutational events. Two questions remain before decisions regarding the potential of the RES mapping strategy for inclusion into a human gene mutation screening protocol can be made. The first is the mutation rate for I/D/R events detected with functional gene probes and the ability to develop cost-effective techniques for generating the requisite data base. The second question is the inducibility of mutations at the hypermutable loci and, if the mutation rate at these loci is responsive to exposure to mutagenic agents, the classes of mutagenic agents that can induce mutational events. The DGGE strategy is capable of detecting most (if not all) nucleotide substitutions in defined DNA segments (Theophilus et al. 1989). The critical question is the mutation rate and the ability to generate sufficient data for detecting a significant number of mutational events in a timely manner. It is also possible that hypermutable regions/sequences for nucleotide substitutions may be identified (Cooper and Youssoufian 1988; Koeberl et al. 1989), as has been observed with the repeat element regions. The identification of hypermutable regions for nucleotide substitutions could change the estimates for the quantity of data necessary for obtaining statistically significant estimates of background and induced mutation rates, although the same caveats as outlined for the tandem repeat regions would apply.

The results from the somatic cell prototype experiment are sufficiently encouraging that these techniques are beginning to be incorporated into mouse germinal gene mutation screening programs (Lewis et al., this volume). The results from these two efforts will be the basis for subsequent decisions regarding the feasibility of undertaking new efforts to estimate background and induced human germinal mutation rates, data that are absolutely critical for the important regulatory decisions regarding allowable human exposures to suspected mutagenic agents.

ACKNOWLEDGMENTS

This work was performed under the auspices of the U.S. Department of Energy, Office of Health and Environmental Research, by the Lawrence Livermore National Laboratory under contract number W-7405-ENG-48. We thank R. Myers for the globin variants.

REFERENCES

Antonarakis, S.E. and H.H. Kazazian, Jr. 1988. The molecular basis of hemophilia A in man. *Trends Genet.* **4**: 233.

Armour, J.A.L., I. Patel, S.L. Thein, M.F. Fey, and A.J. Jeffreys. 1989. Analysis of somatic mutations at human minisatellite loci in tumors and cell lines. *Genomics* **4**: 328.

Asakawa, J. and H.W. Mohrenweiser. 1982. Characterization of two new electrophoretic variants of human triosephosphate isomerase: Stability, kinetic, and immunological properties. *Biochem. Genet.* **20**: 59.

Baird, P.A., T.W. Anderson, H.B. Newcombe, and R.B. Lowry. 1988. Genetic disorders in children and young adults: A population study. *Am. J. Human Genet.* **42**: 677.

Brown, J.R, I.O. Daar, J.R. Krug, and L.E. Maquat. 1985. Characterization of the functional gene and several processed pseudogenes in the human triosephosphate isomerase gene family. *Mol. Cell. Biol.* **5:** 1694.

Byers, P.H. 1988. Osteogenesis imperfecta: An update. *Growth: Genet. Horm.* **4(2):** 1.

Chu, E.H.Y., M. Boehnke, S.M. Hanash, R.D. Kuick, B.J. Lamb, J.V. Neel, W. Niezgoda, S. Pivirotto, and G. Sundling. 1988. Estimation of mutation rates based on the analysis of polypeptide constituents of cultured human lymphoblastoid cells. *Genetics* **119:** 693.

Cooper, D.N. and H. Youssoufian. 1988. The CpG dinucleotide and human genetic disease. *Hum. Genet.* **78:** 151.

Costa, T., C.R. Scriver, and B. Childs. 1985. The effect of Mendelian disease on human health: A measurement. *Am. J. Med. Genet.* **21:** 231.

Cotton, R.G.H. and R.D. Campbell. 1989. Chemical reactivity of matched cytosine and thymine bases near mismatched and unmatched bases in a heteroduplex between DNA strands with multiple differences. *Nucleic Acids Res.* **17:** 4223.

Czeizel, A. and K. Sankaranarayanan. 1984. The load of genetic and partially genetic disorders in man: I. Congenital anomalies: Estimates of detriment in terms of years of life lost and years of impaired life. *Mutat. Res.* **128:** 73.

Czeizel, A., K. Sankaranarayanan, A. Losonci, T. Rudas, and M. Keresztes. 1988. The load of genetic and partially genetic diseases in man: II. Some selected common multifactorial diseases: Estimates of population prevalence and of detriment in terms of years of lost and impaired life. *Mutat. Res.* **196:** 259.

Delehanty, J., R.L. White, and M.L. Mendelsohn. 1986. Approaches to determining mutation rates in human DNA. *Mutat. Res.* **167:** 215.

DeMarini, D.M., H.E. Brockman, F.J. deSerres, H.H. Evans, L.F. Stankowski, Jr., and A.W. Hsie. 1989. Specific-locus mutations induced in eukaryotes (especially mammalian cells) by radiation and chemicals: A perspective. *Mutat. Res.* **220:** 11.

den Dunnen, J.T., E. Bakker, G.K. Breteler, P.L. Pearson, and G.J.B. van Ommen. 1987. Direct detection of more than 50% of the Duchenne muscular dystrophy mutations by field inversion gels. *Nature* **329:** 640.

Evans, H.H., J. Mencl, M.-F. Horng, M. Ricanati, C. Sanchez, and J. Hozier. 1986. Locus specificity in the mutability of L5178Y mouse lymphoma cells: The role of multilocus lesions. *Proc. Natl. Acad. Sci.* **83:** 4379.

Fischer, S.G. and L.S. Lerman. 1983. DNA fragments differing by single base-pair substitution are separated in denaturing gradient gels: Correspondence with melting theory. *Proc. Natl. Acad. Sci.* **80:** 1579.

Grosovsky, A.J., E.A. Drobetsky, P.J. de Jong, and B.W. Glickman. 1986. Southern analysis of genomic alterations in gamma-ray-induced APRT⁻ hamster cell mutants. *Genetics* **113:** 405.

Grosovsky, A.J., J.G. de Boer, P.J. de Jong, E.A. Drobetsky, and B.W. Glickman. 1988. Base substitutions, frameshifts, and small deletions constitute ionizing radiation-induced point mutations in mammalian cells. *Proc. Natl. Acad. Sci.* **85:** 185.

Hanash, S.M., M. Boehnke, E.H.Y. Chu, J.V. Neel, and R.D. Kuick. 1988. Nonrandom distribution of structural mutants in ethylnitrosourea-treated cultured human lymphoblastoid cells. *Proc. Natl. Acad. Sci.* **85:** 165.

Hook, E.B. 1987. Surveillance of germinal mutations for effects of putative environmental mutagens and utilization of a chromosome registry in following rates of cytogenetic disorders. In *Cytogenetics, basic and applied aspects* (ed. G. Obe and A. Basler), p. 141, Springer-Verlag, Berlin.

Jeffreys, A.J., N.J. Royle, V. Wilson, and V. Wong. 1988. Spontaneous mutation rates to new length alleles at tandem-repetitive hypervariable loci in human DNA. *Nature* **332:** 278.

Koeberl, D.D., C.D.K. Bottema, J.M. Buerstedde, and S.S. Sommer. 1989. Functionally important regions of the Factor IX gene have a low rate of polymorphism and a high rate of mutation in the dinucleotide CpG. *Am. J. Hum. Genet.* **45:** 448.

Kovacs, B.W., B. Shahbahrami, and D.E. Comings. 1989. Studies of human germinal mutations by deoxyribonucleic acid hybridization. *Am. J. Obstet. Gynecol.* **160:** 798.

Kronenberg, A. and J.B. Little. 1989. Molecular characterization of thymidine kinase mutants of human cells induced by densely ionizing radiation. *Mutat. Res.* **211:** 215.

Lewis, S.E. and F.M. Johnson. 1986. The nature of spontaneous and induced electrophoretically detected mutations in the mouse. In *Genetic toxicology of environmental chemicals*, part B: *Genetic effects and applied mutagenesis* (ed. C. Ramel et al.), p. 359. A.R. Liss, New York.

Lyon, M.F. 1985. Attempts to estimate genetic risks caused by mutagens to later generations. *Banbury Rep.* **19:** 151.

Mohrenweiser, H.W. 1987. Functional hemizygosity in the human genome: Direct estimate from twelve erythrocyte enzyme loci. *Hum. Genet.* **77:** 241.

Mohrenweiser, H.W. and E.W. Branscomb. 1989. Molecular approaches to the detection of germinal mutations in mammalian organisms, including man. In *New trends in genetic risk assessment* (ed. G. Jolles and A. Cordier), p. 41. Academic Press, New York.

Mohrenweiser, H.W. and I.M. Jones. 1990. Review of the molecular characteristics of gene mutations of the germline and somatic cells of the human. *Mutat. Res.* (in press).

Mohrenweiser, H.W., R.D. Larsen, and J.V. Neel. 1989. Development of molecular approaches to estimating germinal mutation rates. I. Insertion/deletion/rearrangement variants in the human genome. *Mutat. Res.* **212:** 241.

Mulvihill, J.J. and J. Byrne. 1985. Offspring of long-time survivors of childhood cancer. Childhood cancer: Late effects. *Clin. Oncol.* **4:** 333.

Mulvihill, J.J. and A. Czeizel. 1983. Perspectives in mutation epidemiology 6: A 1983 view of sentinel phenotypes. *Mutat. Res.* **123:** 345.

Myers, R.M., V.C. Sheffield, and D.R. Cox. 1988. Detection of single base changes in DNA: Ribonuclease cleavage and denaturing gradient gel electrophoresis. In *Genome analysis: A practical approach* (ed. K. Davies), p. 95. IRL Press, Washington D.C.

Nakamura, Y., M. Leppert, P. O'Connell, R. Wolff, T. Holm, M. Culver, C. Martin, E. Fukimoto, M. Hoff, E. Kumlin, and R. White. 1987. Variable number of tandem repeat (VNTR) markers for human gene mapping. *Science* **235:** 1616.

Neel, J.V., H.W. Mohrenweiser, and H. Gershowitz. 1988a. A pilot study of the use of placental cord blood samples in monitoring for mutational events. *Mutat. Res.* **204:** 365.

Neel, J.V., C. Satoh, K. Goriki, J. Asakawa, M. Fujita, N. Takahashi, T. Kageoka, and T. Hazama. 1988b. Search for mutations altering protein charge and/or function in children of atomic bomb survivors: Final report. *Am. J. Hum. Genet.* **42:** 663.

Office of Technology Assessment (OTA). 1986. Technologies for detecting heritable mutations in human beings (OTA-H-298). U.S. Government Printing Office, Washington, D.C.

Proudfoot, N.J., A. Gil, and T. Maniatis. 1984. The structure of the human ξ-globin gene and a closely linked, nearly identical pseudogene. *Cell* **31**: 553.

Russell, W.L. and L.B. Russell. 1959. The genetic and phenotypic characteristics of radiation-induced mutations in mice. *Radiat. Res.* **1**: 296.

Saiki, R.K., D.H. Gelfand, S. Stoffel, S.J. Scharf, R. Higuchi, G.T. Horn, K.B. Mullis, and H.A. Erlich. 1988. Primer-directed enzymatic amplification of DNA with a thermostable DNA polymerase. *Science* **239**: 487.

Sankaranarayanan, K. 1988. Prevalence of genetic and partially genetic diseases in man and the estimation of genetic risks of exposure to ionizing radiation. *Am. J. Hum. Genet.* **42**: 651.

Searle, A.G. and J.H. Edwards. 1986. The estimation of risks from the induction of recessive mutations after exposure to ionising radiation. *J. Med. Genet.* **23**: 220.

Sheffield, V.C., D.R. Cox, L.S. Lerman, and R.M. Myers. 1989. Attachment of a 40-base-pair G+C-rich sequence (GC-clamp) to genomic DNA fragments by the polymerase chain reaction results in improved detection of single-base changes. *Proc. Natl. Acad. Sci.* **86**: 232.

Stamatoyannapoulos, G. and P.E. Nute. 1982. *De novo* mutations producing unstable HbS and HbM. II: Direct estimates of minimum nucleotide mutation rates in man. *Hum. Genet.* **60**: 181.

Thacker, J. 1985. The molecular nature of mutations in cultured cells: A review. *Mutat. Res.* **150**: 431.

Thacker, J. and A.N. Ganesh. 1989. Molecular analysis of spontaneous and ethyl methanesulphonate-induced mutations of the hprt gene in hamster cells. *Mutat. Res.* **210**: 103.

Theophilus, B.D.M., T. Latham, G.A. Grabowski, and F.I. Smith. 1989. Comparison of RNase, a chemical cleavage and GC-clamped denaturing gradient gel electrophoresis for the detection of mutations in exon 9 of the human acid β-glucosidase gene. *Nucleic Acids Res.* **17**: 7707.

United Nations Scientific Committee on the Effects of Atomic Radiation (UNSCEAR). 1986. Committee report. In *Genetic and somatic effects of ionizing radiation*, publication E.86.1X9. United Nations, New York.

Urlaub, G., P.J. Mitchell, E. Kas, L.A. Chasin, V.L. Funanage, T.T. Myoda, and J. Hamlin. 1986. Effect of gamma rays at the dihydrofolate reductase locus: Deletions and inversions. *Somatic Cell Mol. Genet.* **12**: 555.

Vrieling, H., J.W.I.M. Simons, F. Arweet, A.T. Natarajan, and A.A. van Zeeland. 1985. Mutations induced by X-rays at the HPRT locus in cultured Chinese hamster cells are mostly deletions. *Mutat. Res.* **144**: 281.

COMMENTS

Hecht: When you use probes like actin, do you see any difference between the coding actin genes and the pseudogenes of actin?

Mohrenweiser: We found two or three polymorphic single base substitutions with actin. I didn't dig in deep enough to know whether the bands we were picking up were in pseudogenes. All I know is they were things that were positive with the actin probe.

Hecht: You mentioned with the hypervariable regions there was a much higher mutation rate.

Mohrenweiser: Yes.

Hecht: One wonders whether pseudogenes are under the same or whether pseudogenes would be something in between.

Mohrenweiser: Oh, yes. I would think they would be in between.

Hecht: Has that been established actually?

Mohrenweiser: No. If you go to something like actin, it probably will be. I'm not sure, in hindsight, whether or not actin was a good choice, because actin mutations could well be dominant. For example, after I had gone through all the work to get the collagen gene and do some background work, I said, "What am I doing?" because they're going to be dominant—in fact, they're going to be lethal. A mutation on the collagen gene is unlikely to make it through. It doesn't in the human. When I tell you about what germinal mutations in humans look like—it's Factor VIII, a few HPRTs, and a whole lot of others—using actin was probably not a good choice of probes.

Arnheim: It depends really on what kind of probe you are using. If you are using a cDNA probe, most of the polymorphisms you pick up probably won't be within the coding region anyway; they'll be spread out.

Hecht: Right. However, there are some specific probes for each of the actins, so that you could fish out which ones you want to be looking at.

Mohrenweiser: Yes, I think you're right. What you would see is probably going to be in the flanking region. Our strategy is if we can't monitor 10–15 kb, we don't use the probe. It takes as long to run a blot with 1 or 2 kb as it does with 15. There are plenty of probes out there to use.

Quantitation and Characterization of Human Germinal Mutations at Hypervariable Loci

BRUCE W. KOVACS,[1,2] BEJAN SHAHBAHRAMI,[1] AND DAVID E. COMINGS[2]
[1]Department of Obstetrics and Gynecology
University of Southern California, School of Medicine
Los Angeles, California 90033
[2]Department of Medical Genetics
City of Hope Medical Center
Duarte, California 91010

OVERVIEW

The application of molecular genetic approaches to the study of germ-line mutations contributes much to the understanding of their nature and frequency. Loci in the human genome composed of tandem reiterations of short oligonucleotide core sequences (short tandem repeats, STRs) exhibit exceptional levels of polymorphism (hypervariability) due to variations in the number of repeats. We have used synthetic oligonucleotide probes complementary to several STRs to detect de novo insertion/deletion mutations at multiple loci. In studies of human pedigrees, we have been able to quantitate and characterize de novo germ-line mutations occurring at many loci. Our results indicate that some of the loci detected by multilocus STR probes are hypermutable, and rates as high as 1% have been detected with these probes. Mutations at single hypervariable loci, identified by multilocus probes, have been studied using polymerase chain reaction (PCR) to improve detection of new alleles arising in the germ line. The use of hypervariable STR loci offers significant advantages for study of this type of mutational event in humans. Some hypervariable STR loci may have functional significance, and the processes producing mutations may be intrinsic to the STR sequences motif, emphasizing their usefulness in mutation study.

INTRODUCTION

The genetic material of all living things is neither static nor stable. Spontaneous and induced mutational events in germ-line cells drive evolution and in their wake produce human disease. Dysfunctional alleles arising as a result of new mutations that have occurred in previous generations or that occur de novo are responsible for a significant proportion of premature mortality, incurable morbidity, mental dysfunction, and infertility.

Because of their obvious importance, a variety of methods have been employed to study these events in humans. Estimates of occurrence rates have relied on the incidence of characteristic phenotypic abnormalities, frequency of chromosomal aberrations, and the frequency of electrophoretic variants in proteins (Office of Technology Assessment 1986). These methods have contributed much to our knowledge of germ-line mutations and their role in human disease. However, these approaches detect mutations that produce differences only in phenotype, which limits their usefulness in studying rates. Furthermore, they yield limited information as to the molecular nature of events involved.

The advent of recombinant DNA technologies has now added new dimensions to these studies, allowing direct analysis of mutational events. This methodology has revealed a great deal relative to the nature of mutant alleles resulting in some human diseases. However, there are several major problems in detecting spontaneous and induced germinal mutations in humans that have confounded measurement of mutation rates. One of the primary difficulties in such quantitations is that when single genes or gene products are studied, an enormous number of individuals or gene loci need to be examined. For example, if the spontaneous rate per locus is 10^{-6}/locus/generation, one million parent/child sets would need to be examined to detect one new mutation. Screening 100 loci would reduce the number of individuals required for study to 30,000, still a formidable task. Other impediments include the apparently low frequency of mutation events at a given locus, the need to circumvent the problem of positive and negative selection of mutants, the requirement to establish correct paternity in human pedigrees under study, and the need to be able to detect so-called silent mutations. Therefore, there is a need to examine a large number of loci with high efficiency to maximize information obtained from human pedigrees.

A significant fraction of the eukaryotic genome is composed of repetitive sequence DNA. Much of this DNA has been grouped into various families based on sequence, organization, and size (Singer 1982). In some of these families, variation in sequence and/or in the number of repeat units occurs between and within species. One class of repetitive DNA elements, STRs, are characterized by a motif of short oligonucleotide core sequences reiterated in tandem arrays. These elements have been variously called "minisatellites" (Jeffreys et al. 1985a), "midisatellites" (Nakamura et al. 1987), and "microsatellites" (Litt and Luty 1989). This repetitive DNA has been found to occur at many highly polymorphic (hypervariable) loci dispersed throughout the genome. The exceptionally high levels of polymorphic variation at these loci are due to variation in the number of the tandem repeat (VNTR) core. Familial studies have demonstrated that STR loci are inherited in codominant Mendelian fashion.

These discoveries suggested to us a system of detection that would help resolve the difficulties of study enumerated above and allow direct measurement of germinal mutations in humans. Particularly exciting was the idea that in order to generate the diversity of alleles at these loci, high rates of mutation must also be present. In the technique, molecular probes comprising tandem repeats of STR core sequences would be hybridized to restriction digests of human DNA after electrophoresis. Because variation occurs as a result of insertions or deletions of core repeats, new mutations would appear as novel fragments with many restriction endonucleases. By examining the DNA "fingerprint" patterns of parent/child sets, new mutations could be detected by demonstrating fragments not present in the patterns of the parents. Furthermore, because each person has an individually specific pattern, correct paternity could be simultaneously established (Jeffreys et al. 1985b).

The focus of our research has been to quantitate spontaneous germinal mutations in man at hypervariable loci detected by STR probes and to understand the nature of this genetic instability at the molecular level. In this paper, we summarize progress achieved toward these ends and outline the future directions of research relevant to this topic. In addition, we attempt to integrate specific results of our research into the body of current understanding of mutagenesis as it relates to germ-cell mutations.

RESULTS

Hypermutability in STRs

Shortly after the discovery of minisatellite loci (Jeffreys et al. 1985a), we began preliminary experiments to explore the potential of this approach. Initially, we sought to improve accuracy of detection by designing and synthesizing synthetic DNA probes based on two of the most hypervariable sequences (33.6 and 33.15) reported by Jeffreys. We thought that the use of short oligonucleotide probes (32-mer and 29-mer, respectively) would improve accuracy in that exact specificity of hybridization could be achieved by using high-stringency conditions (Suggs et al. 1981). In addition, the use of synthetic probes allowed for in-gel hybridizations, which would simplify detection. We performed an initial screening of 25 pedigrees, consisting of 190 individuals, using digests produced by several restriction enzymes (*Hin*fI, *Hae*III, and *Pst*I) using these two synthetic DNA probes. We found 3×10^{-3} new mutations/locus/generation in our preliminary survey. Since these spontaneous mutation rates were several orders of magnitude higher than any previously reported, it was clear that at least some of the loci detected were indeed hypermutable, and rates were directly measurable. Encouraged by the validity of our approach, we next screened a larger number of families that contained three generations, in order to demonstrate transmission of mutant alleles to subsequent progeny. In collaboration with the Centre d'Etude Polymorphism Humaine (Paris, France), we obtained a reference panel of three-generation families with large sibships to supplement our own reference panel. During this time, several other STR sequences were identified by other investigators, and we decided to enlarge our battery of probes with some of these sequences. Among these repetitive sequences, the satellite III repeat (Fowler et al. 1987) was of particular interest. The tandem repeat sequence in this DNA consists primarily of a pentamer (TTCCA) core. Polymorphic length variations at the multiple loci detected by the oligonucleotide probe appeared to be detectable only with *Taq*I digestions. This suggested that new length alleles were the result of guanine-cytosine transversions in the tandem repeat pentamer core producing *Taq*I recognition sites (TGCA). Therefore, rates of nucleotide substitution in this classic satellite DNA could be compared to insertion/deletion mutations in other STR loci. Armed with an enlarged battery of multiple-locus probes, we screened our family panel comprising 458 offspring corresponding to 916 gametes (Fig. 1). We found that both base substitution mutations and insertion/deletion mutations occur frequently enough to be directly quantitated and compared at many hypervariable loci in human pedigrees (Table 1). Importantly, the rate of occurrence of insertions/deletions detected with these multilocus STR probes exceeds that of base transversions and approaches 1% at some loci in live-born children (Kovacs et al. 1989).

Expressed STR Loci

Prior to the identification of VNTR loci, a hot spot of insertion/deletion mutations in the *lac-I* gene of *Escherichia coli* was found to be composed of a simple tandem repeated DNA sequence (Farabaugh et al. 1978). Furthermore, these authors suggested that genetic variation was likely to occur in many sites of this type, perhaps as a result of replication slippage or unequal homologous recombination. In an extensive analysis of accumulated DNA sequence data, it was noted that repeated motifs of trinucleotides occurred more frequently than did random motifs (Tautz et al. 1986). Moreover, high levels of allelic length variation have also been demonstrated in several proteins. Some of these proteins contain regions composed of tandem reiterations of an amino acid core, and in some, polymorphic variation is due to unit changes in the number of tandem repeats of the amino acid core (Swallow et al. 1987).

In hopes of detecting new regions of hypervariability in expressed loci, we investigated the occurrence of simple tandem repeats of trinucleotide sequences occurring in the human genome. We used 30 synthetic oligonucleotide probes composed of 10 different simple trinucleotide repeat core sequences in this work. The lengths of the probes used corresponded to 5, 6, and 7 repeats of trinucleotide monomer cores plus an additional 2 bases of the core (general formula $(XYZ)_nXY$ where n = 5, 6, or 7 repeats).

All but one of the 10 simple trinucleotide core repeat motifs hybridized to restriction digests at relatively high stringency afforded by Td-5°C conditions, used for oligonucleotide probes. The patterns produced by all of the other probes revealed the presence of multiple hybridizing fragments, with sizes ranging from 500 to 20,000 bp. High-stringency washes in 3 M tetramethylammonium chloride, at temperatures independent of base composition (Wood et al. 1984) resulted in the

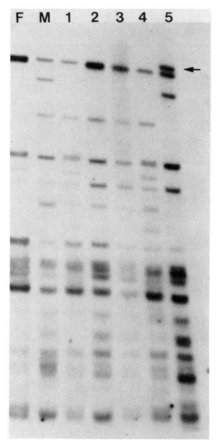

Figure 1
A new length allele (arrow) in a child (lane 5), not present in the patterns of the father (F) or mother (M), detected in a seven-member family by the myoglobin 33.6 probe. New allele is present in restriction digests using several endonucleases (not shown).

loss of some fragments. However, most of the multiple fragments detected by these hybridizations were not removed by this treatment, indicating an absence of nonspecific hybridization.

We used these probes to analyze patterns produced by digesting DNA obtained from 50 unrelated, random individuals. Several of the probes revealed hybridization patterns in which many of the fragments were highly polymorphic, allowing discrimination of individuals by fragment pattern differences. Hybridizations to DNA digests using several other restriction endonucleases revealed different polymorphic fragment patterns, suggesting that the polymorphism is due to length

Table 1
Germinal Mutation Rates

Short tandem repeat family	No. of loci	Mutation rate/gen.	Mutation rate/gen./locus
Satellite III	20	3.2×10^{-4}	1.6×10^{-5}
Zeta globin Int. 1	9	8.7×10^{-3}	9.6×10^{-4}
Myoglobin 33.15	12	6.5×10^{-3}	5.4×10^{-4}
Phage M13	9	5.4×10^{-3}	6.0×10^{-4}
Myoglobin 33.6	11	1.2×10^{-2}	1.1×10^{-3}

variations of the trinucleotide repeat cores. The allelic frequency of the loci detected by these probes was estimated by pairwise comparisons of fragment patterns present in the 50 individuals, in the manner described for minisatellites (Jeffreys et al. 1985b). We also estimated the number of loci detected by these probes by hybridization to DNA digests obtained from several hydatidiform moles (Buroker et al. 1987) previously determined to be completely homozygous at all loci. The results of these studies are summarized in Table 2.

We also used these probes in family studies to examine the inheritance pattern of these fragments and found that most of the fragments detected represent multiple independent loci, inherited as codominant alleles. In these initial family studies, we have detected the occurrence of a few new alleles in offspring, presumably arising as new germinal mutations. These findings indicate that some single loci identified with these multilocus probes will be hypermutable.

We have also screened λgt11 cDNA libraries, generated from human brain, liver, intestine, and lymphocytes, with some of these probes. Approximately 500,000 plaques were screened at high stringency with the most hypervariable probes in these hybridizations. We have isolated 650 positive clones, indicating that at least some of these polymorphic trinucleotide repeats are found in expressed loci.

Table 2
Characteristics of Some Trinucleotide STR Loci

Core repeat sequence	Fragments/indiv. (ave)	No. of loci detected	Mean allele frequency[a]
CCA	24.9	11	0.18
AAG	18.3	9	0.23
CCG	18.3	8	0.43
AGC	11.5	5	0.48
CCT	19.6	10	0.61

[a] Maximum frequency as determined in 50 individuals with *Hin*fI, as described in text.

One of the inserts from a positive isolate was isolated and subcloned. We tested this plasmid subclone to determine if it would identify a single polymorphic locus. The result of these hybridizations demonstrated that this probe identified a single locus with multiple alleles, which was polymorphic using several restriction enzymes, indicating a VNTR type polymorphism. An example of the hybridization pattern produced by this probe is shown in Figure 2. These results suggest that these probes enlarge the spectrum of hypervariability to expressed loci and will be useful in germ-line mutation studies.

Mutations at Single STR Loci

Single hypermutable loci detected by multilocus STR sequence probes offer the potential for more detailed study of germ-line mutations. Furthermore, mutation rates of some single loci are likely to be greater than determined with multilocus probes. With this in mind, we have also studied germ-line mutations at several single VNTR loci to better characterize de novo mutations. Among others, we have used an oligonucleotide to the D1Z2 locus. This large (>100 kb) locus is composed almost entirely of a 40-base core repeat (Buroker et al. 1987). We detected new de novo mutations using *Taq*I digestions; however, repeat studies using other restric-

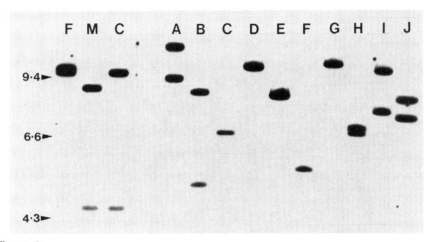

Figure 2
Restriction fragment patterns produced with the highly polymorphic single locus probe isolated from cDNA by multilocus trinucleotide (CCA) probe. A small family in lanes *1–3*, father (*F*), mother (*M*), and child (*C*), demonstrates inheritance of alleles. Hypervariability of locus is demonstrated by patterns in unrelated individuals *A–J*. Molecular weight range is shown in kilobases.

tion enzymes demonstrated that only a portion of these mutations were due to insertion/deletion events. This finding emphasizes that hypervariable loci do not necessarily undergo only a single type of mutation. Furthermore, although many of the loci composed of STR sequences appear to be hypermutable, not all of the hypervariable STR loci detected by the probes have equal mutation rates. Our own work (Kovacs et al. 1987) and that of other investigators (Jeffreys et al. 1988) has shown that some hypervariable loci undergo mutations at higher frequency than do others. Indeed, as might be anticipated, it appears that a direct correlation exists between the levels of heterozygosity and mutation rates. However, some of this difference might relate to the rather gross method of detecting novel alleles using restriction endonuclease digestions, since small changes in repeat length would not be resolved. To examine this possibility, we have recently begun to analyze single VNTR loci using the method of PCR (Saiki et al. 1988). Previously, we had used an oligonucleotide probe to the VNTR sequence at the apolipoprotein B locus and did not detect any de novo mutations in our family studies. More recently, however, using primers immediately flanking this repeat element for PCR amplification, we have begun to rescreen the reference panel families. Thus far, this approach has detected a new allele arising as a new mutation in a member of the panel (Fig. 3) This finding demonstrates the improved sensitivity of the PCR method to detect subtle alterations. Encouraged by this success, we are now developing PCR primers for other VNTR loci, including those identified with the trinucleotide repeat probes. We will then use the PCR method in family studies to detect and characterize de novo germ-line mutations.

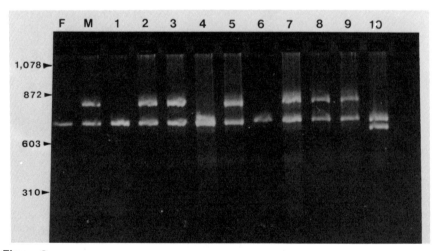

Figure 3
A new length allele at the APO B locus in one child (*10*) detected in a 12-member family by PCR amplification of the STR. Molecular weight range is shown in base pairs.

DISCUSSION

The results of this work have demonstrated that many hypervariable STR loci have exceptionally high spontaneous mutation rates. Furthermore, the results of our studies and those of other workers (Jeffreys et al. 1988) indicate some of the STR motifs are the most unstable sequence elements yet discovered in the human genome. The use of hypervariable loci to identify germ-line insertion/deletion mutations is exceptionally powerful and is not dependent on dominant selectable markers. Indeed, because of the high heterozygosity, wide dispersion, high mutation rate, and presumptive neutral nature of mutations at these loci, they may represent preferred loci for study. Importantly, short sequence tandem repeats are being identified in many areas of the genome, which serves to enlarge the number of potentially hypervariable loci available for study. Increasingly, insertion/deletion mutations are being identified in dysfunctional alleles responsible for many inherited and acquired diseases (Mohrenweiser and Jones 1989). Furthermore, insertions and deletions are often identified in a variety of experimental systems after exposure to γ irradiation and other environmental agents (DeMarini et al. 1989). Therefore, STR loci appear to be markers of choice to examine a variety of questions concerning spontaneous and induced mutations.

The functional significance of STR sequence loci is unknown, but their widespread occurrence in diverse species suggests some functionality. Indeed, their functional role in the genome may be related to their intrinsic instability. Many of these STRs occur in regions of the genome that are not expressed, but this may not be invariably true of all STR sequence elements. We have identified several STR loci that exhibit VNTR polymorphism and are likely to occur as elements in coding regions. These sequence repeats occur as tandem reiterations of trinucleotide cores (codons) and have the same characteristics of hypervariability (and potentially, hypermutability) as other STRs (Kovacs et al. 1988), suggesting that insertion/deletions of the repeat core could result in variable peptides. Recently, for example, this type of polymorphism was found to be responsible for phase variation of an outer membrane protein of *Neisseria gonorrhoeae* (Murphy et al. 1989) and in the signal peptide region of apolipoprotein B (Boerwinkle and Chan 1989).

The mutations demonstrated in our family studies are clearly germinal, since the new alleles are passed to the progeny. Recently, however, it was reported that fragment pattern differences at some STR loci occur in normal versus tumor tissue DNA (Armour et al. 1989). These results indicate that insertion and deletion events at these loci also occur in somatic cells, and that the mechanism producing length variation at these loci operates in somatic as well as germ cells.

The most intriguing question arising from these studies is: What process produces the extraordinary hypermutability at these STR loci? Recently, evidence has accumulated to suggest that unequal recombination occurring by virtue of misalignment of tandem repeats is not responsible for the production of new length alleles

(Wolff et al. 1989). Rather, these new alleles appeared to be generated by either unequal sister chromatid exchange (SCE) or replication slippage.

Replication slippage has long been thought to explain insertions and deletions in tandem repeat sequences (Streisinger et al. 1967). Moreover, slippage occurring in regions of short sequence repeats has been demonstrated in a variety of gene loci (Canning and Dryja 1989). This appears to reflect a susceptibility of repeat motifs to this process. However, it is important to note that mutation rates at STR loci differ greatly, with many short sequence repeats being very stable. This indicates that some characteristic of the repeat sequence itself, or its context within the genome, affects the stability of these loci.

Unequal SCE could also be responsible for the generation of new length alleles. Recently, it was demonstrated in cells exhibiting high levels of SCE (Bloom's syndrome) that a correspondingly increased rate of mutation at STR loci was present (German and Groden 1988). Interestingly, a well-supported mechanistic model of SCE relates the occurrence of SCE to aberrations in replication (Painter 1980), suggesting the reason for this.

Additionally, unusual DNA structures induced by STR sequence motifs may also be related to the hypermutability of these loci. These unusual DNA structures may promote replication infidelity (Lapidot et al. 1989) or misalignment (Glickman and Ripley 1984). A site of unequal SCE has been found to comprise a simple sequence tandem repeat. Notably, this sequence also has the potential to form an alternate DNA secondary structure (Weinreb et al. 1988). Moreover, recent evidence suggests that sequence-induced, unusual tertiary DNA structures can occur at some STR loci (Coggins and O'Prey 1989), which might contribute to their instability.

Taken together, these findings suggest that the phenomenon of hypermutable STR loci, replication slippage, altered DNA structures, and SCE may be interrelated. It may well be that mutations at STR loci are due to misalignments caused by tandem reiterations of sequences, which also promote replication slippage. Such misalignments, produced by some STR sequence motifs, would also cause unusual DNA conformations. Moreover, these conditions might also predispose these loci to frequent breakage events, causing SCEs, which could also contribute to the production of new alleles. Several studies using somatic cell lines are in progress to explore these possibilities. These considerations underscore the importance of using STR loci to understand the processes producing this type of genetic instability in humans.

ACKNOWLEDGMENTS

This work was supported by grants from the American Gynecological and Obstetrical Society and the Department of Energy (FG-0385ER60305). We thank Dr. R. Bruce Wallace for many helpful discussions.

REFERENCES

Armour, J.A.L., I. Patel, S.L. Thein, M.F. Fey, and A.J. Jeffreys. 1989. Analysis of somatic mutations at minisatellite loci in tumors and cell lines. *Genomics* **4:** 328.

Boerwinkle, E. and L. Chan. 1989. A three codon insertion/deletion polymorphism in the signal peptide region of the human apolipoprotein B (APOB) gene directly typed by the polymerase chain reaction. *Nucleic Acids Res.* **17:** 4003.

Buroker, N., R. Bestwick, G. Haight, R.E. Magenis, and M. Litt. 1987. A hypervariable repeated sequence on chromosome 1p36. *Hum. Genet.* **77:** 175.

Canning, S. and T.P. Dryja. 1989. Short direct repeats at the breakpoints of deletions of the retinoblastoma gene. *Proc. Natl. Acad. Sci.* **86:** 5044.

Coggins, L.W. and M. O'Prey. 1989. DNA tertiary structures formed in vitro by misaligned hybridization of multiple tandem repeat sequences. *Nucleic Acids Res.* **17:** 7417.

DeMarini, D.M., H.E. Brockman, F.J. deSerres, H.H. Evans, L.F. Stankowski, Jr., and A.W. Hsie. 1989. Specific locus mutations induced in eukaryotes (especially mammalian cells) by radiation and chemical: A perspective. *Mutat. Res.* **220:** 11.

Farabaugh, P.J., U. Schmeisser, M. Hofer, and J.H. Miller. 1978. Genetic studies of the *lac* repressor. VII. On the molecular nature of spontaneous hotspots in the *lac* gene of *Escherichia coli*. *J. Mol. Biol.* **126:** 847.

Fowler, C., R. Drinkwater, L. Burgoyne, and J. Skinner. 1987. Hypervariable lengths of human DNA associated with a human satellite III sequence found in the 3.4kb Y specific fragment. *Nucleic Acids Res.* **15:** 3929.

German, J. and J. Groden. 1988. Molecular evidence for unequal sister chromatid exchange in Bloom's syndrome. *Am. J. Hum. Genet.* **43:** A730.

Glickman, B.W. and L.S. Ripley. 1984. Structural intermediates of deletion mutagenesis: A role for palindromic DNA. *Proc. Natl. Acad. Sci.* **81:** 512.

Jeffreys, A.J., V. Wilson, and S.L. Thein. 1985a. Hypervariable "minisatellite" regions in human DNA. *Nature* **314:** 67.

―――. 1985b. Individual-specific "fingerprints" of human DNA. *Nature* **316:** 76.

Jeffreys, A.J., N.J. Royle, V. Wilson, and Z. Wong. 1988. Spontaneous mutation rates to new length alleles at tandem repetitive hypervariable loci in human DNA. *Nature* **332:** 278.

Kovacs, B.W., B. Shahbahrami, and D.E. Comings. 1988. Polymorphic variation and dispersion of tandem codon repeats in the human genome. *Am. J. Hum. Genet.* **43:** A190.

―――. 1989. Studies of human germinal mutations by deoxyribonucleic acid hybridization. *Am. J. Obstet. Gynecol.* **160:** 798.

Kovacs, B.W., F.O. Sarinana, R.B. Wallace, and D.E. Comings. 1987. Spontaneous germinal mutations in hypervariable regions of the human genome. *Am. J. Hum. Genet.* **41:** A223.

Lapidot, A., N. Baran, and H. Manor. 1989. (dT-dC)n and (dG-dA)n tracts arrest single stranded DNA replication in vitro. *Nucleic Acids Res.* **17:** 883.

Litt, M. and J.A. Luty. 1989. A hypervariable microsatellite revealed by in vitro amplification of a dinucleotide repeat within the cardiac muscle actin gene. *Am. J. Hum. Genet.* **44:** 397.

Mohrenweiser, H.W. and I.M. Jones. 1989. Review of the molecular characteristics of gene mutations of the germline and somatic cells of the human. *Mutat. Res.* **212:** 241.

Murphy, G.L., T.D. Connell, D.S. Barritt, M. Koomey, and J.G. Cannon. 1989. Phase variation of gonococcal protein II: Regulation of gene expression by slipped-strand mispairing of a repetitive sequence. *Cell* **56:** 539.

Nakamura, Y., C. Julier, R. Wolff, T. Holm, P. O'Connell, M. Leppert, and R. White. 1987. Characterization of a human "midisatellite" sequence. *Nucleic Acids Res.* **15**: 2537.

Office of Technology Assessment. 1986. *Technologies for detecting heritable mutations in human beings.* OTA-H-298. U.S. Government Printing Office, Washington, D.C.

Painter, R.B. 1980. A replication model for sister chromatid exchange. *Mutat. Res.* **70**: 337.

Saiki, R.K., D.H. Gelfand, S. Stoffel, S.J. Scharf, R.G. Higuchi, G.T. Horn, K.B. Mullis, and H.A. Erlich. 1988. Primer directed enzymatic amplification of DNA with thermostable DNA polymerase. *Science* **239**: 487.

Singer, M.F. 1982. Highly repeated sequences in the mammalian genome. *Int. Rev. Cytol.* **76**: 67.

Streisinger, G., Y. Okada, J. Emrich, J. Newton, A. Tsugita, E. Terzaghi, and M. Inouye. 1967. Frameshift mutations and the genetic code. *Cold Spring Harbor Symp. Quant. Biol.* **31**: 77.

Suggs, S.V., R.B. Wallace, T. Hirose, E.H. Kowashima, and K. Itakura. 1981. Use of synthetic oligonucleotides as hybridization probes: Isolation of cloned cDNA sequences for B_2-microglobulin. *Proc. Natl. Acad. Sci.* **78**: 6613.

Swallow, D.M., S. Gendler, B. Griffiths, G. Corney, J.T. Papadimitriou, and M.E. Bramwell. 1987. The human tumor associated epithelial mucins are coded by an expressed hypervariable gene locus PUM. *Nature* **328**: 82.

Tautz, D., M. Trick, and G.A. Dover. 1986. Cryptic simplicity in DNA is a major source of genetic variation. *Nature* **322**: 652.

Weinreb, A., D.R. Katzenberg, G.L. Gilmore, and B.K. Birshtein. 1988. Site of unequal sister chromatid exchange contains a potential Z-DNA forming tract. *Proc. Natl. Acad. Sci.* **85**: 529.

Wolff, R.K., Y. Nakamura, and R. White. 1989. Molecular characterization of a spontaneously generated new allele at a VNTR locus: No exchange of flanking DNA sequence. *Genomics* **3**: 347.

Wood, W.I., J. Gitschier, L.A. Lasky, and R.M. Lawn. 1984. Base composition independent hybridization in tetramethylammonium chloride: A method for oligonucleotide screening of highly complex gene libraries. *Proc. Natl. Acad. Sci.* **82**: 1585.

Analysis of DNA Sequences in Individual Human Sperm Using PCR

NORMAN ARNHEIM
Molecular Biology Section
University of Southern California
Los Angeles, California 90089-1340

OVERVIEW

We have developed a method to study DNA sequences in individual diploid cells and human sperm. Our procedure utilizes the polymerase chain reaction (PCR), which is capable of amplifying, in vitro, specific DNA sequences hundreds of millions to billions of times (Saiki et al. 1985, 1988; Mullis and Faloona 1987). Analysis of DNA sequences in individual cells can provide genetic information heretofore thought unattainable. In addition to allowing fine structure genetic mapping in humans it may also contribute to a further understanding of the molecular nature of germinal mutations.

INTRODUCTION

In a single cell, each chromosome is normally represented by two, or in the case of a sperm, a single DNA molecule. Until the development of PCR, the analysis of DNA sequences in a single cell was impossible. Using PCR, the target DNA sequence must be amplified on the order of 10^{11}-fold to detect the target-specific product with a radioactive probe.

The principle of the PCR method is shown in Figure 1. Two small stretches of DNA of known sequence that flank the target region to be amplified (Fig. 1A) are used to design two oligonucleotide primers. Each primer is a single strand of DNA, usually 20 nucleotides long, and is made using an automated DNA synthesizer. The sequence of each primer is chosen so that one has base-pair complementarity with one flanking sequence, whereas the other is complementary to the other flanking sequence. After denaturation of the double-stranded target DNA into single strands, the primers hybridize to their complementary sequences flanking the target gene (Fig. 1B). The primers are oriented so that when they form a duplex with the flanking sequences, their 3′ hydroxyl ends face the target sequence. DNA polymerases extend DNA chains by adding deoxyribonucleoside triphosphates to the 3′ end of each chain. Therefore, after denaturation of the genomic DNA and hybridization of the primers to the sequences flanking the target, the addition of a DNA polymerase will result in an extension of the primers through the target sequence, thereby making copies of the target (Fig. 1C). DNA denaturation, primer hybridiza-

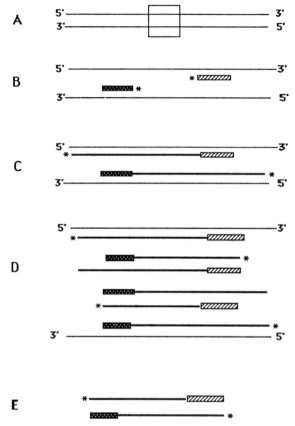

Figure 1
Principle of the polymerase chain reaction. (*A*) Boxed target DNA. The 5′ and 3′ orientations of the two single strands are indicated. (*B*) Two PCR primers annealed to the sequences flanking the target after DNA denaturation. The 3′ end of the primer is denoted by an asterisk. (*C*) Extension of the two primers by DNA polymerase. (*D*) Second cycle of PCR. Each of the four DNA strands shown in part *C* can anneal to a primer and be extended. Note that at the end of this cycle there is four times as much target DNA as was originally present. Two of the eight strands are equal in length to the length of the two primers and the intervening target. Extension products of these molecules (*E*) accumulate exponentially as additional cycles are carried out.

tion, and DNA polymerase extension represent one PCR cycle, and each of the three steps is carried out at an appropriate temperature for a particular length of time. If the first cycle is followed by a second one (Fig. 1D), more copies of the target sequence will be made. The main product of PCR is a DNA fragment exactly equal in length to the sum of the lengths of the two primers and the target DNA (Fig. 1E).

By extending each primer to include the sequences complementary to the other primer, the extension products themselves can act as templates. Thus, production of copies of the target sequence is exponential with respect to cycle number. Since the amount of target doubles with each cycle, as few as 20 cycles will generate about a million times more target sequence than is present initially.

Amplified DNA products can be analyzed by a variety of methods (see Erlich 1989; White et al. 1989). For example, the allelic status of PCR products from a single locus that differ by base substitutions can be determined by the use of allele-specific oligonucleotide (ASO) probes (Conner et al. 1983; Saiki et al. 1986). These short DNA segments can, under the appropriate DNA hybridization conditions, recognize a single base difference between two otherwise identical DNA sequences.

RESULTS

Analysis of DNA Sequences in Single Diploid Cells

Our first attempts to analyze DNA sequences in single cells involved the analysis of the human β-hemoglobin gene locus in two tissue culture cell lines. One was derived from an individual homozygous for the sickle cell mutation (β^S), and the other was homozygous for the normal β^A allele. PCR primers that amplify a β-globin fragment containing the codon with the β^S mutation and ASO probes that are capable of distinguishing between these two alleles have already been described (Li et al. 1988). We cocultivated the cells homozygous for β^A and cells homozygous for β^S in the same tissue culture flask for several days. Individual cells from this mixture were drawn into a thin plastic pipette while being observed under a phase-contrast microscope. Each individual cell was delivered into a PCR tube containing a lysis solution. After lysis, the PCR reaction mixture was added, and 50 cycles of amplification were carried out. Duplicate samples of the amplified product were fixed on nylon membranes and hybridized with either the β^A or β^S ASO probe. Each sample has four possible outcomes: hybridization to (1) the β^A probe, (2) the β^S probe, (3) both probes, or (4) neither probe. The results of this cocultivation experiment were published previously (Li et al. 1988). Of the 37 cells analyzed, 84% hybridized with only one of the two allele-specific probes; 19 hybridized with the β^A probe, and 12 hybridized with the β^S probe. The ratio of the two cell types detected is consistent with the fact that the normal cell line grows faster than the sickle cell line. None of the 12 control tubes that received water in place of a cell was positive in this test, which suggested that DNA contamination was insignificant. That no sample hybridized with both probes suggested that only a single cell was introduced into each tube and that DNA from lysed cells present in the cocultivation mixture did not adhere to individual cells. We estimated that starting with two molecules (the diploid amount of β-globin DNA) or 3.3×10^9 femtomoles of β-globin sequence, between 5 and 500 femtomoles of PCR product was produced in 50 cycles. This was equivalent to an average amplification of 7.6×10^{10}-fold. Considering the extent of amplification, eliminating all sources of possible contamination in the reagents or their introduction into any of the PCR reaction tubes is critical.

Analysis of DNA Sequences in Individual Human Sperm

We next attempted to analyze the genotype of single sperm cells. Sperm were derived from an individual heterozygous at the gene that codes for the low-density lipoprotein receptor (LDLr). PCR primers and probes for an LDLr polymorphism (Hobbs et al. 1987) were prepared with the help of unpublished DNA sequence information kindly provided by David Russell. Individual sperm were drawn into a fine plastic needle while under microscopic observation and delivered to the PCR tube for lysis and amplification. In one series of experiments (Li et al. 1988), we analyzed the LDLr genotype in 80 individual sperm. Approximately 55% of the sperm amplification products gave a hybridization signal. Twenty-two were found to carry one allele and 21 the other. Only one sperm was positive with both probes. Sixteen additional control tubes that received all of the reagents but no sperm did not give any hybridization signal. Typical results are shown in Figure 2. The observation that the distribution of the two amplified alleles obeyed Mendel's law of independent segregation indicates that the PCR reactions were initiated with a single meiotic product and that little if any contaminating diploid DNA sequences were present.

Measuring the Frequency of Genetic Recombination Using Single Sperm Typing

The construction of genetic maps in higher organisms depends on the ability to analyze the progeny of selected matings or to compute linkage relationships by means of pedigree analysis. In humans only the latter is possible. Using restriction fragment length polymorphisms (RFLPs), genetic distances of approximately 1–2 cM encompassing about 1000–2000 kb of DNA are considered to be measurable with statistical reliability using pedigree analysis. The analysis of smaller distances, however, requires such a large number of individuals from informative families that it is impractical. However, the measurement of genetic recombination over shorter physical distances could be accomplished at a resolution far greater than that currently possible, and without the necessity of family studies, if large numbers of individual meiotic products (sperm) could be typed as recombinant or nonrecombinant by DNA analysis using PCR (for a detailed discussion, see Arnheim 1989). We have recently tested this idea by analyzing single sperm from an individual heterozygous at two genetic loci on chromosome 11 (Cui et al. 1989). We attempted

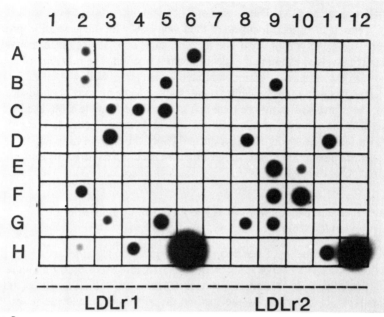

Figure 2
PCR analysis of the LDL receptor locus of individual human sperm. The autoradiographs of two filters are shown side by side. One filter (columns *1–6*) was hybridized with the probe for one LDLr allele (LDLr1) and the other (columns *7–12*) with the allele-specific probe for LDLr2. Two aliquots from each sperm sample were taken to be analyzed, one for each filter. Aliquots from the same sample were placed in comparable positions on the two filters, e.g., A1 and A7 or E3 and E9. Samples 1A–1H (7A–7H) are six water blanks. 6H and 12H are aliquots of amplified DNA from an LDLr1/LDLr2 heterozygote. The remaining samples are from individual sperm. The experimental details are described by Li et al. (1988).

to determine, for each of the two loci, which of the two alleles present in the diploid cells of the donor were present in any one sperm. Dividing the number of recombinant sperm by the total number of sperm examined will estimate the frequency of recombination between the loci. We studied the parathyroid hormone and $^G\gamma$ globin gene loci. Previous studies using pedigree analysis estimated the frequency of recombination between these two loci at between 12% and 17% (for references, see Cui et al. 1989). We studied over 700 individual meiotic products from two men who were heterozygous at both genetic loci for the two DNA polymorphisms. PCR was carried out in two steps. In the first round of PCR cycles, primers for amplifying the polymorphic regions at both loci were present simultaneously. After the twenty-fifth cycle, the sample was divided in half. One half was transferred to a tube where it was diluted with PCR reaction mix that contained only the primers for the PTH locus. This was then subjected to an additional 50 cycles of PCR. The other half was treated identically, except that $^G\gamma$ globin primers were used instead. The PCR products at each locus were then analyzed with allele-specific probes to determine the genotype of each sperm. The data were subjected to a maximum likelihood analysis to estimate the recombination fraction. The maximum likelihood estimate for the frequency of recombination was 16%, well within the limits of earlier estimates using family studies. Our estimate had a narrower confidence interval and, therefore, was more accurate due to our larger sample size. The statistical analysis also showed that, on the average, if a single target molecule was present in a tube, we would detect its PCR products 94% of the time. Extrapolation of these and related parameters led us to suggest that we should be able to measure frequencies of recombination as low as 0.1% with statistical accuracy (Cui et al. 1989).

Ethidium Bromide Detection of PCR Products from Single Sperm on Polyacrylamide Gels

Although PCR is capable of generating highly pure product under most experimental conditions, the very large number of cycles required to amplify a single DNA molecule usually leads to impure product. Thus, when the LDLr locus in single sperm samples is amplified by the method described above and the products are studied by gel electrophoresis, a number of background bands and a smear are observed in addition to the expected product. We attempted to reduce this background by using hemi-nesting, a modification of the PCR "nesting" method (Mullis and Faloona 1987). The concept of hemi-nesting is shown in Figure 3. By carrying out two rounds of PCR using, in the second round of cycles, one old primer and one new primer that is internal to the original two, we find the background to be almost entirely eliminated (Li et al. 1990.).

Because hemi-nesting virtually eliminated background PCR bands, we were able to design a system to allow us to distinguish between alleles at a locus on the basis of the size of the PCR product itself (Fig. 4) (Li et al. 1990). If two alleles differ by a single base substitution, two PCR primers can be designed that contain the polymorphic nucleotide at their 3' ends. Thus, each primer is identical to one allele and differs from the other by a single base substitution at its 3' end. It is expected that under the appropriate conditions, the extension by *Taq* DNA polymerase of the completely matched primer will be more efficient than the extension of the primer with a mismatch at the 3' end. To distinguish between the PCR products from the two alleles, we constructed the two allele-specific primers so that they differed in length by 15 bp, and, therefore, their products also were of a different size and could be distinguished by gel electrophoresis.

We analyzed three independent genetic loci (PTH, $^G\gamma$ globin, and LDLr) simultaneously from six single sperm samples using the system described above. The initial rounds of amplification of each sperm contained three primer pairs; one pair each for each locus. Following this, three aliquots of each PCR mixture were

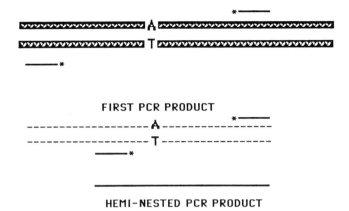

Figure 3
Principle of hemi-nesting. PCR is initiated with two primers that flank the region of interest, which in this case is an A/T base pair. As in Fig. 1, the 3' ends of the primer are indicated by an asterisk. The expected product of the first round of PCR is shown. An aliquot of this product in taken through a second round of PCR cycles with one of the first two primers and a second new primer that is internal to the original ones. The product of this hemi-nesting PCR is shorter than the original PCR product but is less contaminated by nonspecific PCR products.

taken. Each aliquot was further amplified with the hemi-nesting procedure using a pair of allele-specific primers for that locus and one of the primers used in the initial round of amplification. The results are shown in Figure 5. A low-molecular-weight background product is seen in all lanes, including the sample (Fig. 5, lane B) to which no DNA was added. Each of the six single sperm samples (lanes 1–6) shows three main PCR products representing one allele at each of the three loci. Lane 7 and lane D show the mobility difference between the PCR products of the two alleles at each locus.

Because we are using a sperm system where each cell, as a result of meiosis, is expected to contain only one of two alleles present in the somatic tissue of the donor, our data provide the strongest kind of evidence that we are in fact analyzing the amplification products from a single DNA molecule and that our results are not confounded by contamination.

CONCLUSION

The ability to analyze the genotype of single sperm at the DNA level provides a unique tool for the study of problems in human genetics considered intractable using current approaches. We have developed a procedure that allows us to detect PCR products derived from a single target DNA molecule in a human sperm using

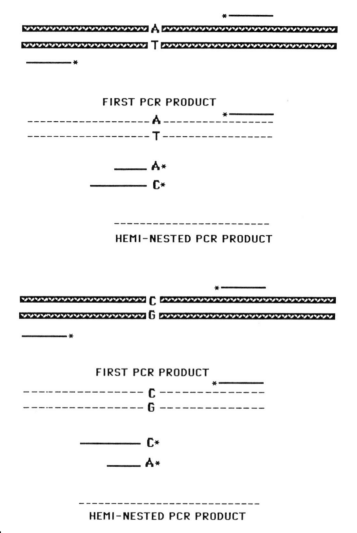

Figure 4
Combination of hemi-nesting and allele-specific PCR. As an example, two possible PCR targets are shown that are identical in sequence except for an A/T to G/C substitution. The products from the first round of PCR cycles will be identical in size for both alleles. In the second round of PCR cycles with hemi-nesting using allele-specific primers, conditions can be found that allow only the primer with the perfectly matched 3′ end to be incorporated into the PCR product. Since the lengths of the two allele-specific primers are not the same (———A* is shorter than ————C*), the PCR products from the different alleles will also differ in size and can be distinguished by gel electrophoresis.

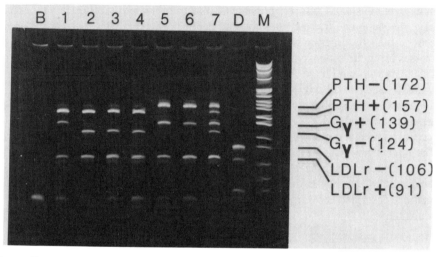

Figure 5
Determination of the allelic status at three loci in single sperm using allele-specific primers and hemi-nesting. (Lane *B*) Negative PCR control, which received all reagents except a sperm. (Lanes *1–6*) Single sperm samples from an individual heterozygous at the PTH and $^G\gamma$ globin loci. (Lane *7*) Products from a mixture of sperm from the same individual showing the expected heterozygosity at PTH and $^G\gamma$ globin and homozygosity at LDLr. (Lane *D*) LDLr PCR products from an individual heterozygous at this locus to indicate the different mobilities of the two LDLr allelic products. (Lane *M*) pBR322 digested with *MspI*. The sizes of the allelic products from each of the three loci are shown to the right of the gel. The + and − notations refer to the restriction enzyme polymorphism that results from the single base substitution. The sizes of each PCR product (in bp) are also given. Additional technical details can be found in Li et al. (1990).

either radioactive probes or, using hemi-nesting, by direct detection using ethidium bromide staining of polyacrylamide gels. At least three genetic loci present in a single sperm can be amplified simultaneously. The amplification procedure is specific as well as efficient. By using allele-specific PCR primers that differ in length, the PCR products of different alleles also differ in size, allowing the allelic state at each locus to be determined directly by gel electrophoresis.

Studies on individual sperm using this procedure should facilitate the measurement of genetic recombination in humans over very small physical distances. To date, no practical method of accurately measuring genetic distances less than 1–2 cM has been developed. A significant advantage to the approach described here is that a large number of meiotic products can be examined from a single individual, allowing determination of the recombination frequency between genetic markers physically very close to each other. Since it is now possible for the first time to obtain statistically significant data on recombination frequencies from a single individual, it should also be possible to determine whether different individuals have the same or different rates for the identical interval. Strategies for making a genetic map by ordering a set of DNA markers on a chromosome by three point crosses using the sperm-typing approach have also been put forward (Boehnke et al. 1989; Goradia and Lange 1990).

PCR analysis of single sperm may also facilitate studies on human germ-line mutations. As described by Wyrobek et al. (this volume), flow cytometric methods are being developed for selecting mutant sperm that were derived from cells incapable of making a specific gene product following the first meiotic division. The mutant population of sperm selected could then be further sorted and individually isolated. Thus, the nature of the mutations in a specific gene causing such lesions could be investigated on a sperm by sperm basis, using the amplification methods described above in conjunction with DNA sequence analysis.

ACKNOWLEDGMENTS

We acknowledge helpful discussions with Dr. Gino Cortopassi and we thank Honghua Li for the figures showing the hemi-nesting protocols. We also thank Cetus Corporation for the gift of a Perkin-Elmer Cetus thermocycler. This research was supported in part by grants from the National Institute of General Medical Sciences, National Institutes of Health.

REFERENCES

Arnheim, N. 1989. A new approach to studying genetic recombination and constructing genetic maps: PCR analysis of DNA sequences in individual gametes. In *PCR technology: Principles and applications for DNA amplification* (ed. H. Erlich), p. 119. Stockton Press, New York.

Boehnke, M., N. Arnheim, H. Li, and F.S. Collins. 1989. Fine structure genetic mapping of human chromosomes using the polymerase chain reaction on single sperm: Experimental design considerations. *Am. J. Hum. Genet.* **45**: 21.

Conner, B.J., A.A. Reves, C. Morin, K. Itakura, R.L. Teplitz, and B. Wallace. 1983. Detection of sickle cell β^S globin allele by hybridization with synthetic oligonucleotides. *Proc. Natl. Acad. Sci.* **80**: 278.

Cui, X., H. Li, T.M. Goradia, K. Lange, H.H. Kazazian, D. Galas, and N. Arnheim. 1989. Single sperm typing: Determination of genetic distance between the G-γ globin and parathyroid hormone loci. *Proc. Natl. Acad. Sci.* **86**: 9389.

Erlich, H.A., ed. 1989. *PCR technology: Principles and applications for DNA amplification.* Stockton Press, New York.

Goradia, T.M. and K. Lange. 1990. Multilocus ordering strategies based on sperm typing. *Ann. Hum. Genet.* (in press).

Hobbs, H.H., V. Esser, and D.W. Russell. 1987. Ava II polymorphism in the LDL receptor gene. *Nucleic Acids Res.* **15(1)**: 379.

Li, H., X. Cui, and N. Arnheim. 1990. direct electrophoretic detection of the allelic state of single DNA molecules in human sperm using PCR. *Proc. Natl. Acad. Sci.* (in press).

Li, H., U.B. Gyllensten, X. Cui, R.K. Saiki, H.A. Erlich, and N. Arnheim. 1988. Amplification and analysis of DNA sequences in single human sperm and diploid cells. *Nature* **335**: 414.

Mullis, K. and F. Faloona. 1987. Specific synthesis of DNA *in vitro* via a polymerase catalysed chain reaction. *Methods Enzymol.* **55**: 335.

Saiki, R.K., T.L. Bugawan, C.T. Horn, K. Mullis, and H.A. Erlich. 1986. Analysis of enzymatically amplified β-globin and HLA-DQα DNA with allele-specific oligonucleotide probes. *Nature* **324**: 163.

Saiki, R.K., S. Scharf, F. Faloona, K.B. Mullis, C.T. Horn, H.A. Erlich, and N. Arnheim. 1985. Enzymatic amplification of β-globin genomic sequences and restriction site analysis for diagnosis of sickle cell anemia. *Science* **230**: 1350.

Saiki, R.K., D.H. Gelfand, S. Stoffel, S.J. Scharf, R. Higuchi, G.T. Horn, K.B. Mullis, and H.A. Erlich. 1988. Primer-directed enzymatic amplification of DNA with a thermostable DNA polymerase. *Science* **239**: 487.

White, T.J., N. Arnheim, and H.A. Erlich. 1989a. The polymerase chain reaction. *Trends Genet.* **5**: 185.

COMMENTS

Favor: Where you were talking about your nested and hemi-nested PCR amplification, what were those junk bands? Are they amplification mistakes that are only confined to the one arm, and why aren't they amplified?

Arnheim: There are two principal reasons why you see junk. One is called primer dimer. If you have two PCR primers and you're amplifying a piece of DNA, and you carry out many, many cycles—and remember, in order to start with a single molecule, you have to carry out many cycles—there is a probability that the two primers can actually prime off each other, producing a PCR product that is a dimer of the primers. That will easily get amplified. So, if you're starting out with a couple of real target molecules and this event happens, there may be a couple of primer dimers around and a couple of your targets and there will be a lot of competition. You can get dimers as well as higher oligomers of it. That's number one. Number two is that the PCR primers are not perfectly specific. Even though they have 100% sequence homology with the region you're interested in, they will occasionally anneal somewhere else and just get extended. Now, that doesn't hurt. That happens every cycle, some fraction of them will anneal somewhere and get extended. The problem is that there is some finite probability that the other primer will anneal to a non-specific extension product and will also become extended. When that happens, you end up with a legitimate PCR product that will now exponentially increase in amount. The more cycles you do, the higher the probability that these two rare events—the original extension plus the other primer extending off of it—occur and you tend to get that junk.

Why the hemi-nesting and nesting work so well is that you start to build junk up in your first rounds of PCR cycling. If you now add a new primer specific to only the internal region, all the random PCR products don't contain that sequence and all you do is further amplify the real PCR products that you are trying to see, and not the PCR products that were generated nonspecifically.

Liane Russell: Do you find any mutations to be induced by PCR and, if so, how often?

Arnheim: Yes. There is a misincorporation rate that is quoted for PCR. It has actually changed over the last few months. It used to be 1 misincorporation per 10,000 nucleotides per PCR cycle. Those numbers came from older data where the concentration of nucleotides used in the reaction was not optimal. Now, under the conditions most people use, which are the optimal conditions, the rate is about one-eighth to one-tenth of that. If you carry out many cycles, a larger and larger fraction of the molecules are going to have a hit somewhere. There is a group of mathematicians in West Germany who published a paper in *NAR* in which they exactly calculated, using the appropriate methods, what fraction of the molecules will have a hit in them due to misincorporation. Of course, the more cycles you do the larger the fraction that is going to have a hit somewhere in it.

Wyrobek: Any thoughts on improving that further?

Arnheim: By fooling around with magnesium concentration, and maybe temperature, that may be possible. I don't know anybody who is systematically trying, because the assay is difficult to do. Most of these assays were done either by sequencing PCR product, estimating the frequency of misincorporation in the final product, and then back-calculating it to what the misincorporation rate was, or by an M13 in vitro DNA replication method. I have a feeling that unless someone really decides to do it carefully, it's not going to get done, because for most things that people want to do they are happy with the conditions already available.

Wyrobek: Do you think it's enzyme dependent?

Arnheim: The Cetus Corporation, for example, and I think probably a lot of other people too, are looking at other thermostable organisms. I don't know of any that has a $3'-5'$ exonuclease—at least all the ones that have been looked at in this thermostable group don't have it—but there might be one somewhere.

Fritz: If you had taken Harvey's [Mohrenweiser] 14 probes and taken a single ejaculate, and then with as many as 300 sperm done an ELISA plate, what would you have found? Would you have found from the single ejaculate the same incidence of insertion/deletion and rearrangements that Harvey found in a bunch of patients? He chose 130 students; each has 130 sperm. Would you get the same kind of variation he found?

Arnheim: I don't see why not.

Fritz: From a single ejaculate, I mean.

Arnheim: Right.

Fritz: Check. So you would predict the same incidence, more or less?

Arnheim: Because of the new data on the misincorporation rate, there are problems looking for insertion/deletion mutations when you do many, many cycles. Bruce Kovacs didn't have to do many, many cycles to use PCR to look at length variation. But if you're starting with a single sperm, you have to do many, many, many cycles. The more cycles you do, the more possible it is, for example, to get slipped mispairing by the *Taq* polymerase. In fact, Alec Jeffreys has tried to do some single sperm analysis of his minisatellite sequences using PCR, and it's tough. Spurious bands can arise just due to the PCR reaction because you can have the slipped mispairing phenomena. Certainly, I would expect that the answer would be exactly the same.

Smithies: It shouldn't matter if you get a mistake after about four or five rounds, should it, because you won't see it? It will be too small a percentage of the total. If you've done, let's say for the sake of argument, five rounds and you have 2^5 copies already, any mistake from now on will be less than $1/2^5$.

Arnheim: Yes, unless the rate of slipped mispairing is exceptionally high. I know people who are looking at these VNTRs and they see lots of bands. So, the rate may be incredibly high, which means that the simple way of thinking about it may not be accurate.

Kovacs: From the standpoint of an early event, it would depend on the sensitivity of your detection. For example, using allele-specific oligonucleotide probes, you can detect as little as 5% mosaicism within a given PCR amplification product. So you may see something that you read as a mosaic that, in fact, is a spurious product.

Hecht: Using the nested procedure with sperm, is there an optimum number of cycles to be carried out before adding a second set of primers?

Arnheim: Yes. I think you have to work that out. You want to generate as much product as you can before you have to split the reaction. You'd like to have maybe 20 or 25 cycles with the hemi-nested primer because that way you won't start building up background material. If you did another 50 cycles with the hemi-nested primers, it's exactly the same as it was in the beginning. If you do 20 or 25 cycles, you don't suffer from this kind of background problem.

Kovacs: What's your limit as far as the allele size that you have been able to go up to, as far as the amplification product?

Arnheim: We try to keep things as small as possible. We try not to amplify very large genes. What we're interested in right now is the polymorphism itself, and so we really only have to amplify one base, literally, although we try to keep things on the order of 100–200 bp. The biggest is, I think, 250 or 300 bp.

Insertional Mutagenesis in Transgenic Mice

RICHARD P. WOYCHIK, BARBARA R. BEATTY, WILLIAM L. MCKINNEY,
DEBRA K. ANDREADIS, ANNE J. CHANG, AND P. EUGENE BARKER
Biology Division, Oak Ridge National Laboratory
Oak Ridge, Tennessee 37831-8077

OVERVIEW

A considerable effort is under way in medicine and biology to characterize the genes associated with development in humans and other species. An important feature of much of this research involves the molecular analysis of mutations, because the alteration in gene expression associated with mutations is, in many cases, the basis for establishing the normal function of individual genes. In fact, the remarkable recent success in characterizing genes with specific functions in yeast, *Caenorhabditis elegans*, and *Drosophila* followed from earlier genetic analyses and from the availability of mutations in these organisms that could be analyzed at the molecular level.

The mouse, with its well-characterized genetic makeup and accessibility to experimental manipulation, is a good system for studying genes with specific functions in mammals. Many spontaneous and chemical- or radiation-induced mutations are being studied at the molecular level to learn more about the structure and expression of genes regulating complex developmental processes. However, to be able to use mutations to characterize genes with specific functions, molecular probes must be available to facilitate the cloning and structural characterization of the mutant loci.

The paucity of well-characterized molecular probes mapping at particular loci throughout the mouse genome has, in many instances, precluded the molecular analysis of existing mutations with developmentally interesting phenotypes. Partly for this reason, increasing attention is now being directed toward an alternative method for generating germ-line mutations in the mouse, namely, insertional mutagenesis. This form of mutagenesis depends on the integration of a foreign fragment of DNA into a host gene, which can cause a disruption in the expression of that gene. When this occurs, the mutant locus is "tagged" with the integrated foreign DNA fragment and can be readily identified and cloned with standard techniques.

PRODUCTION OF INSERTIONAL MUTATIONS IN THE MOUSE

The procedures most often used to generate insertional mutations in the mouse are based on the technology utilized for the production of transgenic mice. The most common methods involve the microinjection of DNA fragments directly into the pronucleus of the zygote (Palmiter and Brinster 1986) or the infection of early-stage embryos or embryonic stem cells with murine retroviruses (Jaenisch 1976; Jenkins and Copeland 1985). In either case, the exogenously added DNA integrates into one of the host chromosomes to become a stable complement of the animal's genome. Insertional mutations occur when the exogenous DNA integrates into, and disrupts the function of, a host gene that is essential for the normal development of the animal. (For genes that have already been cloned, insertional mutations can alternatively be generated by targeting integration to specific sites on the genome by homologous recombination in embryonic stem cells [Koller et al. 1989; Thompson et al. 1989].) Since most insertional mutations are recessive, any phenotype would be revealed by breeding the animal to homozygosity.

The analysis of insertional mutations is proving to be an effective means of identifying genes with specific functions in mice. Many insertional mutations generated in transgenic mice have phenotypes ranging from preimplantation lethality to morphological defects, e.g., limb deformities (Jaenisch et al. 1983; Palmiter et al. 1983; Wagner et al. 1983; Mark et al. 1985; Woychik et al. 1985; Overbeek et al. 1986; Allen et al. 1988; Kothary et al. 1988; McNeish et al. 1988; Beier et al. 1989; Costantini et al. 1989; Krulewski et al. 1989). In fact, a number of these insertional mutations, which are alleles of existing mutations, are providing DNA probes for the characterization of loci that were previously not accessible at the molecular level.

Cloning of the "tagged" mutant locus has been complicated, in some instances, by methylation of the exogenously added DNA (transgene). Many of these problems are likely to be resolved now that *mcr* host strains of *Escherichia coli* are readily available that can be used to propagate clones of highly methylated regions of eukaryotic DNA (Raleigh et al. 1988). In addition, in some mutants, structural rearrangements, such as deletions and duplications of the host DNA, can occur at the integration site. Even under these conditions, flanking-sequence probes can usually be derived and used to characterize the structure of the mutant locus. In fact, it is useful to consider these structural alterations as advantageous because they add to the disruption of the original locus and are more likely than simple integration of the transgene alone to inactivate a host gene and give rise to a detectable phenotype in the animal.

LARGE-SCALE INSERTIONAL-MUTAGENESIS PROGRAM AT OAK RIDGE

We have initiated a large-scale insertional-mutagenesis program at the Oak Ridge National Laboratory. The object of the program is to generate 200 different lines of transgenic mice each year for 5 years, and to systematically screen each line for the expression of insertional mutations. All animals are being derived from the FVB/N inbred line. To facilitate cloning of the mutant locus, each line is being generated by microinjection with either the chloramphenicol acetyltransferase (CAT) gene, which confers resistance to chloramphenicol, or with other bacterial antibiotic-resistance genes. Therefore, the mutant locus in these animals can be cloned, not with tedious standard procedures, but rather by screening genomic libraries for the antibiotic-resistance function provided by the marker gene on the transgene.

The screening procedure developed to identify mutants limits the number of animals that need to be generated within each line and exploits the fact that each animal can be genotyped with the Southern blotting procedure. Initially, the founder is bred to a wild-type FVB/N partner, and transgene carriers are identified in the F_1 offspring. Any mosaicism arising from the integration of the transgene at different sites in the genome, identified with the Southern blotting procedure, is eliminated at this point by breeding the F_1 carriers back to a wild-type partner. Transgenic heterozygotes are then intercrossed to generate homozygotes. Lines that fail to generate a single homozygote in 30 offspring from heterozygous parents are scored as recessive lethals. In all other lines, the homozygous animals are being carefully examined for a number of characteristics, including soft-tissue abnormalities, skeletal defects, reduced fertility or infertility, growth retardation, and decreased life expectancy, as well as any visible or behavioral effects.

The insertional mutation frequency in transgenic mice has been estimated to be about 10% (Palmiter and Brinster 1986). Therefore, we expect that over the course of the next 5 years, at least 100 insertional mutations should be produced from this effort. Since available data suggest that the transgene integrates randomly throughout the genome, we also expect to see a broad spectrum of phenotypes in the resulting mutant stocks. Since each mutant will be directly accessible at the molecular level, this effort will promote the characterization of the structure and expression of the genes associated with a large number of different heritable traits.

ACKNOWLEDGMENTS

This research was jointly sponsored by the Office of Health and Environmental Research, U.S. Department of Energy, under contract DE-AC05-84OR21400 with Martin Marietta Energy Systems, Inc., and by the National Institute of Environmental Health Sciences under IAG No. 22Y01-ES-10067. The submitted manuscript has been authored by a contractor of the U.S. Government under contract No. DE-AC05-84OR21400. Accordingly, the U.S. Government retains a nonexclusive, royalty-free license to publish or reproduce the published form of this contribution, or allow others to do so, for U.S. Government purposes.

REFERENCES

Allen, N.D., D.G. Cran, S.C. Bartonl, S. Hettle, W. Reik, and M.A. Surani. 1988. Transgenes as probes for active chromosomal domains in mouse development. *Nature* **333**: 852.

Beier, D.R., C.C. Morton, A. Leder, R. Wallace, and P. Leder. 1989. Perinatal lethality (ple): A mutation caused by integration of a transgene into distal mouse chromosomal 15. *Genomics* **4**: 498.

Costantini, F.D., G.R. Radice, J.L. Lee, K.K. Chada, W. Perry, and H.J. Son. 1989. Insertional mutations in transgenic mice. *Prog. Nucleic Acid Res. Mol. Biol.* **36**: 159.

Jaenisch, R. 1976. Germ line integration and Mendelian transmission of the exogenous Moloney leukemia virus. *Proc. Natl. Acad. Sci.* **73**: 1260.

Jaenisch, R., K. Harbers, A. Schnieke, J. Löhler, I. Chumakov, D. Jähner, D. Grotkopp, and F. Hoffman. 1983. Germline integration of Moloney murine leukemia virus at the *Mov13* locus leads to recessive lethal mutation and early embryonic death. *Cell* **32**: 209.

Jenkins, N.A. and N.G. Copeland. 1985. High frequency germline acquisition of ecotropic MuLv proviruses in SWR/J-RF/J hybrid mice. *Cell* **43**: 811.

Koller, B.H., L.J. Hagemann, T. Doetschman, J.R. Hagaman, S. Huang, P.J. Williams, N.J. First, N. Maeda, and O. Smithies. 1989. Germ line transmission of a planned alteration made in hypoxanthine phosphoribosyltransferase gene by homologous recombination in embryonic stem cells. *Proc. Natl. Acad. Sci.* **87**: 8927.

Kothary, R., S. Clapoff, A. Brown, R. Campbell, A. Peterson, and J. Rossant. 1988. A transgene containing *lacZ* inserted into the *dystonia* locus is expressed in neural tube. *Nature* **335**: 435.

Krulewski, T.F., P. Neumann, and J. Gordon. 1989. Insertional mutation in a transgenic mouse allelic with Purkinje cell degeneration. *Proc. Natl. Acad. Sci.* **86**: 3709.

Mark, W.H., K. Signorelli, and E. Lacy. 1985. An insertional mutation in a transgenic mouse line results in developmental arrest at day 5 of gestation. *Cold Spring Harbor Symp. Quant. Biol.* **50**: 453.

McNeish, J.D., W.J. Scott, and S.S. Potter. 1988. Legless, a novel mutation found in PHT1-1 transgenic mice. *Science* **241**: 837.

Overbeek, P.A., S.P. Lai, X.R. Van Quill, and H. Westphal. 1986. Tissue-specific expression in transgenic mice of a fused gene containing RSV terminal sequences. *Science* **231**: 1574.

Palmiter, R.D. and R.L. Brinster. 1986. Germ line transformation of mice. *Annu. Rev. Genet.* **20**: 465.

Palmiter, R.D., T.M. Wilke, H.Y. Chen, and R.L. Brinster. 1983. Transmission distortion and mosaicism in an unusual transgenic mouse pedigree. *Cell* **36**: 869.

Raleigh, E.A., N.E. Murray, H. Revel, R.M. Blumenthal, D. Westaway, A.D. Reith, R.W.J. Rigby, J. Elhai, and D. Hanahan. 1988. McrA and McrB restriction phenotypes of some *E. coli* strains and implications for gene cloning. *Nucleic Acids Res.* **16**: 1563.

Thompson, S., A.R. Clarke, A.M. Pow, M.L. Hooper, and D.W. Melton. 1989. Germ line transmission and expression of a corrected HPRT gene produced by gene targeting in embryonic stem cells. *Cell* **56**: 316.

Wagner, E.F., L. Covarrubias, T.A. Stewart, and B. Mintz. 1983. Prenatal lethalities in mice homozygous for human growth hormone gene sequences integrated in the germ line. *Cell* **35**: 647.

Woychik, R.P., T.A. Stewart, L.G. Davis, P. D'Eustachio, and P. Leder. 1985. An inherited limb deformity created by insertional mutagenesis in a transgenic mouse. *Nature* **318:** 36.

COMMENTS

McDonald: Richard, I applaud your enthusiasm at wanting to produce 200 transgenic strains a year. I'm wondering about archives and how you are going to archive these strains; how long will it be before that becomes rate limiting?

Woychik: We're going to screen them for mutations and save the ones that have apparent mutations. The ones that show apparent mutations, or have interesting positions relative to the other mutations, will be kept as frozen embryos. They would be made available to anyone who might be interested. You can do experiments on a scale at the Oak Ridge National Laboratory that you can't do at most other places. That's part of the reason why we are doing them.

Hecht: Rick, what do you mean by an animal being semisterile?

Generoso: It is a rearrangement. The original animal is heterozygous for rearrangements for that insertion. When that male animal goes through meiosis, he produces two types of gametes: One type will be balanced and another type will be unbalanced. The unbalanced gametes are capable of fertilization, but because of the imbalance, they lead to early embryonic mortality. You have the translocation on the average of about 50% of the sperm carrying imbalanced chromosomes, and those will fertilize, but the conceptus will die, so in fact, you have about half as many live births as the normal ones.

Hecht: That Northern blot that you showed that didn't have the 13, the 7, and the 5 kilobases, but had a little small band down at the bottom, what was the history of that?

Woychik: There is another band just above that at about 2 kb, which is uninterrupted in the insertional mutation. However, the breakpoint is downstream from the insertion, and is likely to interrupt that 2-kb fragment. The small band could be a round spermatid or some other testis-specific transcript. I can tell you certain things about what I think the structure would be.

Wyrobek: I wanted to get some sense of how you're going to screen these animals. Will it be phenotype or enzyme activity?

Woychik: Mainly by phenotype. You take the founder animal and simply cross it with a wild type and do Southern blots on the first generation. That's important because if you have multiple integrations—and that happens maybe about 40% of the time—you have a couple of different sites where the integration occurred. Now, it is very important to segregate those out if you're doing genetics. Many people don't do this. That's part of the reason why it would be dangerous simply to follow the inheritance of the transgene with dot blots, or by following, say, CAT expression. If you segregate these independent integration sites out, then you take heterozygotes and cross them and then quantitatively blot the offspring to determine whether you have homozygotes. If you can't make homozygotes in 30 animals, there is a high likelihood that you have a recessive embryonic lethal. Once you have four breeding pairs of homozygotes, you could take all those offspring and give them to Virginia Godfrey, who is our resident pathologist, and do soft-tissue analysis. It's all within an inbred-type genetic background.

Wyrobek: You're basically assessing for incorporation and lethality?

Woychik: We're simply making homozygotes and testing them for any type of defect that may arise. There are a number of things that you could test with this. You might want to test for imprinting, developmental genes, or chromosomal translocations, which are actually marked by the transgene sequence. There are a number of things that would come up in a relatively simple screen.

Genetic Risk Estimation

Problems and Possibilities in Genetic Risk Estimation

WILLIAM L. RUSSELL
Biology Division, Oak Ridge National Laboratory
Oak Ridge, Tennessee 37831-8077

OVERVIEW

Genetic risk estimation has been, and still is, such an inexact process that it is appropriate that society's needs for decisions and regulation have been served by the judgment of committees rather than of individuals. In this paper, I suggest some additional research on mammalian germ-cell mutagenesis, and the choice is perhaps best explained by reviewing certain aspects of how, over the years, the major national and international committees have wrestled with the difficult problem of estimating the genetic hazards of radiation. The treatment is enlarged slightly to serve the additional purpose of an introduction to this symposium's session on genetic risk estimation, particularly for those not familiar with the results of the committee deliberations. The research to be suggested can be done with any mutagen and is, therefore, of potential help in the estimation of genetic risks from both radiation and chemicals. (The national committees specifically cited are the ones convened by the U.S. National Academy of Sciences [NAS], the first of which is usually called the BEAR Committee, after the title of its reports: *The Biological Effects of Atomic Radiation*, and the subsequent ones, called BEIR Committees, after the title of their reports: *The Biological Effects of Ionizing Radiation*. The international committees referred to are the series called United Nations Scientific Committee on the Effects of Atomic Radiation, abbreviated UNSCEAR. The genetic reports of both series of committees have been appearing at intervals from the late 1950s to the present time.)

The BEAR Committee published the first report by a national committee convened to estimate the genetic risks of radiation (BEAR 1956a). At that time, the public was concerned about the possible biological hazards of radioactivity in fallout from above-ground atom bomb testing. With support from the Rockefeller Foundation, the NAS established six committees to study and report to the public on the biological risks of radiation. The sixteen-member genetics committee was one of these. Thirteen of us were research geneticists, one a radiological physicist, one a pathologist, and the chairman, Warren Weaver, a mathematician. Weaver, from the Rockefeller Foundation, was experienced and knowledgeable in many fields of science. He was chosen as chairman by Detlev Bronk, president of NAS, reportedly because Bronk believed that only a nongeneticist could be a neutral chairman for a bunch of obstreperous geneticists. I suspect Bronk's choice was influenced also by Weaver's skill at explaining scientific principles to the lay public. The year after the genetics report appeared, Weaver was awarded the NAS Public Welfare Medal, and the citation read in part that he "was able to fashion the various points of view expressed by geneticists into agreement on most of the fundamental issues." As I remember the meetings, there was only one major disagreement, an argument between H.J. Muller and Sewall Wright, which has been described in an article of personal reminiscences by Crow (1989).

The Committee reported two attempts at estimating genetic damage, and it was one of these that precipitated the dispute between Wright and Muller. The Committee estimated the total induced mutation rate in man for a unit dose of radiation by starting with the radiation-induced specific-locus mutation rate in the mouse, and then multiplying that by the estimated number of gene loci in *Drosophila* and by the ratio of human to *Drosophila* total chromosome length. Muller then applied the principle, suggested by Haldane (1937), that all mutations are equal with respect to population fitness. Thus, the sum of the damages caused by a mutation with a mild effect, which affects many individuals before it is eliminated by selection, is equal to that of a mutation with severe effect, which is eliminated sooner and affects correspondingly fewer individuals. On this view, that all mutations eventually produce equal damage, one needs only to estimate the number of mutations induced by a given dose of radiation in order to appraise the damage.

The objections voiced by Wright to this view were mentioned briefly in the report, but were later expressed in more detail in an appendix to the BEAR Committee's second report (Wright 1960). It will be sufficient here to quote from a few sentences in this interesting 6-page paper: "...the equating of all unequivocally injurious mutations is very unrealistic without consideration of the personal and social impact... the occurrence of a dominant mutation, lethal in the first week of development, will produce no appreciable damage to the population or any one in it... On the other hand, a dominant mutation that gives rise to a distressing and incapacitating but not lethal condition that is usually not manifest until after the family is complete may produce enormous personal and social damage before becoming extinct."

To my knowledge, and perhaps for the above-mentioned reason, the method of estimating risk by treating all mutations as being equally injurious is no longer accepted, and it was not actually used by the BEAR Committee in arriving at a recommended limiting dose of radiation for the population.

The other attempt by the BEAR Committee at estimating genetic damage is now usually called the doubling-dose method. This also was not used by the Committee, which finally decided to recommend a limiting dose of radiation that was considered not to be an unreasonable excess over the background level of radiation to which the population is exposed. However, the doubling-dose method has been used by many other committees up to the present time. (Doubling-dose method is, as Denniston [1982] has rightly pointed out, something of a misnomer, and other terms have been used: relative-mutation-risk method, indirect method, and proportional method.)

The doubling dose is the dose required to produce as many mutations as occur spontaneously in a generation. The doubling-dose *method* requires an estimate of the prevalence of that portion of the physical disorders in the human population that is believed to increase in proportion to the mutation rate. An estimate is then made of the amount of damage expressed if a given dose is repeated generation after generation until an equilibrium is reached at which mutations are eliminated from the population as rapidly as they are induced. The risk in the first generation is then estimated from that at equilibrium by using certain assumptions as to the rate of elimination of mutants. There are considerable uncertainties in all these steps. There is also at least some degree of error, which is discussed in the next paragraph, in the assumption used in the doubling-dose method that spontaneous and induced mutations are qualitatively alike.

There was some evidence from our mouse results against the validity of this assumption when it was first made by the BEAR Committee, and additional data have raised further doubts. For example, for specific-locus mutations in the mouse, which provide the most extensive information on this question, it is quite apparent that radiation-induced and spontaneous mutations are *not* qualitatively alike (Russell and Rinchik 1987). Some committees seem to have ignored or minimized this problem, although the BEIR (1980) Committee did refuse to use the doubling-dose method for chromosomal aberrations. The Committee argued that "chromosomal disorders in human populations result largely from primary and secondary trisomy (resulting from nondisjunction and Robertsonian translocation, respectively), and these are not expected to be increased materially by low-level radiation exposures. A 'doubling dose' determined for reciprocal translocations induced in spermatogonia, however accurate it might be, would have little relevance to the induction of these abnormalities." The UNSCEAR (1986) report now expresses concern about the problem, not with regard to the evidence cited above, but because of the information that transposable elements may cause a sizable proportion of the spontaneous mutations, and because of lack of evidence that transposable-element integration can be induced by radiation.

Uncertainties with the doubling-dose method have led the UNSCEAR (1977 and later reports) and the BEIR (1980) Committee to use an additional approach to risk estimation, usually called the direct method, the development of which has involved my participation. To my knowledge, it was first advocated in simple form by Sewall Wright in 1947. Elsewhere (Russell 1989) I have mentioned that, in 1947, Alexander Hollaender organized an informal conference of Sewall Wright, H.J. Muller, and myself to discuss the mouse radiation genetics program being planned at Oak Ridge National Laboratory. Wright strongly recommended that the genetic hazards of radiation could be best estimated in the mouse by a simple empirical measure of the effects of radiation on vital traits in the offspring of irradiated parents. I agreed to do such experiments after first undertaking the measurement of induced mutation rates with a specific-locus test. From what Wright said much later, I am sure he guessed that the radiation-induced effects would be at least as great as the marked deleterious effects of inbreeding in guinea pigs that he had reported. Such was not the case. For example, the percentage reduction in litter size observed in the offspring of male mice whose spermatogonia had been exposed to a massive dose of X rays was much less than the percentage reduction resulting from inbreeding in guinea pigs, and it was not until additional data had been obtained that we were really confident that we had observed a statistically significant effect. Many other experiments at our laboratory and elsewhere also had difficulty in demonstrating significant effects of the irradiation of parents on the vital traits of offspring. In one study, we were able to show apparent shortening of life in the offspring of male mice exposed to neutron radiation from an atomic bomb (Russell 1957), but there remained the major difficulty, because of the vast differences in living conditions between mice and humans, of translating this finding into human detriment, even if one had known the cause of death of all individuals.

A slightly different approach to measuring phenotypic damage in first-generation offspring of irradiated parents had also been tried, but had likewise proved to be unrewarding for the effort expended. This was the scoring of dominant visible mutations, defined as those whose effects in the living mouse can be seen with the naked eye. My first published report on the frequency of radiation-induced specific-locus mutations (Russell 1951) lists the dominant visibles observed in the same population. The frequency was so low, and the time taken in careful observation for these effects interfered so greatly with the rapid scoring for specific-locus mutations, that a detailed observation protocol was abandoned. Studies by several members of the Radiobiology Unit of the Medical Research Council Laboratory at Harwell, England (for review, see Searle 1974), have provided data from more careful observation for dominant visibles. Again the frequencies have proved to be very low. However, results reported by Searle and Beechey (1986) indicate that a significant addition to the frequency of dominant visibles can be obtained if growth retardation is added to the effects scored. The study of congenital malformations has also proved of limited use for risk estimation, but Lyon (1983) has pointed out that since this method "uses an endpoint that can also be studied epidemiologically in man, it may at some future date be possible to get direct comparisons."

A new approach seemed worth trying, namely, instead of looking for effects on vital traits or externally visible characteristics, to concentrate on recording all disorders detectable in one of the major body systems in the offspring of control and irradiated parents. We had found the skeleton to be a useful endpoint for the detection of experimentally induced abnormalities in mice (Russell and Russell

1954). The skeleton *is* one of the major body systems, it is formed over a considerable period in development, it has basically the same features in mouse and human, and it is subject to similar disorders in these two species. Permanent preparations for observation with a dissecting microscope are obtained by skinning, eviscerating, clearing, and staining the specimens. Starting in 1959, we encouraged Udo Ehling, who had joined us on a fellowship from Germany for postdoctoral work in our laboratory, to find out if a comparison of the offspring from irradiated male mice with those from untreated parents would reveal significant differences in skeletal disorders. His experiments were successful (Ehling 1966), and UNSCEAR was urged to use the results for risk estimation. I argued that extrapolation to damage in all body systems might be approximated by using a round figure of 10 as a multiple of the damage in the skeleton. It was thought, by me at least, to be unlikely that this could be wrong in either direction by more than a factor of 2, which was considered not to be a large range of uncertainty in risk estimation. However, the Committee decided not to use the data, arguing that, without further evidence on the transmission of the skeletal defects, there was not adequate proof that these defects were genetic in origin. I thought that the data had provided adequate proof. The Committee would accept, without further genetic evidence, an empirical difference between the offspring of control and irradiated parents as proof of the mutational origin of endpoints such as congenital abnormalities or vital traits, including those being studied in the children of atom bomb survivors in Japan. I could not understand why the Committee would not accept the same evidence for skeletal abnormalities. The objection persisted, and when Paul Selby consulted me about the choice of a problem for his postdoctoral research, I suggested repeating the skeletal work, but also raising litters from the first-generation offspring so that if any skeletal defect was found when a first-generation animal was killed for study, its offspring would be on hand to test for a mutational cause. Ehling's work had included a few such cases, and, when Selby decided to tackle this onerous task, Ehling kindly provided support in his laboratory for the work. The results were presented to UNSCEAR in manuscript form before their publication (Selby and Selby 1977, 1978a,b) and, 30 years after Sewall Wright had proposed a simple form of the direct method, the UNSCEAR (1977) report adopted the direct method for risk estimation.

UNSCEAR had requested that Victor McKusick be asked to provide his estimate of a multiplication factor to be used in extrapolating from inherited skeletal defects to inherited damage in all body systems in humans. Selby and I conferred with McKusick, and, after making adjustments for pleiotropy and ease of detection, McKusick informed UNSCEAR of his agreement that a factor of 10 was reasonable. He doubted that it would be higher than that. UNSCEAR used the estimate of 10, without a range of uncertainty. Selby and Selby (1977) had given a range of 5–20, and the BEIR (1980) report, which also adopted the direct method, used a range of 5–15. UNSCEAR had also requested McKusick's estimate of the proportion of the skeletal defects scored in the mouse that would be clinically important if they occurred in humans. McKusick found the data of particular interest because of the seeming validity of extrapolation to man. He concluded that an estimate of 1/2 would appear to be a valid one for the fraction of abnormalities found in the Selby study which, in humans, would impose a serious handicap. UNSCEAR accepted this estimate. Selby and Selby (1977) and BEIR (1980) used a range of 1/4 to 3/4.

Ehling and colleagues have subsequently reported on the frequency of radiation-induced cataracts in the offspring of treated mice, and, using McKusick's list of human genetic disorders, Ehling (1980) arrived at an estimate of damage in all body systems similar to that based on skeletal disorders. No adjustment has been reported for the proportion of cataracts in the mouse that would be considered clinically important if they occurred in humans.

This brief excursion through a limited part of the history of risk estimation has not mentioned many other aspects, other committees, and the papers on risk estimation by individuals. As was stated at the outset, the purpose was to provide some background for choosing particular suggestions for future research on mouse germ-cell mutagenesis that might help in risk estimation. Since these will be my personal suggestions, the background given has included more of my involvement than would otherwise have seemed warranted.

It appears that the mouse cannot contribute much more to the doubling-dose method. The method is based primarily on the incidence of human disorders of both simple and complex genetic origin (estimates of which vary widely), on the mathematics of population genetics, and on various assumptions, including one on the mutational component in the causes of disorders of complex etiology. In the absence of statistically reliable estimates of the doubling dose in humans, the mouse has contributed estimates of the doubling dose for various endpoints. More precise data might be desirable on some of these, but acquiring them seems unnecessary at this time in view of the large uncertainties mentioned above. Data obtained from the mouse could, however, provide more information on the degree of qualitative difference between spontaneous and induced mutations, a key issue for the validity of the doubling-dose method.

Turning to the direct method, I think there are many possibilities by which additional research in the field of mouse germ-cell mutagenesis could provide information useful in risk estimation. One obvious set of experiments needed is to test the validity of the current assumption made for risk estimation that the effects of such factors as dose rate, dose fractionation, sex, and germ-cell stage will be the same for dominant skeletal mutations as they are for specific-locus mutations. Another obvious need is the investigation of one or more phenotypic endpoints in addition to the skeleton and cataracts. Perhaps disorders in the hematopoietic system are worth exploring for this purpose. More than one endpoint could be scored in the same population of offspring, thus reducing effort and expense.

There are ways in which the design of mouse experiments aimed at contributing to the direct method of risk estimation can be improved. Selby (1990 and this volume) has presented improvements that I find convincing. He suggests that a simple empirical scoring of the frequency of disorders in the first generation after treatment of the parents, and subtracting from this the frequency in the controls, will give a more complete estimate of risk than going through the process of identifying the variants that transmit their defects to subsequent generations (the procedure that he used in his original skeletal studies and the one customarily used in work on cataracts). Selby (this volume) points out that reliance on breeding tests of the offspring to identify mutations will fail to score the following mutants: (1) those dying before maturity or before the breeding test is complete, (2) those that are sterile or produce too few offspring for a conclusive breeding test, and (3) those carrying a mutation with penetrance too low to be recognized in the sample size of the breeding test. It seems likely that these three categories, together, might comprise a significant proportion of the total mutants and perhaps include many with serious defects. None of these need be missed in the simple empirical comparison of experimental and control offspring.

Category 3 is of special interest. By far the largest class of inherited disorders in man is made up of those of complex etiology, in which there is evidence of a genetic component, but the identification of which is obscured, presumably by interaction with other genetic or environmental factors. These comprise 84–98% of the classified inherited disorders in man, according to BEIR (1980) and UNSCEAR (1988), respectively, and they have been a major problem in the use of the doubling-dose method for risk estimation. Committee assumptions as to the mutational component of these disorders have varied over a tenfold range from 5% to 50%. Because of these uncertainties, UNSCEAR (1986) states that "using the doubling-dose method...the Committee is not in a position to provide risk estimates for these disorders." In contrast, the direct method is not faced with the problem of guessing the mutational component of this large class of disorders. The complexity of the etiology might be due mainly to incomplete penetrance. If so, then the empirical approach, using the direct method, would score the correct average proportion of these happening to be expressed in the first descendant generation. However, as Selby (this volume) has stated, some of them would be missed if use of the direct method relied solely on a breeding test.

It is ironic to me that Selby, who did the breeding tests on skeletal mutations to satisfy a committee that would not accept the simple empirical approach that seemed adequate to some of us, is now pointing out that the empirical approach actually gives a better estimate of risk. Neither Selby nor I recommend that breeding tests be dropped completely. Much useful information on the nature of the mutations can be obtained from them. For example, mutations with incomplete penetrance are quite common, and, contrary to an opinion sometimes expressed, mutations with very low penetrance can have serious deleterious effects (Selby and

Selby 1977). Therefore, in addition to the empirical comparison of experimental and control offspring, a sample of breeding tests is recommended for any new body system or endpoint under investigation.

Two general concepts concerning genetic risk estimation seem worth mentioning. First, it is now usually accepted that damage to the offspring in the first generation after exposure is the most important effect to evaluate. The damage will be greatest in that generation, and the less urgently needed estimation of damage to later generations can wait for the rapid advances being made in the field of genetics. Therefore, the direct method, which measures first-generation damage directly, seems more applicable than the doubling-dose method, which arrives at an estimate of first-generation damage circuitously. This view is supported by the BEIR (1980) Committee, which recommends "that the direct method be used for first-generation estimates and the relative-mutation-risk method be used for equilibrium estimates."

Second, although it is frequently stated that all mutations are deleterious, except for the very rare ones that make evolution possible, this extreme view may not be true. In fact, Sewall Wright's concept of the process of evolution, derived originally from his study of genetic improvement in livestock, argues that it is not usually the occurrence of a rare mutation that promotes evolutionary advance, but a shifting in the relative frequencies of mutations already in the population. Wright (1960) makes the following statement: "It is possible that the optimal state of any population is one in which many alleles with slight differential effect are carried at a large proportion of all loci at more or less equal frequencies. Even conspicuously unfavorable effects of mutations in particular combinations may be balanced by favorable effects in others." Allowance for this in risk estimation requires judgment, such as was used for skeletal mutations by basing the estimate of risk only on those mutations whose effects would be considered clinically important if they occurred in humans.

So far the discussion has been limited to radiation-induced mutations. Although current research is devoted much more to chemical mutagenesis than to radiation mutagenesis, there have been relatively few comprehensive committee reports, and most of the research has dealt only with the first step in risk estimation, namely, with the determination of whether a substance is mutagenic. This is commonly done first in bacteria or other lower organisms, followed by the use of a tier of tests, ending with the mammal, to judge the likelihood that the chemical is mutagenic in man. This focus of attention has undoubtedly been dictated by the vast number of chemicals to which man is exposed. The limited facilities available for studies of mammalian germ-cell mutagenesis have been used mainly to search for basic principles of chemical mutagenesis in mammals, using a few model compounds. When an estimate of the genetic hazard of a potentially dangerous mutagenic chemical is needed, it seems to me that the research with radiation indicates that the direct method, in the form recommended by Selby (this volume) will provide the most useful contribution that a mouse germ-cell test can offer.

As a final comment, I should like to draw attention to the change in attitude, over the past three decades, regarding the relative importance of the risk from mutagenesis and the risk from carcinogenesis. Of the six committees involved in the 1956 BEAR report, the Genetics Committee was considered by the Rockefeller Foundation, by the National Academy of Sciences, and by the media to be the most important. The conclusions of the Genetics Committee were given first place in that report and in a companion "Report to the Public" (BEAR 1956b). This latter report stated: "There are at least two good reasons for beginning with genetics. All of us tend to be more concerned for our children than for ourselves. And, as we shall see, the inheritance mechanism is by far the most sensitive to radiation of all biological systems."

The Genetics Committee had concluded that any dose of radiation, no matter how small, would cause some mutations, and it recommended that the mutational damage could be kept at an acceptable level if the dose to the public from man-made sources of radiation were restricted to 10 roentgens total accumulation during the period from conception to age 30. The report of the Pathology Committee (BEAR 1956a), which dealt with cancer and all other somatic effects of radiation, was approximately only one-quarter the length of the Genetics Committee report. Many people today would be surprised if they read the conclusion reached by this Committee of 14 well-known leaders in the field. As stated in the "Report to the Public," the conclusion reads: "Therefore, if the general level of exposure is held down to genetically acceptable levels, there would be no noticeable effects on the bodies of the persons exposed."

At the present time, the opposite view seems to prevail. The media echo or amplify this view. For example, as this paper was in final preparation, the latest BEIR report was released (unfortunately too late for study and discussion here). It includes a 70-page genetics section, but the public radio program, "All Things Considered," and the lengthy front-page *New York Times* article that appeared the following day failed even to mention the existence of the genetics report, although both discussed fully the conclusions on radiation induction of cancer.

In the original BEAR (1956a) report, the minimizing of the cancer risk relative to the genetic risk may have been too extreme. It is still difficult to make an accurate comparison of these two kinds of risk, but the mere fact that cancer is a group of related diseases, occurring predominantly in later life, whereas germ-cell mutations can act from conception to old age to cause major disorders in every bodily structure and function, suggests that the current relative inattention to genetic risk may be out of balance.

The organizers of this meeting are trying to redress the imbalance, as they have demonstrated not only by their choice of the general topic for the symposium, but also by the speakers and subjects chosen for this session on genetic risk estimation. Vicki Dellarco has organized, and presents the results of, an estimable attempt by the U.S. Environmental Protection Agency to demonstrate an example of what can be done, for at least one hazardous chemical, to make a real quantitative genetic risk assessment. Michael Cimino explains the goals of EPA as set forth by Congress and interpreted by the Agency to regulate chemicals that pose genetic hazards. Jack Bishop describes the major support of, and results achieved by, the research on mammalian heritable effects funded by the National Toxicology Program. Paul Selby explains, in more detail than I have presented here, ways to improve the direct method of risk estimation. All of these, and the other participants of this symposium who dealt with the specifics of mutation induction in mammalian germ cells, or with related biological topics, have emphasized that it should not be forgotten that germ-cell mutagens *are* mutagens, i.e., they *do* cause genetic damage as well as, some of them, being carcinogens.

ACKNOWLEDGMENTS

This research was jointly sponsored by the Office of Health and Environmental Research, U.S. Department of Energy, under contract DE-AC05-84OR21400 with Martin Marietta Energy Systems, Inc., and by the National Institute of Environmental Health Sciences under IAG No. 22Y01-ES-10067. The submitted manuscript has been authored by a contractor of the U.S. Government under contract No. DE-AC05-84OR21400. Accordingly, the U.S. Government retains a nonexclusive, royalty-free license to publish or reproduce the published form of this contribution, or allow others to do so, for U.S. Government purposes.

REFERENCES

Biological Effects of Atomic Radiation (BEAR). 1956a. *The biological effects of atomic radiation: Summary reports.* National Academy of Sciences, National Research Council, Washington, D.C.
———. 1956b. *The biological effects of atomic radiation: A report to the public.* National Academy of Sciences, National Research Council, Washington, D.C.
Biological Effects of Ionizing Radiation Committee (BEIR). 1980. Genetic effects. In *The effects on populations of exposure to low levels of ionizing radiation,* p. 71. National Academy Press, National Research Council, Washington, D.C.
Crow, J.F. 1989. Concern for environmental mutagens: Some personal reminiscences. *Environ. Mol. Mutagen.* (suppl. 14) **16:** 7.
Denniston, C. 1982. Low level radiation and genetic risk estimation in man. *Annu. Rev. Genet.* **16:** 329.
Ehling, U.H. 1966. Dominant mutations affecting the skeleton in offspring of X-irradiated male mice. *Genetics* **54:** 1381.
———. 1980. Strahlengenetisches Risiko des Menschen. *Umsch. Wiss. Tech.* **80:** 754.
Haldane, J.B.S. 1937. The effect of variation on fitness. *Am. Nat.* **71:** 337.
Lyon, M.F. 1983. Problems in extrapolation of animal data to humans. In *Utilization of*

mammalian specific locus studies in hazard evaluation and estimation of genetic risk (ed. F.J. de Serres and W. Sheridan), p. 289. Plenum Press, New York.

Russell, L.B. and E.M. Rinchik. 1987. Genetic and molecular characterization of genomic regions surrounding specific loci of the mouse. *Banbury Rep.* **28:** 109.

Russell, L.B. and W.L. Russell. 1954. An analysis of the changing radiation response of the developing mouse embryo. *J. Cell. Comp. Physiol.* (suppl. 43) **1:** 103.

Russell, W.L. 1951. X-ray-induced mutations in mice. *Cold Spring Harbor Symp. Quant. Biol.* **16:** 327.

———. 1957. Shortening of life in offspring of male mice exposed to neutron radiation from an atomic bomb. *Proc. Natl. Acad. Sci.* **43:** 324.

———. 1989. Reminiscences of a mouse specific-locus test addict. *Environ. Mol. Mutagen.* (suppl. 16) **14:** 16.

Searle, A.G. 1974. Mutation induction in mice. *Adv. Radiat. Biol.* **4:** 131.

Searle, A.G. and C. Beechey. 1986. The role of dominant visibles in mutagenicity testing. In *Genetic toxicology of environmental chemicals, part B: Genetic effects and applied mutagenesis* (ed. C. Ramel et al.), p. 511. A.R. Liss, New York.

Selby, P.B. 1990. Experimental induction of dominant mutations in mammals by ionizing radiations and chemicals. *Issues Rev. Teratol.* **5:** 181.

Selby, P.B. and P.R. Selby. 1977. Gamma-ray-induced mutations that cause skeletal abnormalities in mice. I. Plan, summary of results and discussion. *Mutat. Res.* **43:** 357.

———. 1978a. Gamma-ray-induced dominant mutations that cause skeletal abnormalities in mice. II. Description of proved mutations. *Mutat. Res.* **51:** 199.

———. 1978b. Gamma-ray-induced dominant mutations that cause skeletal abnormalities in mice. III. Description of presumed mutations. *Mutat. Res.* **50:** 341.

United Nations Scientific Committee on the Effects of Atomic Radiation (UNSCEAR). 1977. Committee report. In *Sources and effects of ionizing radiation*, p. 425. United Nations, New York.

———. 1986. Committee report. In *Genetic and somatic effects of ionizing radiation*, p. 27. United Nations, New York.

———. 1988. Committee report. In *Sources, effects and risks of ionizing radiation*, p. 375. United Nations, New York.

Wright, S. 1960. On the appraisal of genetic effects of radiation in man. In *The biological effects of atomic radiation. Summary reports*, p. 18. National Academy of Sciences, National Research Council, Washington, D.C.

Quantification of Germ-cell Risk Associated with the Induction of Heritable Translocations

VICKI L. DELLARCO AND LORENZ R. RHOMBERG
Human Health Assessment Group
Office of Health and Environmental Assessment
U.S. Environmental Protection Agency
Washington, D.C. 20460

OVERVIEW

In this paper, we examine how quantitative risk methods might be extended to analyses of risks associated with the induction of heritable translocations in the human germ line. The chemical used in our modeling is ethylene oxide (EtO). It is hoped that this risk analysis will help define key types of data needed to quantify genetic risk, the biological considerations that go into such an analysis, and the inferences that must be drawn from mouse dose-response data when making predictions about human germ-cell risk.

INTRODUCTION

EtO was chosen for analysis because it is a relatively simple alkylating agent that is a direct-acting mutagen. It is efficiently absorbed into the blood via the respiratory system, and from the blood it is rapidly distributed to all tissues. Moreover, there is a tremendous amount of information regarding its mutagenicity, including some human somatic cell data and data from mouse heritable germ-cell tests (Dellarco et al. 1990). Given all the available data on EtO, a mouse heritable translocation study (Generoso et al. 1990) provided the best data to establish a mathematical expression relating the dose of this chemical to the probability of effects occurring at low exposures. This study was the basis for our risk analysis.

Mouse Heritable Translocation Test on EtO

In the experiment of Generoso et al. (1990), (C3H x 101)F_1 male mice were exposed to inhaled EtO 6 hours per day for up to about 8 weeks, at exposures that ranged from 165 ppm to 300 ppm. It is evident from Figure 1 that the frequency of

---

This paper is adapted from Rhomberg et al. (1990). The views expressed in this document are those of the authors and do not necessarily reflect the views and policies of the U.S. Environmental Protection Agency.

Banbury Report 34: Biology of Mammalian Germ Cell Mutagenesis
Copyright 1990. Cold Spring Harbor Laboratory Press. 0-87969-234-0/90.$1.00 + .00

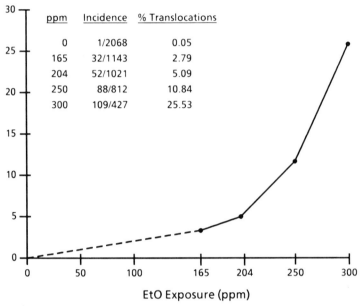

Figure 1
The dose-response curve for EtO-induced heritable translocations in male (C3H × 101)F$_1$ mice (Generoso et al. 1990). Males were exposed to inhaled EtO for 6 hr per day, 5 days per week for 6 weeks and then daily beginning at week 7 for 2.5 weeks. During the last 10 days and for 1 day after treatment, exposed males were mated with untreated T stock and (SEC × C57BL)F$_1$. The data from the two female stocks were pooled.

heritable translocations shows a clear dependence on the EtO concentration experienced by the treated mice. It is also readily apparent from this figure that several types of extrapolations are necessary to quantify genetic risk. Since we are using mouse data, estimation of human risk will require a species-to-species extrapolation. If it is presumed that the biological mechanisms tying different levels of EtO exposure to ensuing risks of heritable translocations are common in mouse and humans, then the dose response from the mouse study can be applied to humans if the doses that result in similar risks in the two species can be established. Thus, in essence, the species-to-species extrapolation becomes a determination of equivalency of dose across species. The mouse data were obtained at exposure levels much higher than those usually encountered by humans; thus, estimates of low-dose risk will require a consideration of how the mouse dose-response pattern can be extrapolated to lower exposures. Finally, in order to determine the amount of human disease and disability associated with a particular translocation frequency, a mutation-to-disease extrapolation will be necessary. In this paper, our intention is to consider these various extrapolations and the biological considerations and assumptions made for each.

Equivalency of Dose across Species

In this paper, we do not delineate the pharmacokinetic considerations (see Rhomberg et al. 1990) necessary for the interspecies dose extrapolation but will simply point out that the simplest dose metric—ppm concentration in air times a given duration of exposure in hours—was chosen. We think that ppm·hour is a reasonable dose metric to use for EtO when assessing the risks of chronic and constant types of environmental exposures, since external exposure is likely to be proportional to internal target-tissue exposure for this chemical.

A major assumption under the ppm·hour dose metric is that the response is a function of concentration times time, i.e., no dose-rate effects. We will not discuss the dose-rate issue for EtO (for discussion, see Rhomberg et al. 1990) but only point out that the mice in the Generoso et al. (1990) study experienced prolonged exposure to constant air concentrations of EtO, so that all sperm would be equally and fully exposed.

The length of exposure must be defined for the ppm·hour dose metric, which involves determining the time period during which damage or critical adducts are formed. This brings us to an important consideration in genetic risk assessment, the relative susceptibility of the various germ-cell stages to the induction of heritable or transmitted damage. It is important to determine what germ-cell stage(s) heritable lesions are derived from in order to define the critical time over which exposure should be integrated. For instance, if a mutagen produces heritable damage in stem cells, then it is presumed that such cells would become permanent sources of translocation-bearing sperm because of their long life span. Heritable damage would accumulate throughout the exposure history of the individual. The individual would be at some level of increased risk for his entire reproductive life even after the termination of exposure. On the other hand, post-stem cells have a much shorter life span. Risk to such cells would be a function only of recent exposures, and thus damage would accumulate only over that period. Such risk would decline following cessation of exposure as affected cells are cleared out of the epididymis.

A mouse dominant lethal study on EtO was used to determine from what germ-cell types induced transmissible damage was derived (Generoso et al. 1980). As shown in Figure 2, induced heritable genetic damage as a result of EtO treatment occurs in early spermatozoa and late spermatids. The life span of these cell types is about 10 days in the mouse, which corresponds to approximately a 21-day window of susceptibility to transmissible damage in humans. Thus, to calculate exposures in ppm·hours in the mouse translocation study, one would only accumulate exposures during the last 10 days before mating, not over the 8 weeks of exposure actually

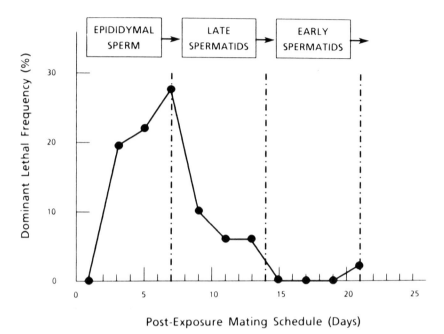

Figure 2
Frequency of dominant lethal mutations induced in different postmeiotic stages of males after intraperitoneal injection of 150 mg/kg EtO (Generoso et al. 1980). Treated T stock males were mated with untreated (SEC x C57BL)F_1 females.

given in the Generoso et al. (1990) experiment. For humans, one would accumulate dose over the last 21 days of exposure.

Postmeiotic Risk

If induced heritable damage is yielded only in postmeiotic cells, then it can be estimated how long it would take to clear the EtO-damaged sperm from the epididymis after exposure had ceased. The life span of the susceptible germ-cell stage is about 21 days in humans. Once spermatids are released from the seminiferous tubules, the transit time through the epididymis is about 6 days for a male at sexual rest (Bedford 1983). Although a few affected sperm might linger, the risk should be greatly reduced by the dilution with unaffected sperm in about 27 days. However, this estimate is confounded by the time that sperm might reside in the epididymis. Although there is no clear information on residence time, it is probably prudent to add an additional 10 days to the 27-day estimate. Since EtO has a short biological half-life, this estimate should not be confounded by persistence of the chemical at the target tissue.

Dose-response Relationships Associated with Induction of Heritable Translocations at Low Exposures

To determine the risk associated with exposure or immediately following exposure requires a high- to low-dose extrapolation, which basically involves selecting or developing a mathematical model to predict incidences at low exposures (where there is no information) based on the observed responses at higher exposures. As shown in Figure 1, the translocation frequency or risk rises sharply with increasing exposure. A convex relationship is observed, with the incidence of translocations in male offspring increasing from approximately 3% at 165 ppm to about 25% at 300 ppm. Although the highest exposure is not quite double the lowest, the risk it produces is over eightfold higher. The key challenge is to predict how trends of the experimental region should be extended to low exposures, i.e., whether risk drops just as precipitously below the lowest tested dose, which would indicate little or no low-dose risk, or whether the dose response diminishes more gradually.

Extrapolation of apparent dose-response trends can be explored on a purely empirical basis. One approach is to connect a straight line segment between the background and lowest tested dose producing a significant response. This tactic is a very conservative one and surmises that risk is proportional to dose at all exposures below the observed responses. A simple linear extension of the curve is naive, however, because it disregards information provided by the higher exposure points. It is preferable to use an extrapolation technique that permits the observed curvature in the experimental region to influence the shape of the curve in the low-dose region. By fitting a curve to the experimental data, one presumes that there is an underlying functional relationship of dose to response that dictates the shape of the whole curve. In other words, the curvature in the experimental region is not due to mechanisms operating solely at high doses, e.g., a perturbation of a protective mechanism.

Empirical curve fitting uses a quite general mathematical model that embodies the general shape expected of a dose-response curve and yet grants sufficient flexibility to fit the observed data. The extrapolation is conducted by tracing the low-dose continuation of the fitted curve. The general equation that is used in the multistage model fulfills this specification (Guess and Crump 1976; Portier and Hoel 1983):

$$P(d) = 1 - \exp - (q_0 + q_1 d + q_2 d^2 + ... + q_n d^n)$$

where $P(d)$ is the probability that an individual experiencing dose d will manifest the toxic response. The coefficients of the polynomial in the dose (the qs) are the parameters to be chosen by fitting the curve to the data using the methods of maximum likelihood and are constrained to be positive or zero.

The multistage model is used in cancer risk assessment. There is some thinking relevant to the cancer process that led to the development of this equation. However, it can be applied empirically without reflecting any particular biological rationale. This particular fitted curve is indicated by an M in Figure 3. As shown in

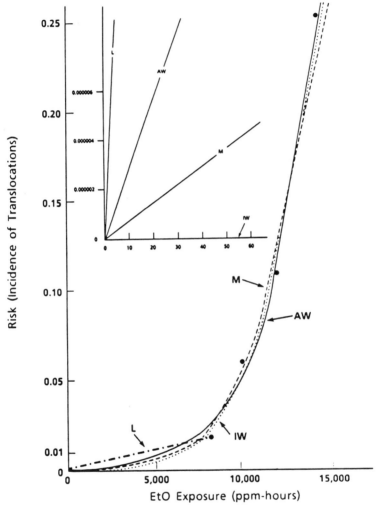

Figure 3
Mouse heritable translocation test data (from Fig. 1) for EtO fitted by different extrapolation models. (L) Straight line segment connecting lowest tested dose with background response; (M) multistage model; (AW) Weibull model assuming additive background; and (IW) Weibull model assuming independent background. The upper left inset is a magnification of the low-dose region for these various curve-fitting models.

Figure 3, the multistage equation fits the experimental data quite well. However, it is not satisfying to simply establish a dose-response trend on a purely empirical basis with no biological information being used to guide what the shape of the curve should look like at low exposures.

To obtain more meaningful quantitative dose-response assessments, one should employ or develop models that are biologically reflective, requiring an understanding of the mechanisms leading to translocations. For reciprocal translocations, it is generally accepted that they are the product of chromosomal "breaks" in two nonhomologous chromosomes and the mutual exchange through reunion of the chromosome fragments. Although mechanisms for conversion of alkylated DNA to mutations have been advanced, and relatively well-described pathways have been depicted for repair of alkylated DNA, knowledge is still incomplete regarding the mechanisms for conversion of alkylated DNA or chromosomes to a reciprocal translocation. Nonetheless, knowing that the mechanism must involve strand breaks occurring in close proximity on two different chromosomes provides a starting concept for modeling. One can think of a nucleus as being divided into boxes with different parts of DNA or chromosomes being contained in each box. For a reciprocal translocation to occur, two nonhomologous chromosomes must have repairable lesions that give rise to DNA strand breaks occurring in close proximity, i.e., within the same box. This critical event need happen only in one box to result in a translocation-bearing sperm. The question is, how should the probability of two DNA strand breaks falling close together diminish with dose?

An extrapolation model that would embody this concept is the Weibull equation (Gumbel 1958):

$$P(d) = 1 - \exp - (\exp[A + B \ln d])$$

where d is dose and A and B are parameters that are found by optimizing the fit to the data using the method of maximum likelihood.

This model is used to describe the pattern of occurrence of the worst outcome among many units, all of which are equally subjected to a certain pattern of risk. It has been used in cancer risk assessment, and the rationale is that several rare events must coincide within a single cell to cause that cell to be transformed (Pike 1966). There are many cells at risk, but only one needs to be affected in the appropriate manner to result in neoplasia. Whether or not an individual will develop cancer depends on the fate of its most damaged cells, i.e., the extreme of the distribution of damage. So the concept applied in cancer is similar to our thinking regarding heritable translocation occurrence. The translocation event may be viewed as the extreme or worst outcome in any local region among many such regions in the nucleus or boxes at risk. It should be stressed that the Weibull equation is still a general one.

Figure 3 depicts two Weibull plots. One (additive-Weibull) assumes that spontaneously arising lesions, which can potentially lead to reciprocal translocations, and the chemically induced lesions form a common pool of chromosomal damage on which the heritable translocation incidence depends. The other approach (independent-Weibull) assumes that the chemically induced lesions act independently and separately from the spontaneous lesions. Thus, the EtO-induced heritable translocations arise by a different mechanism than the background response.

Although the multistage, additive-Weibull, and independent-Weibull models provide similar fits to the experimental data, they provide dissimilar projections of risk in the low-dose region (Fig. 3). The low-dose regions for the multistage and additive-Weibull models are linear, whereas the curve fitted by the independent-Weibull model goes down steeply, producing a convex (sublinear) relationship in the low-dose region. At very low doses, the risks implied by the independent-Weibull extrapolation are so low that the curve becomes essentially indistinguishable from the x axis. This trend can be interpreted to mean that as chemically induced chromosome breaks become rare, the likelihood that two such breaks are in close proximity becomes extremely low.

It is evident from Figure 3 that one cannot apply a goodness-of-fit test to designate the appropriate extrapolation model. All the models depicted in Figure 3 provide a good fit to the EtO dose-response data. Selection of the appropriate extrapolation procedure will be based on our fundamental understanding of the biological mechanisms leading to heritable genetic damage. As shown by the other papers in this volume, the biological processes involved in or influencing genetic risk are complex.

As discussed above, dose-response relationships for heritable translocations induced by EtO can be determined by several alternative means. Estimates of low-dose risk can be predicted by the applied models. The exposure in ppm·hours must be calculated first, however. For example, EtO exposure, 8 hours per day, 5 days per week, for 21 days, at 10 ppm (i.e., presumed human window of sensitivity) would be accumulated over a total exposure dose of 1200 ppm·hours (i.e., ppm concentration was multiplied by 120 hours). Since the window of susceptibility in humans is presumed to be 21 days, exposures prior to this window would have no effect on current risk.

For extrapolations that are linear at low doses (i.e., multistage and additive-Weibull models), calculations of risks at other exposures are easy, e.g., an exposure twice as high carries twice the risk. On the other hand, the slope changes with dose for extrapolations that are nonlinear (i.e., independent-Weibull model), and the full equation for the model, as estimated by the curve-fitting procedure, must be used for each exposure to estimate its ensuing risk. For exposures near or in the experimental region of the dose-response curve, all applied models will project similar results, since they were all estimated by fitting curves to these data.

Risk can be expressed as the excess risk (i.e., over and above background due to spontaneous causes) of fathering a translocation carrier in the time period closely following a particular EtO exposure. For example, the excess risk associated with a 1200 ppm·hour to EtO is estimated to be 1×10^{-4} by the multistage model and

5.6 × 10^{-4} by the additive-Weibull model. In other words, the excess risk would be 10–60 children with heritable translocations of 100,000 offspring born to fathers who were exposed in this fashion, in addition to the background frequency of translocation carriers. The background incidence of heritable translocation carriers in the human population has been estimated at 190 translocation carriers per 100,000 live

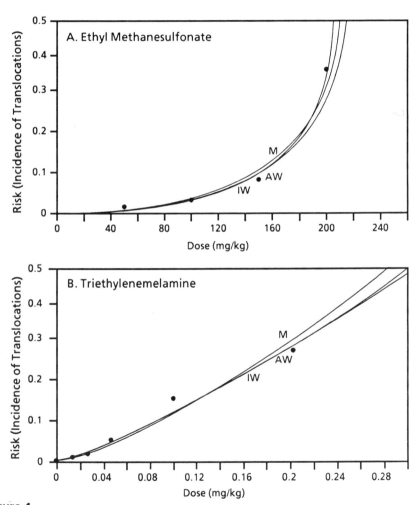

Figure 4
Mouse heritable translocation test data (from Generoso et al. 1978) for (A) EMS and (B) TEM. (C3H × 101)F_1 males were injected with the mutagen and mated with untreated (SEC × C57BL)F_1 females 6.5–7.5 days after the EMS treatment and 11.5–15.5 days after the TEM treatment.

births (ICPEMC 1983). Thus, this exposure would result in an increase of about 5–32% over background.

Figure 4 illustrates that the multistage and Weibull models applied to the EtO heritable translocation data do not uniquely fit this particular mouse data set. The convex dose-response curve for induced heritable translocations by ethylmethane sulfonate (EMS) is similar to that for EtO, and thus it is not surprising to observe a good fit with the applied models. However, unlike EtO and EMS, the dose response for triethylenemelamine (TEM) does not appear to deviate significantly from linearity in the experimental region, yet the TEM data still fit the multistage and Weibull extrapolation models.

Relationship between Translocations and Disease/Disability/Death

Heritable translocations can lead to a number of health consequences, such as infertility, death of the fetus, and serious medical disorders. Although the consequences of heritable translocations are serious, extrapolation of translocation frequency measured in the mouse to determination of frequencies of specific human diseases cannot be made at present because the fraction of heritable translocations that are associated with human disease and disability is not known. Use of a factor of one (i.e., all translocations produce a disease state) would err on the side of safety. This approach might not be overly conservative, however, since a large number of birth defects and other medical disorders are associated with heritable translocations. Nonetheless, the determination of health impact is a difficult step in genetic risk assessment and a step where science will have to merge with risk assessment science policy.

SUMMARY

We have presented a quantitative risk analysis based on the induction of heritable translocations. Although our analysis is limited to a particular genetic endpoint, namely heritable translocations, we view the basic structure of our analysis as applicable to other types of genetic lesions. We think it is critical to go through such examples of risk analysis in order to define the necessary information or research needed to formulate appropriate strategies for minimizing genetic risk.

REFERENCES

Bedford, J.M. 1983. Considerations in evaluating risk to male reproduction. *Adv. Mod. Toxicol.* **3**: 44.

Dellarco, V.L., W.M. Generoso, G.A. Sega, J. Fowle, and D. Jacobson-Kram. 1990. A review of the mutagenicity of ethylene oxide. *Environ. Mol. Mutagen.* (in press).

Generoso, W.M., K.T. Cain, S.W. Huff, and D.G. Gosslee. 1978. Inducibility by chemical mutagens of heritable translocations in male and female germ cells of mice. *Adv. Mod. Toxicol.* **5**: 109.

Generoso, W.M., K.T. Cain, M. Krishna, C.W. Sheu, and R.M. Gryder. 1980. Heritable translocation and dominant-lethal mutation induction with ethylene oxide in mice. *Mutat. Res.* **73**: 133.

Generoso, W.M., K.T. Cain, C.V. Cornett, N.L.A. Cacheiro, and L.A. Hughes. 1990. Concentration-response curves for ethylene oxide heritable translocations and dominant lethal mutations. *Environ. Mol. Mutagen.* (in press).

Guess, H.A. and K.S. Crump. 1976. Low-dose-rate extrapolation of data from animal carcinogenicity of experiments—Analysis of a new statistical technique. *Math. Biosci.* **32**: 15.

Gumbel, E.J. 1958. *Statistics of extremes.* Columbia University Press, New York.

International Commission for Protection Against Environmental Mutagens and Carcinogens (ICPEMC). 1983. Estimation of genetic risks and increased incidence of genetic disease due to environmental mutagens. Committee 4 final report. *Mutat. Res.* **115**: 255.

Pike, M.C. 1966. A method of analysis of a certain class of experiments in carcinogenesis. *Biometrics* **22**: 192.

Portier, C. and D. Hoel. 1983. Low-dose extrapolation using the multistage model. *Biometrics* **39**: 897.

Rhomberg, L., V.L. Dellarco, C. Siegel-Scott, K.L. Dearfield, and D. Jacobson-Kram. 1990. A quantitative estimation of the genetic risk associated with the induction of heritable translocations at low-dose exposure: Ethylene oxide as an example. *Environ. Mol. Mutagen.* (in press).

COMMENTS

William Russell: Thank you, Vicki, for a very detailed attempt at analysis of a very difficult problem. Questions?

Wyrobek: Of all the translocation-carrying animals that have come out of all of these studies, has anybody looked at the phenotypes systematically, semi-systematically, or to any extent?

Generoso: I think there is a program now supported by the NTP to study in more detail for phenotypic effects translocations that are produced in mutagenesis experiments. It is realized that risk quantification is a valuable resource in efforts to study the molecular structure of some of the mammalian genome.

Wyrobek: And also, as Vicki pointed out, you get a sense of what the factor is.

Generoso: Yes. Another thing that may be pointed out here is that if you look at what we might call true environmental mutagens: ethylene oxide, acrylamide, methylene-*bis*-acrylamide, trimethyl phosphate, and so on, the major transmitted genetic changes produced by these are heritable translocations.

Lonnie Russell: You brought up something that struck me as an answer to a question I asked a few days ago concerning cells getting through meiosis. In the rat and a number of other species, a lot of cells fail to get through meiosis even in a normal animal, and animals in the same strain differ. It reminds me of Monty Moses's talk about synaptonemal complex, in which he said that if there is a problem with the spermatocytes, or even spermatogonia, as they go through meiosis they have to do a careful pairing up, and if it doesn't work, the cells can degenerate. Instead of repairing themselves, they have the option of dying. Maybe that's part of why as a male you have a lot of gametes, so that you can self-select during that process of meiosis.

Dellarco: In fact, if you look at all the heritable translocation data available in the mouse, the stem cells don't seem to play a major role in risk; it's the post-stem cells, and mostly the postmeiotic ones. Although that conclusion is based on limited data, it's probably correct.

Handel: There has been a lot of speculation in the literature about some kind of quality assessment going on in meiosis that has not been investigated in terms of biological mechanisms. One form that speculation has taken is the model proposal put forth by Miklos in the early 1970s that somehow a cell can assess "pairing sites" (whatever they are) and determine whether or not they are "saturated" (whatever that means), and if they are not saturated, meiosis does not proceed. Nobody has investigated this in terms of biological mechanisms. There is evidence from a number of different species that if meiosis goes awry, or the recombination mechanisms go awry, meiosis isn't completed. I have a question I'll direct to anybody who has any information. Has there been any study of the interaction between agents that will induce heritable translocations and aging of sperm during sexual rest? Is there any evidence on whether or not sperm aged in the epididymis express more cytogenic abnormalities?

Bishop: My understanding is that the mouse is not an appropriate model to look at the issue of aged sperm. I was told that they are dribblers and they just don't store sperm.

Dellarco: They don't really store sperm.

Handel: And humans do?

Lonnie Russell: Humans definitely store sperm for quite a lengthy period. After a vasectomy, there is a period of time that an individual is told to wait. Someone has looked at the longest period of time you can find a few sperm present, and that's years, but they won't fertilize an egg.

Bishop: I know Dr. C. Auerbach has been interested in sperm storage effects with radiation, with chemicals, and a lot of things like that for many years. They do those kinds of studies with *Drosophila* and other insects.

Handel: I believe Patricia DeLeon is doing some work on that now, too.

Dellarco: We would be grateful for any reference that will give us a handle on this question of residence, because every reproductive biologist we talk to about how long sperm can be stored gives a different answer.

Generoso: The chemical induction of heritable translocations depends on the placement of that adduct on the sperm and the conversion of that adduct. The aging may be responsible, or the time between adduct formation and fertilization may be responsible, for the creation of sites either by hydrolysis or by whatever mechanism. The main point is that this rearrangement is completed in the egg. It is the egg that is doing the processing of this lesion into chromosome rearrangement. I think a more important question is what is the role of the aging egg.

Fritz: Several comparative aspects would be quite interesting here. First of all, zoologically comparative: Bedford has analyzed the residence time of sperm in the epididymis of different species, and by and large, the better the chilling effect in the epididymis, the longer the sperm can stay. The very tail of the epididymis from which the vas emanates is usually the coldest part in terms of temperature difference. That's a very interesting phenomenon. There are some bats who become functional hermaphrodites in the sense that sperm introduced in the winter live in crypts of the uterus during the winter and then fertilize for the first ovulation the following year. That's at least a one-year period. Bats are very successful as far as I know. So, at least a one-year period with a favorable environment for a sperm is just great.

If you can't predict the future, you sure can predict the past. You've got all these sperm samples, along with other cell types, that have been frozen in different ways. You now have all this marvelous methodology to be able to look at deletions, insertions, translocations, whatever. Why not do a study on sperm that have been frozen for varying periods of time? The epididymis is just an inefficient freezer compared to what liquid nitrogen is doing.

Hecht: There actually are some data on that. In the dairy breeding industry, farmers or breeders don't like to use sperm that has been stored for a very, very long period of time even in liquid nitrogen. You start seeing chromatin changes, whatever that means. They're just uncomfortable with it. You certainly can fertilize with sperm that has been stored in liquid nitrogen for 30 or 40 years and get progeny, but in practice they prefer to turn it over relatively quickly. Now, I don't know what the actual data are.

Fritz: Maybe it's in the *Farmer's Almanac*.

Kovacs: There's plenty of data for humans from the standpoint of in vitro fertilization, just what you're seeing with cryopreservation of eggs, cryopreservation of sperm, delayed fertilization, and the treatment of some male infertility. There is a whole body of knowledge to draw on; for example, the studies that have been done on epididymis after vasectomy. Knowledge has been accumulated that demonstrates clearly that stored sperm are in many regards inadequate. They have lower incidences of penetration and much greater incidences of morphological anomaly. In using the hamster penetration assay, there was a report, I think, from a British Columbia group that showed a higher incidence of chromosomal anomalies.

Moses: I'd like to introduce another subject. It indicates the tremendous amount of basic information that we lack. It's a semantic problem that has to do with something that was quoted from Lee's (Russell) talk: the sensitivity of a stage to damage. The problem that I have, and I think I have resolved, is that what we mean by sensitivity is really endpoint-measured effectiveness of a dose. I think as long as we define that as what sensitivity means, we are all right. Sensitivity is something that does need to be studied, in the true sense of the word—where you hit a cell and as quickly as possible assess the damage as far as the technology will allow you to do so. What happens to it afterwards, how it is modulated by cell death, by enhancement of the damage, by repair of the damage, all affects the quantity of the damage at the end—in the gamete—which concerns us because that's where the risk is.

Another part of the question concerns the concept of the blood-testis or rather the Sertoli cell-Sertoli cell barrier, which has for a long time been accepted as the reason that meiotic prophase appears to be less affected, for example, than the gonial stem cell in terms of heritable damage. I would like to get on the record just a quick statement from the Sertoli cell experts about transmissibility through that junction, past that junction, or at least accessibility of molecules that are presented on the basal side to cells that are on the adluminal side, namely the spermatocytes. I invite some brief comment about the kinds of drugs that are being tested and how easily they can pass into the adluminal side.

Lonnie Russell: In my first presentation I wanted to say there are four ways that things can get into the tubule. First, they can go between cells, Sertoli cells, to get into the center of the tubule; that seems to be limited mainly to hydrophilic substances of small molecular weight. The second way is that they can go by receptor-mediated means into the Sertoli cell, where the Sertoli cell has a plan to take them up to begin with. The third way is that they are lipid-soluble and can go through the membrane of the Sertoli cell; it seems the more lipid-soluble they are, the better they enter the Sertoli cells themselves. The fourth way is that they could be carried up by germ cells as the germ cells move up from the basal compartment to the adluminal compartment. What are the data in the field as to what gets in? What gets in has best been studied by Okamura et al., who looked at a variety of chemicals and came up with the lipid-soluble versus the water-soluble data. Virtually no other studies have looked at a variety of compounds getting in. It seems to me that we are left at state of the art with empirically testing fluids from the lumina of the seminiferous tubule versus the lymphatics. To make a prediction about a chemical for purposes of risk assessment might be premature.

Wyrobek: Gary Sega gave us some data on adducts, so we have another set of data on adducts that are formed with radiolabeled compounds; e.g., EMS, the acrylamides, ethylene oxide. These data say very clearly that they get into the spermatids.

Use of Germ-cell Mutagenicity Testing by the U.S. Environmental Protection Agency Office of Toxic Substances

MICHAEL C. CIMINO AND ANGELA E. AULETTA
U.S. Environmental Protection Agency, Office of Toxic Substances,
Health and Environmental Review Division, Washington, D.C. 20460

OVERVIEW

The Toxic Substances Control Act provides the U. S. Environmental Protection Agency (EPA) with the authority to regulate chemical use by requiring testing and use restrictions as appropriate to protect human health and the environment. It controls both the acquisition of information to identify and evaluate potential chemical hazard, and the regulation of chemical production, use, distribution, and disposal. Regulation on the basis of heritable mutation induction is specifically mentioned in the Test Rule section of the law and has also been pursued for new chemicals. A tiered scheme of mutagenicity testing has been employed to assess hazard. The final tier contains tests that permit risk assessment for a chemical.

INTRODUCTION

The Toxic Substances Control Act (TSCA; Public Law 94-469) was signed into law by President Ford on October 11, 1976, and went into effect on January 1, 1977. TSCA provides the EPA with the authority "to regulate commerce and protect human health and the environment by requiring testing and necessary use restrictions on certain chemical substances... ." This authority supplements other existing laws, such as the Federal Insecticide, Fungicide and Rodenticide Act (FIFRA), the Clean Air Act, the Water Act, and the Occupational Safety and Health Act. These other laws, however, either deal with chemicals only when they enter the environment as wastes, or deal with only one phase of a chemical's existence. TSCA is designed to fill the gap in the government's authority to test and regulate chemicals.

In TSCA the term "chemical" encompasses a wide variety of organic and inorganic substances manufactured or imported for industrial uses, such as dyes, pigments, lubricant additives, chemical intermediates, synthetic fibers, structural polymers, coatings—essentially any commercial chemical except those used as drugs, food additives, cosmetics, pesticides, and certain other uses. These are controlled by other statutes.

TSCA has two main regulatory features: (1) acquisition of sufficient information by EPA to identify and evaluate the potential hazards from chemical substances and (2) regulation of the production, use, distribution, and disposal of such substances when necessary.

The primary focus of this report is heritable mutagenicity testing and the role of in vivo assays in that testing, primarily under TSCA Section 4 (Test Rules) and, more briefly, under Section 5 (new chemicals).

DISCUSSION

The Agency has published guidelines on risk assessment for mutagenicity (USEPA 1986). These guidelines deal with heritable mutation as a regulatory endpoint. They present a weight-of-evidence scheme for determining if an agent may be a potential human mutagen. Emphasis is placed on a chemical's intrinsic mutagenic potential, its ability to reach the gonad and interact with germ-cell DNA, and its ability to induce heritable mutations in a mammalian species.

There are two categories of evidence for chemical interaction in the gonad (USEPA 1986):

1. *Sufficient evidence* of chemical interaction is given by the demonstration that an agent interacts with germ-cell DNA or other chromatin constituents, or that it induces such endpoints as unscheduled DNA synthesis (UDS), sister chromatid exchange (SCE), or chromosomal aberrations in germinal cells.
2. *Suggestive evidence* includes the finding of adverse gonadal effects such as sperm abnormalities following acute, subchronic, or chronic toxicity testing, or findings of adverse reproductive effects such as decreased fertility, which are consistent with the chemical's interaction with germ cells.

There are eight categories of evidence that contribute to the weight-of-evidence for potential human germ-cell mutagenicity. These are, in order of decreasing strength-of-evidence:

1. Positive data derived from human germ-cell mutagenicity studies.
2. Valid positive results from studies on heritable mutational events (of any kind) in mammalian germ cells.
3. Valid positive results from mammalian germ-cell chromosome aberration studies that do not involve transmission from one generation to the next.
4. Sufficient evidence for a chemical's interaction with mammalian germ cells, together with valid positive mutagenicity test results from two assay systems, at least one of which is mammalian (in vitro or in vivo). The positive results may both be for gene mutation or both for chromosome aberrations; if one is for gene mutations and the other for chromosome aberrations, both must be from mammalian systems.
5. Suggestive evidence for a chemical's interaction with mammalian germ cells, together with valid positive mutagenicity evidence from two assay systems as described under 4, above. Alternatively, positive mutagenicity evidence of less

strength than defined under 4, above, when combined with sufficient evidence for a chemical's interaction with mammalian germ cells.
6. Positive mutagenicity test results of less strength than defined under 4, combined with suggestive evidence for a chemical's interaction with mammalian germ cells.
7. Although definitive proof of nonmutagenicity is not possible, a chemical could be operationally classified as a nonmutagen for human germ cells if it gives valid negative test results for all endpoints of concern.
8. Inadequate evidence bearing on either mutagenicity or chemical interaction with mammalian germ cells.

Section 4: Testing Requirements

TSCA Section 4(a) gives EPA authority to require the testing of chemicals if unreasonable risk to health or the environment is suspected. TSCA specifies that data may be developed in the areas of "carcinogenesis, mutagenesis, teratogenesis, ... and any other effect which may present an unreasonable risk of injury to health or the environment." To require testing, EPA must find that (1) the chemical may present an unreasonable risk or a significant potential for exposure; (2) there are insufficient data available with which to perform a reasoned risk assessment; and (3) testing is necessary to generate such data (and is not already under way).

A testing requirement is promulgated in a Test Rule, which must (1) identify the substance to be tested and the tests to be conducted; (2) provide (or reference) guidelines for performance of the tests; and (3) specify a reasonable period of time for completion of testing.

As presently constituted, the testing scheme for each endpoint is composed of three tiers. In the gene mutation test scheme (Fig. 1A), testing for a chemical for which no known data exist begins with the *Salmonella* (Ames) assay. An Ames negative triggers a test for gene mutation in mammalian cells in culture. A positive in either assay triggers a *Drosophila* sex-linked recessive lethal (SRL) assay. A positive in the SRL triggers a specific-locus assay (SLT).

In the chromosomal effects test scheme (Fig. 1B), testing begins with an in vitro cytogenetics assay. A negative in vitro assay triggers an in vivo bone marrow cytogenetics assay (either aberrations or micronuclei). A positive in either assay triggers a rodent dominant lethal (RDL) assay. A positive RDL triggers a rodent heritable translocation (RHT) assay.

The first four tests in these schemes (Ames, in vitro gene mutation, and in vitro and in vivo cytogenetics assays) are designed to detect intrinsic mutagenic potential. If all are negative, no further testing is required.

The next two tests (SRL and RDL) demonstrate the ability of the chemical to reach the gonad and to interact with germ-cell DNA. If these two assays are negative, no further testing is required. Agents positive in the SRL are further tested

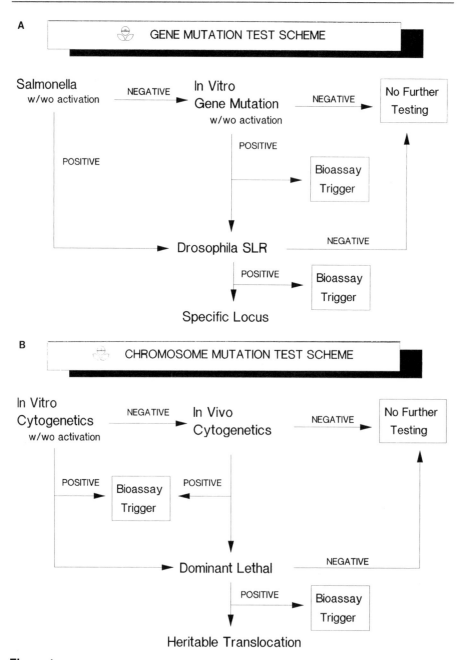

Figure 1
Existing mutagenicity test scheme for (A) gene mutations and for (B) chromosomal effects.

in the mouse SLT; those positive in the RDL assay are tested in the RHT. Agents positive in either of these two assays (SLT and/or RHT) will be presumed to be potential human mutagens as defined in the risk assessment guidelines, category 2, mentioned earlier.

Several factors entered into the choice of tests in this scheme. First, it was deemed necessary to test for both gene mutation and chromosomal effects in the event that agents existed that were specifically gene or chromosomal mutagens.

Second, it was decided to use only those tests that measured defined genetic endpoints. Tests such as the SCE or UDS assays, that either do not measure a defined genetic endpoint or for which the specific endpoint of concern is not known, were not included in the generic test scheme. Such tests have been included in specific Test Rules if existing data suggested that they might be sensitive indicators of genotoxicity for the chemical in question.

Test Rule Chemicals

To date, the Agency has issued final Test Rules for 14 chemicals and proposed Rules for five others. A consent order (a mutual agreement between EPA and industry) has been issued for one chemical, and another one is in preparation.

The final Test Rules are for the C_9 hydrocarbon fraction of petroleum distillates, commercial hexane, diethylenetriamine (DETA), cresols (ortho-, meta-, and para-), 1,2-dichloropropane, fluoroalkenes (vinyl fluoride, vinylidene fluoride, tetrafluoroethane, and hexafluoropropene), hydroquinone, 2-mercaptobenzothiazole, and oleylamine. Mutagenicity testing requirements for both hydroquinone and diethylene glycol butyl ether were completed before their Test Rules were finalized, thus indicating industry's willingness to support this type of testing. Consent orders have been prepared for aniline and seven substituted anilines; one is in preparation for triethylene glycol monomethyl ether. Proposed Rules have been issued for meta-phenylenediamine, cyclohexane, tributylphosphate, isopropanol, and methyl ethyl ketoxime.

Of the 14 chemicals subject to final Test Rules, five (C_9 hydrocarbon fraction, DETA, and three fluoroalkenes) have reached stop points at either the first or second tier level in the mutagenicity test scheme.

One criticism of the Section 4 test schemes is that the mouse visible SLT is not available commercially and is performed in only one laboratory in the United States. The Agency has responded to this criticism by determining that the biochemical SLT, which *is* commercially available, may be substituted for the visible. It is anticipated that in the future, industry will be given a choice between these two assays for those chemicals that reach final tier testing for gene mutation.

A second criticism directed at the final tier questions the ability to generate quantitative risk assessments using data from either the SLT or the RHT. Recently, Agency scientists have used the published data for ethylene oxide to generate a risk assessment using heritable translocation data (Dellarco and Rhomberg, this volume). A risk assessment using mouse visible specific-locus data on procarbazine was published by Ehling and Neuhäuser (1979). These two efforts demonstrate that quantitative risk assessments can be performed using data generated from these tests. Furthermore, the Agency believes that data generated under a test rule will be more conducive to quantitative risk assessment than some of the data that has been available in the scientific literature, since the intended use of the data for risk assessment will be borne in mind when the test protocols are designed. Therefore, factors such as sample size and adequately spaced doses, which have limited attempts at quantitative risk assessment in the past, should not be an issue for chemicals tested under Section 4.

To date, the data generated on Section 4 chemicals have related to intrinsic mutagenic potential and not to the ability to induce heritable mutations in mammals. EPA also uses short-term genotoxicity tests as indicators of potential oncogenicity. Under the present scheme, a positive response in certain key tests will trigger a two-year bioassay:

1. A positive response in both Ames and SRL assays.
2. A single positive response in any one of the following: in vitro mammalian gene mutations; in vitro mammalian chromosome mutations; or in vivo chromosome mutations.

Since the publication of the first Test Rule, which defined these tests as triggers to a 2-year bioassay, new information has become available. This information prompted review of the mutagenicity tests used as bioassay triggers under Section 4. These data were the subject of discussion at an Agency-sponsored Workshop on the Relationship Between Short-term Test Information and Carcinogenicity at Williamsburg, Virginia in January, 1987.

As a result of data presented at this workshop and subsequently published in the scientific literature (Tennant et al. 1987), EPA is in the process of reassessing its position on the use of short-term tests in a regulatory context. Revisions to the first tier of the Section 4 test scheme for mutagenicity and to the tests that serve as triggers to a 2-year bioassay have been proposed, as diagramed in Figure 2.

The major change found in the proposal occurs in the first tier: The test schemes for gene mutation and chromosomal aberrations are now combined. The revision proposes three tests for the first tier: the Ames assay, an in vitro assay for gene mutation, and an in vivo assay for chromosomal effects (bone marrow chromosomal aberrations or micronucleus).

The proposed trigger for a bioassay is then as follows:

1. There is no longer a single test trigger to a bioassay.
2. An automatic trigger to a bioassay is dependent on a minimum of two positive responses, at least one of which must be in an in vivo assay.

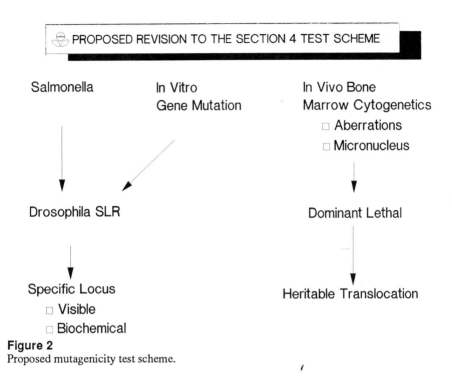

Figure 2
Proposed mutagenicity test scheme.

3. In vitro cytogenetics is no longer part of the scheme. (At one time, OTS also considered removing the in vitro gene mutation assay from the test scheme. However, that position has been reconsidered in light of unofficial comment on the proposal and as a result of work done since the Williamsburg meeting. For now, these assays remain part of the test scheme, but they no longer serve as single test triggers to a bioassay.)
4. The SRL assay no longer serves as a trigger to a bioassay.
5. A positive response in all three first tier tests, or positives in the Ames and the in vivo cytogenetics assays, or positives in the in vitro gene mutation and the in vivo cytogenetics assays, would lead directly to a bioassay.
6. Any other combination of responses, including a single positive response in any one assay, or positives in both the Ames and the in vitro gene mutation assays, would result in a program review. This review, which would occur before a decision were made to require further testing, would consider all available data, including results from tests in other endpoints, structure activity relationships, production volume, and exposure figures.

It is anticipated that no further testing would be required for the majority of chemicals that are negative in all three first tier tests. However, if exposure data, structure activity relationships, or other factors warrant, these agents may also be subject to a program review and subsequent testing in a bioassay.

For now, there are no changes in the second tier of the scheme. The third tier of the scheme for chromosomal effects also remains unchanged; a positive response in the RDL will trigger a heritable translocation assay. The third tier of the scheme for gene mutations has been modified to include a mouse biochemical SLT as an alternative to the mouse visible SLT. Agents positive in the SRL may be tested in either assay; the choice will be left to the regulated industry.

It is anticipated that a program review, similar to that mentioned above for the cancer bioassay, will be performed for those agents that may require testing in either the SLT or the RHT. The first such program review was conducted this past summer for the fluoroalkene vinyl fluoride, which was positive in the SRL assay. As a result of that review, the Agency is presently considering evaluating additional test data (some combination of germ-cell UDS, germ-cell DNA-binding, and RDL assays) before triggering the specific-locus assay.

As part of its ongoing process of reviewing the test schemes used under Section 4 of TSCA, OTS is reconsidering the contents of the second tier of the mutagenicity test scheme. At the same time, the Office of Pesticide Programs (OPP) has been examining its mutagenicity testing requirements under FIFRA. It is expected that the OTS and OPP schemes will closely correspond (allowing for the differences necessitated by regulatory mandate) and that both offices will publish revised schemes in the future.

Section 5: New Chemicals

Section 5 of TSCA requires that manufacturers and importers of "new" chemicals submit a Premanufacture Notification (PMN) to the EPA 90 days before beginning the manufacture or import of the new substance. New TSCA chemicals are those that do *not* appear on the Inventory of Existing Commercial Chemicals, which was compiled in 1977 and is continually being updated (and which currently contains more than 60,000 entries).

Section 5 requires that certain information be provided in the PMN, including chemical identity, production volume, proposed uses, estimates of exposure and release, and any health or environmental test data available to the submitter. It does *not* require that submitters conduct toxicity testing prior to submission of the notification to the Agency, only that they submit whatever such data as they have.

In response to this lack of a specific requirement that mutagenicity data be generated, EPA has developed techniques for hazard assessment that can be used in the presence of little or no test data on the substance itself. Briefly, this approach involves four major components:

1. Evaluation of available toxicity data on the chemical itself.
2. Evaluation of test data available on substances that are "analogous" to the PMN chemical or data available on key potential metabolites or analogs of the metabolites.
3. Use of mathematical indicators of biological activity, or quantitative structure activity relationships (QSARs). The use of QSARs is limited to the estimation of physical chemical properties such as water solubility, log P, and vapor pressure for estimation of aquatic toxicity and bioconcentration factors.
4. The knowledge and judgments of scientific assessors in the interpretation and integration of the information developed in the course of the assessment.

With regard to heritable effects, because the PMN assessment process relies so heavily on the use of analog data, and because of limitations in the size of the data base of chemicals tested for heritable genetic effects, concern for a chemical's ability to induce heritable gene or chromosomal mutations is rarely supportable under Section 5. However, in some instances, where an analog or potential metabolite is a known heritable chromosomal mutagen, EPA has required testing for heritable effects. In these instances, the test most often required has been the RDL assay. This test is usually required with the understanding that positive results may necessitate further testing, most likely a RHT. Data from this assay may then be used in a risk assessment. To date, Section 5 testing has not reached this stage.

REFERENCES

Ehling, U.H. and A. Neuhäuser. 1979. Procarbazine-induced specific-locus mutations in male mice. *Mutat. Res.* **59:** 245.

Tennant, R.W., B.H. Margolin, M.D. Shelby, E. Zeiger, J.K. Haseman, J. Spalding, W. Caspary, M. Resnick, S. Stasiewicz, B. Anderson, and R. Minor. 1987. Prediction of chemical carcinogenicity in rodents from in vitro genetic toxicity assays. *Science* **236:** 933.

United States Environmental Protection Agency (USEPA). 1986. Guidelines for mutagenicity risk assessment. *Fed. Reg.* **51 (185):** 34006.

COMMENTS

Wyrobek: How do you deal with interspecies differences in response in terms of the weight-of-evidence philosophy that you're using here? For example, the thing that comes to mind is the difference in response with dichlorobromopropane (DCBP), where the human responds very sensitively to DCBP in terms of germ-cell killing, but the mouse does not seem to respond at all. I understand the genetic tests in the mouse are negative. Yet the rat responds, and there is a positive dominant lethal result. That's the only example I know, but I imagine that there will be many compounds that differ dramatically in their species responsiveness and mutagenicity. Working only with the mouse could generate some wrong information.

Cimino: In those cases in which we have some reason to suspect that might be the case, on the basis of structural or physical analogs of the chemical, we would have to take that into consideration. We obviously can't test the chemicals that are positive in humans and negative in mice because we can't test them in humans. We would normally utilize the things that Vicki (Dellarco) has said about species-to-species extrapolation. We feel that in vivo tests are more meaningful as far as risk or hazard assessment is concerned than in vitro studies, and mammalian in vitro studies are more telling than nonmammalian in vitro studies. Other than that, I really can't respond to that very well.

Dellarco: If you have information that something is not positive in the mouse, what do you do, because people want to do a quantitative risk assessment and the mouse is our only model?

Wyrobek: We have at least one such compound. If you have a dominant lethal effect in the rat that is very strongly positive, and yet there is no effect in the mouse, how do you do a quantitative risk assessment?

Smithies: As a sort of layman in this area, but a chemist at the same time, I am concerned about what seems to be a much bigger problem than the problem we are addressing in the fine-tuning of the tests. That is the incredible discrepancy between the number of chemicals that are in the inventory—60,000 or thereabouts, as you said—and the number, some 54, if I heard you rightly, that are being thoroughly investigated.

Cimino: Actually, it's fewer than 54 chemicals.

Smithies: Well, let's assume it's 54. How does a chemical ever get on to the 54 list? As a chemist, I would have thought some of the chemicals that were mentioned in your list were very low on the list of likelihood. For example, isopropanol and cyclohexane: My chemistry says cyclohexane is inert, but I may be completely wrong. I could see meta-phenylenediamine being high on our list, but I don't know how they get on that list and how they come out of the 60,000.

Cimino: I can give you two answers to that, one long and one short. I will give you a brief summary of the long one, which I took out of my talk. TSCA allows for the existence of the Interagency Testing Committee, ITC, which is composed of representatives from eight agencies and observers from other agencies of the government, which twice a year meets to decide what chemicals we are worried about today—or this year. They create a list of chemicals of concern and present it to the Agency. The Agency must review the available data and decide whether there is reason for concern; e.g., if analogs of the chemical are dangerous, based on observed effects in the environment. There might be a very large amount of human exposure to this chemical and insufficient data to assess whether there should be a concern.

Lonnie Russell: What role do the media play in that?

Cimino: These things are dealt with in the order in which they come from the ITC. It is a long, laborious process. Some of these test rules have been in existence since 1979, and we are still working them through. As far as new chemicals are concerned, the original Inventory contained, I think, some 37,000 or 38,000 chemicals that were "grandfathered" in. It was decided that, as of 1977, any chemical that already existed would be in the Inventory, and anything produced after that would have to be looked at before we let it out. The ones that are in the Inventory are the ones that the ITC looks at. Granted, there are an incredible number of chemicals, and we are testing them very, very slowly.

Moses: Are there any instances of agents that have been shown to be mutagenic to a mammalian system that have escaped the *Salmonella* test?

Several Participants: Acrylamide.

Handel: Procarbazine is a good example.

Favor: If it should come to the point where a heritable mutation assay is required, must all spermatogenic stages be adequately tested, or would dominant lethal data be used to identify postspermatogonial stages that are at risk and only those stages?

Cimino: The dominant lethal is in the second tier. It would not be necessary to look at all of the spermatogenic stages.

Favor: No. I'm saying if it comes to a heritable mutation assay.

Cimino: Then all stages would be looked at.

Dellarco: But the dominant lethal isn't in the second tier for the specific locus trigger. It's the *Drosophila* assay.

Cimino: The dominant lethal triggers the translocation.

Favor: Dominant lethal data are very predictive for those stages that might be sensitive to mutation induction.

Dellarco: Are you going to include the dominant lethal test in the specific locus half?

Cimino: Well, currently it isn't. This is part of what we are discussing in our modification of the second tier. We are considering using the dominant lethal as a generic second tier test for both endpoints. Once we have decided what we would like to do, we will propose that in the *Federal Register,* at which point everybody can respond to that.

Favor: In any event, the dominant lethal data can identify stages that might be sensitive to mutation induction in postspermatogonial cells.

Mammalian Heritable Effects Research in the National Toxicology Program

JACK B. BISHOP AND MICHAEL D. SHELBY
National Institute of Environmental Health Sciences
Division of Toxicology Research and Testing
Experimental Carcinogenesis and Mutagenesis Branch
Research Triangle Park, North Carolina 27709

OVERVIEW

The National Toxicology Program (NTP), through the Experimental Carcinogenesis and Mutagenesis Branch (ECMB) at the National Institute of Environmental Health Sciences (NIEHS), conducts a program to investigate the mammalian germ mutagenicity of chemicals. The four basic tests used to detect the induction of mutations are the dominant lethal test (DLT), the heritable translocation test (HTT), the morphological specific-locus test (MSLT) and the electrophoretic specific-locus test (ESLT). Mice are the study organisms in these tests. In addition to mutagenicity tests, work is conducted to investigate the processes and characteristics of germ-cell mutagenesis.

This program has made numerous contributions to the understanding of mammalian germ-cell mutagenesis by substantially increasing the number of chemicals tested for germ-cell mutagenicity, providing data for genetic risk assessments, characterizing numerous aspects of germ-cell mutagenesis through the use of ethylnitrosourea, investigating mutation induction in often ignored cells such as oocytes and zygotes, and providing animal models of human genetic diseases.

Future studies will include additional testing of chemicals for mutagenicity, molecular investigation of induced and spontaneous mutations, expansion of tests by combining endpoints, development of methods to detect induced aneuploidy, and molecular studies of the mouse genome. The program also anticipates supporting studies to investigate the frequency and origin of transmitted mutations in humans.

INTRODUCTION

The Division of Toxicology Research and Testing (DTRT) of the NIEHS is the major operating component of the Department of Health and Human Services NTP. The NTP is responsible for the toxicologic characterization of chemicals and other environmental agents through the conduct of long- and short-term toxicology and carcinogenesis studies and for the development and validation of appropriate methodologies to perform and interpret these studies.

As part of its mission to provide the NTP with information regarding the genotoxic activity of potentially toxic or carcinogenic substances in the environment, the ECMB of the DTRT has conducted numerous in vitro and in vivo genetic toxicology tests on hundreds or, for some assays, thousands of chemicals. For the most part, these tests are conducted through contracts with investigators at private and university research facilities or through interagency agreements with other governmental laboratories.

The Heritable Effects Group (HEG) within the ECMB supports a research program in mammalian germ-cell chemical mutagenesis that is one of the largest in the world and the primary one in the United States. Four mouse germ-cell systems have been used in studies conducted in this program; the established MSLT and the more recently developed ESLT are being used in a variety of approaches to detect and study the effects of germ-cell mutagens throughout spermatogenesis, and chemical induction of chromosomal effects, primarily in postmeiotic stages of spermatogenesis and in female oocytes, is being studied using both the DLT and HTT. More than 20 compounds have been studied in one or more of these mouse germ-cell assays through this research program.

The ultimate objective of this program, as with most programs in mammalian germ-cell mutagenesis, is to protect the health of future generations. Our first tasks toward reaching this goal are (1) to determine whether environmental mutagens reach the germ cells and produce genetic damage and (2) to provide data that permit quantitative assessments of the health impact resulting from this damage. These challenges extend beyond simply identifying germ-cell mutagens. Therefore, the research sponsored by the NTP strives to enhance the field of germ-cell mutagenesis by increasing our understanding of the mutation process as well as by increasing the numbers of chemicals studied.

In this paper, we present an overview of the NTP research in germ-cell mutagenesis and its accomplishments; we discuss some of the challenges that remain and initiatives that will address these challenges. The research described has been conducted through contracts with Research Triangle Institute (RTI) and Interagency Agreements with the Department of Energy, Oak Ridge National Laboratory (ORNL). The principal investigators are Dr. S.E. Lewis, RTI; and Drs. W.M. Generoso and L.B. Russell, ORNL.

RESULTS AND DISCUSSION

What Has Been Accomplished through NTP Research in Germ-cell Mutagenesis?

Testing efforts by the NTP have substantially increased the number of chemicals that have been evaluated for germ-cell mutagenicity in mammals. The number of chemicals evaluated in the MSLT for effects in post-stem-cell stages has been more than doubled (Russell 1990). MSLTs have been conducted on dibromochloropropane, ethylene dibromide, adriamycin, platinol, hexamethylphosphoramide, urethane, 6-mercaptopurine, acrylamide, chlorambucil, and melphalan. Of these, only chlorambucil, melphalan, and acrylamide have given evidence of inducing specific-locus mutations in germ cells. None induced mutations in spermatogonial stem cells. Chlorambucil and melphalan appear to induce specific-locus mutations in both pre- and postmeiotic germ-cell stages and, of those mutagens known to induce specific-locus mutations in postmeiotic cells, chlorambucil is the most effective (Russell et al. 1989). With the increased number of chemicals tested in the MSLT and shown to induce specific-locus mutations in one or more stages of spermatogenesis, it is now possible to compare their patterns of mutagenic activity (Table 1).

For those chemicals that have yielded positive test responses to date, all are positive in postmeiotic stages, but the patterns of response and/or the relative effectiveness at a specific cell stage differ. Two chemicals, ethylnitrosourea and methylnitrosourea, produce the highest mutation frequency in premeiotic stages, probably preleptotene spermatocytes or late differentiating spermatogonia. In contrast, chlorambucil and melphalan are most effective in inducing specific-locus mutations in early spermatids or possibly secondary spermatocytes, whereas late- or mid-spermatids appear to be the stages most mutationally responsive to cyclophosphamide, ethylene oxide, methylmethanesulfonate, ethylmethanesulfonate, diethylsulfate, acrylamide, and triethylenemelamine. For most chemicals, the pattern of stage specificity for induction of specific-locus mutations in postmeiotic stages does not differ substantially from that for induction of dominant lethals. Interestingly, the more environmentally relevant chemicals (e.g., ethylene oxide, acrylamide) are positive in postgonial stages but negative in spermatogonia. Although of short duration, these later stages cannot be ignored in assessing mutagenic risk for germ-cell mutagens. Chlorambucil, whose effectiveness in inducing mutations in treated spermatids exceeds that of the stem-cell "supermutagen" ethylnitrosourea, promises to be of similar value for use in in-depth studies of the mutagenesis process in postmeiotic cells.

Numerous in-depth MSLT studies have been conducted with ethylnitrosourea, including investigation of germ-cell stage sensitivities, sex differences, age effects, effects of sequential exposure to other mutagens, and detailed dose-response information. Studies have demonstrated that the dose response for ethylnitrosourea in spermatogonial stem cells does not follow a linear interpolation from high doses. At low doses and with split doses, the curve appears to drop significantly below linear (W.L. Russell et al. 1982; Favor, this volume). Studies of specific-locus mutations induced by treatment of zygotes have shown that a very high proportion of mutants induced are mosaics (Russell et al. 1988); such mosaic mutants should prove useful in studies of embryonic development. The numbers of mutants produced by ethyl-

Table 1
Patterns of In Vivo Genetic Toxicity in Rodents

Chemicals	Bone marrow micronucleus or aberrations	Dominant lethal mutations	Heritable reciprocal translocations	Specific locus mutations[a]
4-Acetylaminofluorene	−	−		
Pyrene	−	−		
Tetrahydrocannabinol	−	−	−	
Caffeine	−	−	−	/−
Ethylene dibromide	−	−		?/−
Dibromochloropropane	−	−		?/−
2-Acetylaminofluorene	+	−		
Urethane	+	−		−/−
Vincristine	+	−		−/−
Adriamycin	+	−[b]		−/−
Platinol	+	−[b]		−/−
Hexamethylphosphoramide	+	−[b]		?/−
Hycanthone		−[b]		?/−
Benzo(a)pyrene	+	+	−	?/−
6-Mercaptopurine	+	+	−	−/−
Myleran	+	+		?/−
Ethylene oxide	+	+	+	?/−
Acrylamide	+	+	+	+/−
Methylenebisacrylamide	+	+	+	
Ethylenebisacrylamide		+	+	
Ethylmethanesulfonate	+	+	+	+/−
Cyclophosphamide	+	+	+	+/−
Chlorambucil	+	+	+	+/−
Melphalan	+	+		+/−
Diethylsulfate		+		+/−
Chlormethine		+		+/−
Mitomycin C	+	+	+	?/+
Ethylnitrosourea	+	+	+	+/+
Methylnitrosourea		+	+	+/+
Procarbazine	+	+	+	+/+
Triethylenemelamine	+	+	+/+[a]	+/+

[a] Specific-locus test responses are indicated as postmeiotic/premeiotic; single entries for the DL and HTT are for postmeiotic germ-cell stages.
[b] Positive DL response in female mice demonstrated to be genetic via egg transfer experiments and/or analysis of first cleavage metaphase chromosomes.

nitrosourea and chlorambucil continue to provide valuable resources for in-depth molecular, genetic, and cytogenetic studies.

In the electrophoretic specific-locus project, emphasis has been placed on acquiring data from the known germ-cell mutagens, ethylnitrosourea and X rays, in order to provide a data base that will allow comparison of this system with other methods of detecting germ-cell mutagens. The dose-response curve for induction of electrophoretic mutations with ethylnitrosourea peaked at a lower dose than that seen in the MSLT and did not reach as high a mutant frequency. A substantial data base on the spontaneous frequency of electrophoretic mutations has also been accumulated by examining control populations. Two chemicals, ethylene oxide and ethylene dibromide, have been tested for mutagenicity by the ESLT.

Mouse models for two human diseases have also been found through the ESLT screen of protein electrophoretic mobility. A mutant deficient in carbonic anhydrase-2, which was discovered in the progeny of an ethylnitrosourea experiment (Lewis et al. 1988), provides a model for carbonic anhydrase deficiency syndrome in humans. A second mouse model for a human genetic disorder was detected among the control population. This mutation, which produces a hemoglobin disorder known in humans as β-thalassemia, has been characterized through extensive genetic and molecular analysis (Skow et al. 1983).

More than 20 compounds have been studied in the standard DLT or HTT assays, and effects on female germ cells and on zygotes have been studied with selected compounds. Ethylene oxide, acrylamide, and methylenebisacrylamide, which are widely used environmental chemicals, were found to be effective inducers of both dominant lethal mutations and heritable translocations. The EPA has made extensive use of these data, especially the dose-response and dose-rate data on ethylene oxide, in their hazard assessment efforts (Rhomberg et al. 1990).

The induction of late fetal death and fetal malformation has been observed following exposure of zygotes to various chemical mutagens (Generoso et al., this volume). Through cytogenetic analysis of first-cleavage metaphase chromosomes, the fetal death induced by the tubulin inhibitor, nocodazole, was shown to be associated with induced numerical chromosome aberrations (Generoso et al. 1989). However, for mutagens such as ethylmethanesulfonate and ethylene oxide, which also are effective in inducing malformations in treated zygotes, the classic causes of mutations (e.g., structural or numerical chromosome damage and gene mutation) seem unlikely (Katoh et al. 1989). This has led to investigation of other possible causes, such as the induction of transposable elements (W.M. Generoso, pers. comm.).

Certain duplication-deficiency products of meiotic segregation in translocation-carrying animals may also result in fetuses with malformations. Rutledge et al. (1990) have recently described associations between specific malformations and methylenebisacrylamide-induced translocations. Some of these malformations are associated with specific duplication deficiencies that have been described cytogenetically (Cacheiro et al. 1990). In addition to providing models of human disease, and hence a direct link to health impact, these findings provide exciting new material for studying genetic control of embryological development.

Most mutagenesis experiments have concentrated on males; assessment of damage to female germ cells has received little attention. One reason for this disparity is that the DLT, as typically used in the initial evaluation of chemicals for mutagenicity, cannot distinguish embryonic mortality that is due to genetic effects in the oocyte from that caused indirectly by maternal or nongenetic oocyte toxicity. Assessment of chemicals in both males and females, including egg-transfer experiments, has led to the identification of two chemicals, adriamycin and platinol, that induce dominant lethals in females but not in males (Katoh et al. 1990). These female-specific dominant lethal effects, which have been shown through cytogenetic analysis of first-cleavage metaphase chromosomes to be associated with the induction of chromosome aberrations, have been hypothesized to occur because of the less condensed nature of chromatin in the oocyte compared to the chromatin in most stages of spermatogenesis.

What Are the Major Gaps in Our Knowledge Regarding Estimation of Genetic Risk?

In general, our ability to detect germ-cell mutagens through the use of animal models is good. Through the use of these animal models, chemicals to which humans are exposed, like ethylene oxide and acrylamide, have been demonstrated to induce mutations in mammalian germ cells. Although there is no direct evidence for induction of germinal mutations in man by such chemicals, the consistency of mutagenicity test results in a wide range of experimental organisms leaves little doubt that humans are likewise susceptible to their effects (National Research Council 1983).

We must improve our ability to extrapolate from specific genetic damage to an impact that is of concern to human health, and to convert these data into meaningful assessments of risk. Through studies such as those of Drs. W.L. and L.B. Russell using their newly discovered supermutagens, questions such as those of dose-rate effects for chemicals can finally be addressed. The advent of molecular technologies within the field of germ-cell mutagenesis promises to provide improved methods for detecting and characterizing mutations in both humans and animal models (Office of Technology Assessments 1986). This should give us greater insight into the nature of induced mutations as well as solutions to a host of other questions.

However, there still remain several gaps in our knowledge that constitute major obstacles to estimating genetic risk. The following, although not intended to be all-inclusive, provides an example of four such areas requiring additional research initiative:

1. There is almost no information on relative effects among species. Acquisition of data on molecular dosimetry and the pharmacokinetics of metabolism in humans and in animal models would be helpful in meeting this challenge. In fact, development of models for extrapolation between rodent species would be a desirable prelude to the development of animal-human extrapolation models.
2. Some rodent mutational tests having endpoints detected by assays used to monitor human disease states have been, or are being developed; for example, altered phenylalanine metabolism (McDonald et al., this volume) and cataracts (Favor, this volume). Additional endpoints need to be identified for which assays may reasonably be developed that can be applied both in rodent mutation tests and in monitoring human populations. Development of such parallel models would provide a better basis for estimating the health impact associated with germ-cell mutagenesis.
3. We have now identified at least three chemicals to which female germ cells are uniquely vulnerable. A greater awareness and understanding regarding the characteristics of such chemicals are essential to improving estimates of genetic risk. Additional effort needs to be expended toward identifying and characterizing chemicals that are particularly more damaging to female germ cells.
4. There are no chemicals that have been demonstrated to be human germ-cell mutagens. This fact is probably attributable to the relative insensitivity of methods available to ascertain an increase in germ-cell mutations in humans and the limited effort that has been directed to this task. Our understanding of the induction of mutations in mammalian germ cells, the mutagenicity of chemicals, and the exposure of humans to germ-cell mutagens all suggest the existence of a tangible genetic risk. It is important that the latest technologies and developments be applied in our efforts to determine the frequency and origins of genetic diseases in humans.

Planned NTP Initiatives for Improving Our Ability to Estimate Genetic Risk

Our program will continue the animal model studies in mice, which not only contribute to regulatory agency efforts in assessing genetic risk, but have also led to development of animal models for human genetic disease and to a deeper understanding of the mutation process in mammalian germ cells. Greater emphasis will be placed on the use of molecular techniques to detect and characterize mutants and to elucidate the structure and function of the genome.

The ESLT is being expanded to combine endpoints, including tests for dominant cataract mutations, morphological specific-locus mutations, and alterations in the lengths of DNA restriction fragments. Testing of chemicals will continue using the MSLT, but additional emphasis will be placed on molecular studies such as the characterization of spontaneous and induced mutations, evaluation of the biological effects of different classes of DNA lesions, and the relationship between the frequencies of induced dominant and recessive mutations.

Additional chemicals are scheduled for evaluation in female DLT tests with concomitant egg-transfer and cytogenetic studies to confirm genetic effects. Studies of the chromosomal and molecular bases for the unique effects resulting from mutagen treatment of pronuclear stage zygotes will continue. An increased effort will also be made to develop methods that will permit the detection and quantification of induced, transmitted aneuploidy.

The large numbers of new translocation stocks being produced through HTT studies involving NTP chemicals are a valuable resource in efforts to molecularly characterize the mammalian genome and in characterizing adverse health effects associated with translocation heterozygosity. Progeny of balanced translocation carriers will continue to be evaluated for developmental anomalies. Some of the stocks have transgenes, for which there are readily available probes, incorporated near the translocation breakpoint. These stocks provide opportunities for detailed molecular analysis of genomic regions surrounding the breakpoints, and of short chromosomal segments, which can facilitate study of genomic structure/function correlations.

Two areas that have not been addressed specifically but that constitute major gaps in our knowledge of estimating genetic risk are (1) our lack of confirmed human germ-cell mutagens and (2) the mutational component of human genetic disorders. These questions will be answered, in part, through studies involving humans. New methods for detecting heritable mutations in humans, such as those described by Wyrobek et al., Mohrenweiser et al., Kovacs et al., and Arnheim (all this volume), should be helpful in identifying de novo changes in the human genome. An additional level of effort involving parallel animal model tests will be required to demonstrate the association of these changes with environmental exposures; this is an effort in which NTP animal model studies will unquestionably play a major role.

ACKNOWLEDGMENTS

The authors are grateful to Ms. Sherry Bolen for typing this paper and to Mr. Stanley Stasiewicz for his assistance in preparation of the table. We are also indebted to Dr. James Mason and Ms. Kristine Witt, whose many helpful suggestions and comments through prolonged discussions contributed significantly to the clarity of concepts presented in this paper; remaining lack of clarity is the accepted sole responsibility of J.B.B. NTP data presented in this paper were produced through work conducted under National Institute of Environmental Health Sciences contracts N0-1-ES-25012 and N0-1-ES-55078, and NIEHS/DOE Interagency Agreements Y0-1-ES-20085 and Y0-1-ES-10067.

REFERENCES

Cacheiro, N.L.A., K.T. Cain, C.V. Cornett, and W.M. Generoso. 1990. Chromosomal imbalance in malformed mouse fetuses. *Mol. Environ. Mutagen.* (in press).

Generoso, W.M., M. Katoh, K.T. Cain, L.A. Hughes, L.B. Foxworth, T.J. Mitchell, and J.B. Bishop. 1989. Chromosome malsegregation and embryonic lethality induced treatment of normally ovulated mouse oocytes with nocodazole. *Mutat. Res.* **210:** 313.

Katoh, M., N.L.A. Cacheiro, C.V. Cornett, K.T. Cain, J.C. Rutledge, and W.M. Generoso. 1989. Fetal anomalies produced subsequent to treatment of zygotes with ethylene oxide or ethyl methanesulfonate are not likely due to the usual genetic causes. *Mutat. Res.* **210:** 337.

Katoh, M., K.T. Cain, L.A. Hughes, L.B. Foxworth, J.B. Bishop, and W.M. Generoso. 1990. Female-specific dominant lethal effects in mice. *Mutat. Res.* (in press).

Lewis, S.E., R.P. Erickson, L.B. Barnett, P. Venta, and R. Tashian. 1988. Ethylnitrosourea-induced null mutation at the mouse Car-2 locus: An animal model for human carbonic anhydrase II deficiency syndrome. *Proc. Natl. Acad. Sci.* **85:** 1962.

National Research Council, Committee on Chemical Environmental Mutagens. 1983. Identifying and estimating the genetic impact of chemical mutagens. National Academy Press, Washington D.C.

Office of Technology Assessments, U.S. Congress. 1986. Technologies for detecting heritable mutations in human beings, OTA-H-298. U.S. Government Printing Office, Washington, D.C.

Rhomberg, L., V.L. Dellarco, C. Siegel-Scott, K.L. Dearfield, L. Valcovic, and D. Jacobson-Kram. 1990. A quantitative estimation of the genetic risk associated with the induction of heritable translocations at low-dose exposure: Ethylene oxide as an example. *Environ. Mol. Mutagen.* (in press).

Russell, L.B. 1990. Patterns of mutational sensitivity to chemicals in poststemcell stages of mouse spermatogenesis. In *Mutation and the environment,* part C: *Somatic and heritable mutation, adduction, and epidemiology* (ed. M.L. Mendelsohn and R.J. Albertini). A.R. Liss, New York. (In press.)

Russell, L.B., J.W. Bangham, K.R. Stelzner, and P.R. Hunsicker. 1988. High frequency of mosaic mutants produced by N-ethyl-N-nitrosourea exposure of mouse zygotes. *Proc. Natl. Acad. Sci.* **85:** 9167.

Russell, L.B., P.R. Hunsicker, N.L.A. Cacheiro, J.W. Bangham, W.L. Russell, and M.D. Shelby. 1989. Chlorambucil effectively induces deletion mutations in mouse germ cells. *Proc. Natl. Acad. Sci.* **86:** 3704.

Russell, W.L., P.R. Hunsicker, G.D. Raymer, M.H. Steele, K.F. Stelzner, and H.M. Thompson. 1982. Dose-response curve for ethylnitrosourea-induced specific locus mutations in mouse spermatogonia. *Proc. Natl. Acad. Sci.* **79:** 3589.

Rutledge, J.C., K.T. Cain, J. Kyle, C.V. Cornett, N.L.A. Cacheiro, K. Witt, M.D. Shelby, and W.M. Generoso. 1990. Increased incidence of developmental anomalies among descendants of carriers of methylenebisacrylamide-induced balanced reciprocal translocations. *Mutat. Res.* (in press).

Skow, L.C., B.A. Burkhart, F.M. Johnson, R.A. Popp, D.M. Popp, S.Z. Goldberg, W.F. Anderson, L.B. Barnett, and S.E. Lewis 1983. A mouse model for β-thalassemia. *Cell* **34:** 104.

COMMENTS

Smithies: I had a question regarding the time of treatment of the adriamycin on the female [see Katoh et al. 1990]. You said it was 4 days prior to mating. When it's such a short time before mating, can you be sure that you're not just seeing some sort of toxicity that is going through it as distinct from a mutagenic effect, or at this point do you not distinguish that? If the treatment had been done, for example, 4 days after mating, what would have happened?

Bishop: Well, the ones that are done after mating are the ones that Generoso described, where you are in fact treating the zygote. In this particular case, the interest was to treat the egg before ovulation. The treatments were just 4 days prior to mating.

Smithies: What I really wondered is are any of the results the same or are they completely different when the chemical is tested before and after when it's such a short time before mating?

Generoso: In fact, that's the reason we did the experiment in this way, to see whether the embryonic effect is due to maternal toxicity or direct effect in the egg. So, to first prove that it's not maternal toxicity, we performed an egg transfer study with the adriamycin; the result is that the effect is not on maternal toxicity. Then we did further cytogenetic analysis on first cleavage, and we showed that the number of aberrations corresponded to the amount of lethality that we had observed. Jack [Bishop] has made the important point that from the beginning when we studied gametogenesis to the end, risk assessment, the methods that have been proposed by the EPA are all male-oriented. I think it's time for women.

Wyrobek: When you mate the heterozygous animals, some progeny will be normal translocation carriers, some will be heterozygotes balanced, and then some will be the unbalanced ones. Which ones show these morphological defects?

Bishop: As you know, a quarter of them would have a duplication deficiency of one chromosome, another quarter would have duplication of the other chromosome. In this case [see Cacheiro et al. 1990], it turns out with the 6 Gso, the ones with the frontal nasal dysplasia are the ones that, indeed, have the small loss of chromosome 19. They happen to have this big piece of Chr 15 stuck on there. The reciprocal of that, the animal that has a big loss of Chr 15 and a small piece of 19 stuck on it, doesn't survive long enough to show any kind of malformation. It's an early death, but it does contribute to the 50% loss that you would normally see among the progeny for that heterozygote.

Wyrobek: How many translocations have been looked at this way or screened?

Generoso: First we see them sporadically, and then when we have interesting ones, we look at them more carefully. Jack mentioned a total of about 200 in the program that were done with two different chemicals over the past year. Beginning with them, we are doing a more systematic study of the nature of the break, the homozygous effects, and the effect of unbalanced segregants from the heterozygotes. In addition to the partial trisomy that leads to these multiple matings, we are also screening them for possible trisomies that may result from a disjunction involving especially the small products of translocations.

Liane Russell: Do you have additional cases where unbalanced segregants had morphological effects?

Generoso: Yes [see Cacheiro et al. 1990].

The Importance of the Direct Method of Genetic Risk Estimation and Ways to Improve It

PAUL B. SELBY
Biology Division, Oak Ridge National Laboratory
Oak Ridge, Tennessee 37831-8077

OVERVIEW

The direct method of genetic risk estimation provides a much more straightforward way of estimating first-generation genetic risk in humans than does the doubling-dose method. Furthermore, committees involved in risk estimation have found that they cannot apply the doubling-dose method to estimate the risk from the large class of genetic disorders with complex etiology. With the direct method, data can be collected in such a way that serious mutations of all types are detected, including those with very low penetrance, which probably relate to many disorders with complex etiology. This goal can be met if the primary emphasis is given to identifying all first-generation offspring with clinically important phenotypes, in the body system being studied, and if the induced mutation frequency is calculated by subtracting the frequency of such offspring in the control group from that in the experimental group. The proposed assessment of dominant damage would provide the necessary information on skeletal damage for use in the direct method of genetic risk estimation as well as additional information for a more sensitive general test of the possible mutagenicity of the treatment. Because only a fraction of induced mutations cause clinically important effects in heterozygotes, it is essential, when estimating genetic risk, to know the health consequences of the mutations induced as well as the mutation frequency.

INTRODUCTION

W.L. Russell (this volume) describes the doubling-dose and direct methods of genetic risk estimation and discusses the history of their development, some of their problems, and possibilities for their improvement. Committees usually make estimates of genetic risk for both the first generation following exposure and for genetic equilibrium, which is assumed to be eventually reached if exposure is continued at a constant level over subsequent generations. The first-generation estimate seems more useful for addressing questions of immediate interest to society. Of the two methods, only the direct method is based on an estimate of the actual damage seen in the first generation. In sharp contrast, in order to estimate first-generation risk by the doubling-dose method, it is necessary to use several assumptions to estimate risk at genetic equilibrium and then to use additional assumptions to estimate first-generation risk from the equilibrium estimate.

Ideally, methods used for genetic risk estimation should apply to all important categories of genetic disorders, but they often do not approach doing so. For example, UNSCEAR (1986, 1988) felt that it was valid to apply the doubling-dose method to only 1.9% of the current incidence of genetic disorders in the human population. Almost all of the disorders *not covered* by UNSCEAR's doubling-dose estimate were those of complex etiology, which make up 97.6% of the current incidence used by UNSCEAR (1986).

An additional advantage of the direct method over the doubling-dose method is that its calculation does not require an estimate of the current incidence of genetic disorders in the human population. Such estimates have varied by more than sixfold over the last decade (UNSCEAR 1988) and are still in dispute.

In this paper, I suggest several ways in which the direct method of genetic risk estimation can be improved to give more reliable estimates of the amount of serious phenotypic damage expected in the F_1 generation following exposure of humans to mutagens. In view of the shortcomings mentioned for the doubling-dose method, it seems imperative that data be collected in such a way that all important categories of disorders are included. Disorders of complex etiology must not be overlooked. The mouse provides many examples of such disorders, probably the most important ones being serious disorders caused by dominant mutations with low penetrance. Such mutations have been found in mutation experiments designed to detect induced skeletal malformations (Selby and Selby 1977, 1978a) and cataracts (Kratochvilova 1981; Favor 1984).

DISCUSSION OF SUGGESTED IMPROVEMENTS IN THE DIRECT METHOD

Applications of Direct Method to Date

When the direct method of genetic risk estimation for radiation was first applied by a committee (UNSCEAR 1977), it was based on frequencies of induction of dominant skeletal mutations found in a large breeding-test experiment (Selby and Selby 1977, 1978a,b) and a smaller non-breeding-test experiment (Ehling 1966). Those two data sets are still used for risk estimation (UNSCEAR 1988), and breeding-test data on dominant cataract mutations have been added (UNSCEAR 1982). W.L. Russell (this volume) described the way in which the mutation frequency for the skeleton was extrapolated to risk for all body systems by UNSCEAR and the BEIR Committee (1980).

Two types of endpoints already used in collecting data for the direct method are presumed dominant mutations (Ehling 1966; Selby and Selby 1978b) and proved mutations (Selby and Selby 1977; Kratochvilova 1981), that is, those for which transmission of abnormalities has been demonstrated. For the skeleton, various morphological presumed-mutation criteria have been used to distinguish true mutants from nonmutational variants with a fairly high probability of being correct, and these are described in detail elsewhere (Selby 1990). As an example, F_1 mice are considered to be presumed dominant mutants if they have two or more rare (i.e., less that 1 in 400 in the control) and easily seen (with a dissecting microscope) skeletal malformations that do not seem to result from one accident of development. Use of such a criterion eliminates the need for an expensive breeding test, and data are not biased against finding those true mutations that, for various reasons discussed later, cannot be shown to transmit in a breeding test. The disadvantage of using such a criterion is that some malformed mice will not be counted as mutations because they have only one malformation or because all of their malformations are in close proximity. Some of those mutations could be recognized using a breeding test. Other experiments have relied exclusively on proved mutations (Kratochvilova 1981; Favor 1983). Breeding-test data are, of course, extremely useful in many ways and essential if the primary goal is to establish mutant stocks.

When our skeletal breeding-test data have been used for genetic risk estimation, committees (UNSCEAR 1977, 1988; BEIR 1980) have followed our suggestion (Selby and Selby 1977) to include the 6 presumed mutations, for which F_1 mice had no progeny, with the 31 proved mutations, many of which would still have been counted based on presumed-mutation criteria even if they had not been proved mutations (Selby and Selby 1978b). By doing so, they eliminated some of the shortcomings of using either endpoint alone.

Even though the direct method has many advantages and much potential, its experimental foundation is still at a rather early stage of development. It thus seems worth while to suggest some major improvements that could be made in it based on my experience in studying dominant mutations and in serving on committees involved in genetic risk estimation and on knowledge gleaned in preparing an extensive review (Selby 1990) of research on the induction of dominant mutations by radiation and chemicals.

The Most Important Type of Data Needed for the Direct Method

It is extremely important for risk estimation not to require proof of transmission of presumed mutations but to identify all first-generation offspring with clinically important phenotypes, in the body system being studied, and to calculate the induced mutation frequency by subtracting the frequency of such offspring in the control group from that in the experimental group. A contrary view, argued by Ehling (1984), is that proof of transmission is essential. He stated that "An excess of variants in the experimental group over the control group ... cannot be used ... for the estimation of the genetic risk ... ." In our response (Selby and Niemann 1984b) to his letter to the editor stating this, we pointed out that Ehling has not been consistent in his position because he has continued to use presumed mutations and to calculate the induced skeletal mutation rate by subtracting control from experimental when using his own skeletal data for risk estimation by the direct method. We also pointed out that both of our sets of morphological criteria used for identifying presumed mutations, to which Ehling objected, were strongly supported by breeding-test results.

When studying induction of dominant mutations in experimental mammals, the choice of the endpoint studied greatly influences the breadth of application of any risk estimate made using the direct method. As long as risk estimates are based on presumed mutations, proved mutations, or a combination of both, it will be unknown whether there is an important component of serious mutational damage that remains undetected. The only way to be sure of including all serious genetic disorders is to make the endpoint of primary importance *the frequency of mice with clinically serious malformations in the part of the body under study*, and then, of course, to subtract the control frequency from the experimental frequency.

The inadequacies of counting only mutations proved to transmit or of counting only variants meeting particular presumed mutation criteria are underlined by the following comparisons, which relate only to the part of the body being studied, for example, the skeleton or lens. If only proved mutations are counted, the following mutations are overlooked: those that cause sterility or death before breeding age (some of which would be among the most malformed), many of those that cause death before being completely tested, many of those with low penetrance, and many of those that are mosaics. If mice are counted as mutants only if they meet specific presumed-mutation criteria, all mutations would be overlooked that failed to cause malformations that meet those particular criteria. In sharp contrast, *no serious mutations would be overlooked* if all mice are counted that show clinically important malformations and if the induced mutation frequency is calculated by subtracting control from experimental.

For the skeleton, malformations would be found by careful examination of alizarin-stained preparations of intact skeletons of young adult mice using a dissecting microscope. It would probably be wise to have a group of clinical geneticists decide which mice had effects that would be likely to cause a serious handicap if the same damage occurred in a person. The mouse skeleton is well suited for making evaluations concerning the possible severity of particular malformations in people because of marked similarities in structure of many parts of the skeletons of the two species. Vast numbers of different kinds of skeletal malformations have been observed in both species, and there are many similarities regarding types of malformations. Large numbers of mutations are known to have incomplete penetrance in both species, and, as a result, many of the disorders in both species have complex etiology rather than simple Mendelian patterns of inheritance.

For chemically induced damage, it would seem straightforward to extrapolate from serious dominant effects in the skeleton to those in the entire body in the same way that committees (UNSCEAR 1977, 1988; BEIR 1980) have done this for radiation. Extrapolation from the experimental dose to the dose for the risk estimate would require information obtained using other methods.

Studies using serious malformations as the critical endpoint must give high priority to examining all animals in the sample, since some of the seriously malformed mice will die early and could be lost in the absence of close surveillance of the colony. For the skeleton, it is impractical to examine mice thoroughly that die at an early age. However, it seems reasonable to attempt to screen every F_1 mouse living at least 3 weeks.

Useful Ancillary Data for Evaluation of Genetic Risk

Many dominant mutations cause malformations that are indistinguishable from normal variation and probably cause no handicap. Such mutations can be used, however, to demonstrate that a particular treatment does induce dominant mutations. A high correlation seems likely between treatments that induce dominant mutations causing innocuous malformations and those that induce serious malformations. By combining these categories, a much larger number of mutations can be detected, thus making tests of mutagenicity more powerful. Endpoints to be described in this section could easily be incorporated into an experiment studying induction of serious skeletal mutations, and some of them could just as easily be applied to studies of induced damage in other body systems. If no significant increase was found for serious skeletal malformations in F_1 offspring, but there was strong indication that dominant mutations causing other effects were being induced, it would appear that the agent under study is a weak mutagen. Evaluation of these other endpoints could help to determine whether a larger study would be warranted to estimate genetic risk.

Skeletal mutation frequencies derived using presumed-mutation criteria could be used in this way. Another method shown to be effective for demonstrating induction of dominant mutations is the mutational-index method (Selby 1983; Selby and Niemann 1984a), which is based to a large extent on innocuous skeletal malformations.

Our results for ethylnitrosourea (ENU) (Selby et al. 1988) and results of Searle and Beechey (1986) and A.G. Searle (pers. comm.) for radiation suggest that strong mutagens induce high frequencies (as high as 7% for ENU) of mutations causing stunted growth. Although some of these mutations are undoubtedly serious clinically, it seems much more difficult to extrapolate to total damage from stunted growth than from serious skeletal malformations or cataracts. Besides providing extra information about whether dominant mutations are being induced, study of this endpoint, if broadened to include weighing of mice every week or so starting at weaning age, would improve the surveillance of the health of the animals and thereby help to ensure that no mice were lost between 3 weeks of age and the age at which their skeletons would be prepared.

Another experimental endpoint useful for demonstrating induction of serious dominant mutations is litter-size reduction (LSR). Comparison of the mean number of offspring weaned per litter at 3 weeks of age in the experimental and control groups provides an estimate of the frequency of induced mutations that cause death between conception and 3 weeks of age. Statistically significant LSR has been demonstrated for radiation (Selby and Russell 1985) and ENU (Russell and Hunsicker 1988). As mentioned earlier, it is impractical to examine mice routinely for skeletal malformations during the first few weeks of life. The estimate of LSR becomes especially useful for determining the extent of induced death in this period. We already have estimates of the LSR up to 3 weeks based on 14 experiments involving 158,490 F_1 litters from experiments with X- or gamma-irradiation (Selby and Russell 1985). Those data, together with data from smaller experiments of Lüning (1972) and A.G. Searle and D.G. Papworth (pers. comm.), provided the basis for an estimate of genetic risk of death in early life (UNSCEAR 1986), and it has been suggested that the estimate of early death could be combined with the risk estimate based on skeletal damage to provide a total estimate of induced dominant damage (Selby and Russell 1985).

By recording litter size at birth, LSR can be divided into that occurring before and after birth. If significant LSR is found, it might prove useful to carry out additional experiments, perhaps congenital-malformation (see Selby 1990) and dominant lethal experiments, to learn more about the time of induced death. Because the number of mice in a litter affects the extent of mortality, any estimate of LSR for agents causing extensive dominant lethality must be restricted to a comparison of death between birth and 3 weeks in litters containing the same number of progeny at birth.

Precautions Always Needed When Studying Induction of Dominant Mutations

For any endpoints that involve subjectivity, an obvious precaution is to code all offspring until they have been classified. It is essential that precautions also be taken to guard against misinterpretations caused by preexisting mutations. The importance of such precautions is underscored by our finding that in some skeletal experiments a high proportion of the presumed mutations were preexisting mutations. In one series of experiments, 78 of 107 presumed dominant mutations were thought to be preexisting mutations (Selby 1990). Almost all of the 78 were either one mutation with high penetrance for a specific syndrome occurring in one sibship in the control, or one of two widespread mutations with low penetrance and variable expressivity. Some carriers of one of these latter two mutations were extremely malformed. Despite the large numbers of preexisting mutations, almost all of our conclusions were the same both before and after correction for preexisting mutations, probably because we had randomized parents between the different groups (Selby 1990).

The experiments in which so many preexisting mutations were found involved crosses between two inbred lines. It has been well recognized for years (Haldane 1936) that inbred lines are not pure lines because of the occurrence of new spontaneous mutations. It would not be surprising for a spontaneous mutation occurring in one animal to become so widespread throughout a subsequently expanded colony that it would have a reasonably good chance of persisting for many generations. Corrections could easily be made for any such mutation with high penetrance for some distinct abnormality. However, if its penetrance was low and/or its expressivity was variable, such a mutation could easily confound results of experiments in which precautions were not taken. The cataract method has also provided a good example of a spontaneous mutation of this type in that the only control mutation found so far had penetrance of only 8% (Favor 1986). In view of the obvious need to take preexisting mutations into consideration when studying induction of dominant mutations, it is troubling that few papers on induction of dominant mutations make any mention of randomizing parents between experimental and control groups (Selby 1990).

Two of the most useful precautions for guarding against complications from preexisting mutations are as follows. Parents can be randomized between experimental and control groups to distribute any preexisting mutations between the groups as evenly as possible. In addition, parents can be selected from several (usually about six) somewhat distantly related sublines of the same inbred stock. If a mutation with low penetrance and/or variable expressivity is widespread in one of these sublines, the expected outcome is that many more mutations will be found than expected in the control, and both the control and experimental groups will have far too many malformation syndromes that seem to be variations on the same theme instead of showing the expected variety of different malformations. If the recurring malformation syndromes all trace back to one or the other of the distantly related sublines, it will be apparent that that particular subline contains a preexisting mutation causing it.

Whenever induced mutation frequencies are calculated, it is essential to assume that certain causes of the endpoint under study occur at the same frequency in both the experimental and control groups. If the endpoint is, for example, clinically important skeletal malformations, it must be assumed that (1) preexisting mutations, (2) new spontaneous mutations, and (3) nonmutational variants occur equally in both groups, and that, after the control is subtracted from the experimental group, only the induced frequency of true mutations is left. Even for proved mutations, it must be assumed that the first two of these causes occur equally in both groups. There would, of course, be much less reason to assume that the other causes are at equal frequency in both groups if the precautions mentioned in this section are not taken.

The Assessment of Dominant Damage

The improvements suggested in the above paragraphs will be incorporated in experiments in which F_1 offspring will be scored for the following types of genetic damage: (a) clinically serious and innocuous skeletal malformations, (b) stunted growth, (c) dominant visibles, and (d) litter-size reduction. Although dominant visibles would probably occur at much lower frequencies than the other three types of damage (Selby 1990), they would add some useful and easily attainable information. These experiments will be referred to as using the assessment of dominant damage (ADD) approach. An essential feature of this approach will be to estimate induced mutation frequencies by subtracting control frequencies from experimental frequencies of disorders appearing in the F_1 generation after exposure of the parental generation to the mutagen. In some cases, breeding tests of limited numbers of especially interesting variants, which showed their effects externally or in radiographs, will be performed to determine the nature of the mutations, for the obtaining of mutant stocks of interest as models of human disorders, and for other purposes.

Application of a Non-breeding-test Approach to the Cataract Test

In view of the strong position taken by Ehling (1984) and Favor (1986) as to the need for using a breeding test, it is instructive to mention a reanalysis of some cataract results in which a non-breeding-test approach is much more effective than a breeding-test approach. This reanalysis, described in detail elsewhere (Selby 1990), was of a 3.0 + 3.0 Gy X-ray experiment (Graw et al. 1986) in which the breeding test showed no statistically significant effect (all one-tailed p values >0.09). Our reanalysis of the same data, using a non-breeding-test approach, revealed clear-cut induction of dominant cataract mutations (both p values for spermatogonial exposure < 0.0001). As described in detail elsewhere (Selby 1990), another cataract experiment with X radiation, in which no effect was shown by the breeding test, was also reanalyzed in the same way and then demonstrated clear-cut induction of dominant cataract mutations. Very few cataracts found in the F_1 generation are shown to be transmitted. This is probably because of serious problems in the cataract breeding test that make it much less efficient than had been thought for detecting mutations with low penetrance (Selby 1990).

Accordingly, a non-breeding-test analysis should probably be made when presenting results from cataract experiments to guard against the danger of concluding that an effective mutagen is ineffective. It should be noted that one earlier cataract experiment using radiation did demonstrate statistically significant induction of dominant cataract mutations in stem-cell spermatogonia (Ehling et al. 1982); however, four other such experiments have not (Ehling et al. 1982; Graw et al. 1986). A disadvantage sometimes mentioned for the cataract test is that mutation frequencies are so low. However, the frequencies of induction of dominant cataracts were so high in some of our non-breeding-test analyses (Selby 1990) that it might even be worth while to add cataracts to the ADD approach. Although many cataracts detected using the non-breeding-test analysis probably have no clinical importance, they might be useful for providing additional information about whether dominant mutations are being induced.

My reanalysis provides a clear illustration of the disadvantage of insisting on proof of transmission even when trying to determine if dominant mutations are being induced, let alone the disadvantage already discussed of requiring proof of transmission for risk estimation.

GENERAL COMMENTS

It is now clear that there must be an astounding number of mutations present in many of the mice that we examine following certain treatments, and yet the great majority of them appear completely normal and have normal offspring. To illustrate, four weekly 100 mg/kg injections of ENU induce, on the average, a specific-locus mutation at one or the other of the seven specific loci in every 93 offspring derived from exposed spermatogonial stem cells (Hitotsumachi et al. 1985). From this, it appears that there are probably anywhere from about 10 to perhaps more than 100 mutations of similar type present in every F_1 mouse. Even so, the great majority of the F_1 mice survive to weaning, and fewer than 10% of those show any effects that appear to be serious. It thus becomes especially important to know whether induction of those types of mutations that are used to make decisions regarding human genetic risk has any correlation with induction of dominant mutations that cause clinically serious disorders. Very little is known about this at present.

By its very nature, the ADD approach is a multiple endpoint approach, and it could be modified to incorporate several other types of mutational assays, including protein or DNA methods, to provide many more sorts of useful information from the same mice. The comparisons that would then become possible between different endpoints could provide valuable information about the correlations between the results of different methodologies, including those useful for quantitative genetic risk estimation. Eventually, such information might lead to confidence that some more quickly applied test or tests could be used to estimate quantitative genetic risk from particular mutagens in humans.

ACKNOWLEDGMENTS

This research was sponsored by the Office of Health and Environmental Research, U.S. Department of Energy, under contract DE-AC05-84OR21400 with Martin

Marietta Energy Systems, Inc. The submitted manuscript has been authored by a contractor of the U.S. Government under contract DE-AC05-84OR21400. Accordingly, the U.S. Government retains a nonexclusive, royalty-free license to publish or reproduce the published form of this contribution, or allow others to do so, for U.S. Government purposes.

REFERENCES

Biological Effects of Ionizing Radiation Committee (BEIR). 1980. Genetic effects. In *The effects on populations of exposure to low levels of ionizing radiation*, p. 71. National Academy Press, National Research Council, Washington, D.C.

Ehling, U.H. 1966. Dominant mutations affecting the skeleton in offspring of X-irradiated male mice. *Genetics* **54**: 1381.

———. 1984. Variants and mutants. *Mutat. Res.* **127**: 189.

Ehling, U.H., J. Favor, J. Kratochvilova, and A. Neuhäuser-Klaus. 1982. Dominant cataract mutations and specific-locus mutations in mice induced by radiation or ethylnitrosourea. *Mutat. Res.* **92**: 181.

Favor, J. 1983. A comparison of the dominant cataract and recessive specific-locus mutation rates induced by treatment of male mice with ethylnitrosourea. *Mutat. Res.* **110**: 367.

———. 1984. Characterization of dominant cataract mutations in mice: Penetrance, fertility and homozygous viability of mutations recovered after 250 mg/kg ethylnitrosourea paternal treatment. *Genet. Res.* **44**: 183.

———. 1986. A comparison of the mutation rates to dominant and recessive alleles in germ cells of the mouse. In *Genetic toxicology of environmental chemicals*, part B: *Genetic effects and applied mutagenesis* (ed. C. Ramel et al.), p. 519. A.R. Liss, New York.

Graw, J., J. Favor, A. Neuhäuser-Klaus, and U.H. Ehling. 1986. Dominant cataract and recessive specific locus mutations in offspring of X-irradiated male mice. *Mutat. Res.* **159**: 47.

Haldane, J.B.S. 1936. The amount of heterozygosis to be expected in an approximately pure line. *J. Genet.* **32**: 375.

Hitotsumachi, S., D.A. Carpenter, and W.L. Russell. 1985. Dose-repetition increases the mutagenic effectiveness of N-ethyl-N-nitrosourea in mouse spermatogonia. *Proc. Natl. Acad. Sci.* **82**: 6619.

Kratochvilova, J. 1981. Dominant cataract mutations detected in offspring of gamma-irradiated male mice. *J. Hered.* **72**: 301.

Lüning, K.G. 1972. Studies of irradiated mouse populations. IV. Effects on productivity in the 7th to 18th generations. *Mutat. Res.* **14**: 331.

Russell, W.L. and P.R. Hunsicker. 1988. Dominant mutagenic effect of ENU on first-generation litter size in mice following treatment of spermatogonial stem cells. *Environ. Molec. Mutagen.* (suppl. 11) **11**: 90.

Searle, A.G. and C. Beechey. 1986. The role of dominant visibles in mutagenicity testing. In *Genetic toxicology of environmental chemicals*, part B: *Genetic effects and applied mutagenesis* (ed. C. Ramel et al.), p. 511. A.R. Liss, New York.

Selby, P.B. 1983. Applications in genetic risk estimation of data on the induction of dominant skeletal mutations in mice. In *Utilization of mammalian specific locus studies in hazard evaluation and estimation of genetic risk* (ed. F.J. de Serres and W. Sheridan), p. 191. Plenum Press, New York.

———. 1990. Experimental induction of dominant mutations in mammals by ionizing radiations and chemicals. In *Issues and reviews in teratology* (ed. H. Kalter), vol. 5, p. 181. Plenum Press, New York.

Selby, P.B. and S.L. Niemann. 1984a. Non-breeding-test methods for dominant skeletal mutations shown by ethylnitrosourea to be easily applicable to offspring examined in specific-locus experiments. *Mutat. Res.* **127**: 93.

———. 1984b. Response to "Variants and mutants" by U.H. Ehling. *Mutat. Res.* **127**: 191.

Selby, P.B. and W.L. Russell. 1985. First-generation litter-size reduction following irradiation of spermatogonial stem cells in mice and its use in risk estimation. *Environ. Mutagen.* **7**: 451.

Selby, P.B. and P.R. Selby. 1977. Gamma-ray-induced dominant mutations that cause skeletal abnormalities in mice. I. Plan, summary of results and discussion. *Mutat. Res.* **43**: 357.

———. 1978a. Gamma-ray-induced dominant mutations that cause skeletal abnormalities in mice. II. Description of proved mutations. *Mutat. Res.* **51**: 199.

———. 1978b. Gamma-ray-induced dominant mutations that cause skeletal abnormalities in mice. III. Description of presumed mutations. *Mutat. Res.* **50**: 341.

Selby, P.B., G.D. Raymer, and P.R. Hunsicker. 1988. High frequency of dominant mutations causing stunted growth is induced in spermatogonial stem cells by ENU. *Environ. Molec. Mutagen.* (suppl. 11) **11**: 93.

United Nations Scientific Committee on the Effects of Atomic Radiation (UNSCEAR). 1977. Committee report. In *Sources and effects of ionizing radiation*, p. 425. United Nations, New York.

———. 1982. Committee report. In *Ionizing radiation: Sources and biological effects*, p. 425. United Nations, New York.

———. 1986. Committee report. In *Genetic and somatic effects of ionizing radiation*, p. 27. United Nations, New York.

———. 1988. Committee report. In *Sources, effects and risks of ionizing radiation*, p. 375. United Nations, New York.

COMMENTS

William Russell: Paul has obviously started a controversy. Jack should be given equal time, which I'm afraid is impossible.

Favor: I'll make one or two comments. You have always said that breeding confirmations are sort of negative, but I would just say it another way around: The only way you can identify steriles, animals that die, animals with low penetrance, or mosaics, is if you have breeding information.

Selby: I agree that that's the only way you can be positive that a variant transmits. Some of what I have presented would suggest that we are paying an awfully high price to have proof of transmission.

Favor: No. If you impose a breeding test on your variants, you also have the variant information, and you can do anything you want with it. If you like to work with variants, that's fine. If we want to work with mutants, that's also fine. But we also provide information on variants.

Selby: I think it's obvious from my past work that I am also interested in getting data from breeding tests. With the ADD approach, there is the possibility of doing all kinds of things with that, for instance, making radiographs, picking up some variants, and doing breeding tests on those.

Lewis: We also have preexisting mutations. How do you identify your preexisting mutations in your system? I'm not at all clear about it.

Selby: The basic clue that you have a preexisting mutation is if you have something that is extremely stable in your experimental system, and suddenly you start seeing many modifications of it that are very similar. Then you can trace back and see where they came from in the pedigrees of your animals.

Wyrobek: If the skeletal abnormalities you see have low penetrance, and if one just thinks probabilistically, there might exist among those F_1 animals some animals that have those mutations. And then, just normal-looking animals would carry those mutations—low penetrance, that animal didn't express it.

Selby: I agree with you. I used to worry about that.

Wyrobek: Can you make a correction factor?

Selby: I used to worry about that until it occurred to me that if you are looking at first generation damage, you're really not concerned with those mice that don't express abnormalities. It appears from ENU studies now, in which 1 out of every 93 animals has a mutation at just those 7 loci, that animals must be filled with specific-locus type mutations scattered throughout the genome. Despite that, the great majority of the animals appear to be completely normal and have normal offspring. If we were to use some of the molecular techniques, we would probably see vastly more types of damage than that. I think one of the huge problems that we have to face in years to come is to try to tie together those tremendous possibilities we have with molecular techniques and build bridges from them to what the damage is. It may require methods such as I have described to make that tie. I'm not sure what we can try to do with the human population, because as far as genetic effects are concerned in the first generation, no one has even been able to see any transmissible genetic effects in Hiroshima and Nagasaki after 40 years of intensive study—which is consistent with the low-level effect we predict from the risk estimates based on the mouse skeleton. That's a major problem. What we have to do is try to figure out ways of getting from measures of molecular damage to what matters—what causes critical health problems in the first generation.

William Russell: There will be available in January a 74-page review of all the studies on dominant effects, and some of what he has said will be in that. It's a very thorough analysis of all the dominant effects that have been studied up to the date when he wrote it. Unfortunately, it's in a book and it has taken two years to come out, so it is not quite as up-to-date as we would like, but that's not our fault.

Selby: It will be in *Reviews in Teratology,* Volume 5.

William Russell: Unfortunately, it's in a teratology book, so I'm afraid unless we draw attention to it, it might be missed. As you can see, it is still a lively field and there are still a lot of problems in risk estimation.

Aspects of Germ Cells Relevant to Mutagenic Risk Evaluations: Some Concluding Remarks

BRYN A. BRIDGES
MRC Cell Mutation Unit
University of Sussex
Falmer, Brighton, United Kingdom

As was expected, this meeting has revealed a wealth of information about the physiology and genetics of mammalian germ cells. In my concluding remarks, however, I want to pass over much excellent and stimulating work and pick out only a few points. All of them are relevant to the question posed at the beginning of the meeting, namely, what are the specific characteristics of germ cells (distinguishing them from somatic cells) that have to be taken into account when attempting to estimate risks arising from exposure to mutagens.

Perhaps the most common fallacy in this area is the myth of the blood-testis barrier. It was apparent from L.D. Russell's talk that the real barrier in the testis is not the blood-testis interface, but is composed of the junctions of Sertoli cells. Beyond this barrier, remote from the lymph, are all but the earliest spermatocytes, and all the spermatids and spermatozoa. The differentiating spermatogonia and stem cells, however, are outside this barrier and are essentially bathed by lymph. Thus, the barrier can only be considered to be effective for spermatocytes and subsequent stages. Even so, there is little evidence that variations of different germ-cell stages in sensitivity to mutagens are due to the Sertoli cell barrier. In the specific-locus test, every mutagen that mutates spermatogonia has, when tested, also been found to mutate one or more of the later (supposedly protected) stages. One is tempted to conclude that screening the male germ cells from chemical attack may not be a strategy employed to any great extent for protecting the mouse germ line from mutagenesis. Indeed, the spermatogonia (including stem cells), although unprotected by the Sertoli cells, are generally the germ-cell stages most resistant to induction of inherited damage.

There is often no more telling evidence to the vulnerability of mammalian germ cells to chemicals than the effects observed in the synaptonemal complexes described by Allen et al. and Moses et al. Synaptonemal complexes are unique to meiotic cells and may represent targets, as well as analytical endpoints, for mutagen effects on germ cells—effects that are not a concern for somatic cells. Many types of damage seen when synaptonemal complexes are visualized are similar to those seen in condensed spermatocyte chromosomes, but the sensitivity for revealing damage

All references cited without a date refer to papers in this volume.

is often much higher. It would appear, therefore, that mechanisms must exist for the elimination of much of this damage in meiotic cells. Whether this is repair or (more likely) selective elimination is a topic that needs to be addressed.

A feature that distinguishes spermatozoa and late spermatids from somatic cells is substitution of a protamine-like basic protein for histone as the protein component of the chromosomes. Sega noted that the period during which protamine was present coincides with the period of peak sensitivity for dominant lethal and heritable translocation induction by small alkylating agents. It also coincides with a period during which these agents bound most strongly to germ cells, binding which is chiefly with protamine. Sega argued persuasively that protamine (particularly its cysteines) might be the target for the chromosomal damage and might represent a unique target peculiar to germ cells. I call to mind the excellent correlation between the number of pairs of brooding storks and the number of newborn babies in West Germany over the period of 1965 and 1980 (Sies 1988). Any child could draw from this a logical strategy to be followed to achieve any desired birthrate.

I suggest that we should be careful about accepting the conclusion that alkylation of protamine is responsible for the chromosomal damage seen when late spermatids and spermatozoa are exposed to small alkylating agents. We know that DNA is the mutagenic target for such agents in most cells and that DNA is alkylated in late spermatids and spermatozoa. We also know that DNA repair is likely shut down during this period; indeed, this may be in part due to the presence of protamine. As far as I can see, the available data are as consistent with DNA being the target as with protamine. Fortunately, it should be possible to distinguish between these hypotheses experimentally. There are a number of chemicals capable of reacting with cysteines in proteins that leave DNA unscathed, iodoacetamide and N-ethylmaleimide being two of the most avid thiol reagents. Compounds such as these (or perhaps others less avid and less toxic) should, if protamine is indeed an important target for chromosome damage, be extremely effective in inducing dominant lethals and heritable translocations in late spermatids and spermatozoa.

Before turning to earlier germ-cell stages, this may be an appropriate place to mention that the final germ-cell stage is the fertilized egg. Before a functionally diploid zygote exists, there is a period during which the sperm genome has to be reactivated and during which many things may happen.

As far as mutagenesis is concerned, spermatogonia appear to be more resistant than other germ-cell stages, but there is much still to be learned about their response to mutagens. There was something unexpected about the findings of de Rooij et al. that gonia that were cycling the slowest were the most sensitive to the lethal effect of ionizing radiation. This contrasts not only with the earlier conclusions from Oakberg's laboratory, but also with our own experience with primary human fibroblasts in culture where slowing cell cycle progression results in a substantial increase in resistance. It would be interesting to know about the sensitivity to chromosomal damage in spermatogonia as a function of cell cycle time. Does the hypersensitivity in slowing cycling cells reflect a mechanism for the elimination of chromosomal damage rather than its repair?

Finally, I would like to share some thoughts on the action of ethylnitrosourea (ENU), by now the best-studied mutagen in mouse germ cells. The specific-locus test in the mouse has from time to time been criticized as using unrepresentative and abnormally mutable loci. It is therefore of major importance for risk estimation that McDonald's three-generation system (McDonald et al.) for screening for forward mutations at the hph locus, besides being a model for human disease, has revealed a rate of ENU-induced mutation that is approximately double that expected on the basis of the specific-locus test. It is also becoming clear that the point mutations induced in germ cells by ENU are largely or all at A:T base-pair sites. This contrasts with the data from cellular systems, where the changes are all at G:C base pairs due to the mispairing potential of O^6-alkyl guanine. To achieve a preponderance of changes at A:T sites, it is necessary to explain both why there are no changes at G:C sites and how the changes at A:T sites arise. Extremely effective repair of O^6 alkylations might explain the former, and relative poor repair of damage at A or T sites is a necessary precondition for the latter. In fact, all the data shown, persistence of the premutational lesion through several replication cycles and mutational mosaicism, are consistent with a lesion at an A or a T that is not detected by error-free repair systems.

A clue to the way in which the mutations arise may come from the apparent threshold in the dose response curve. Given the time that is available for regeneration of repair systems following exposure, this threshold does not seem likely to be due to saturation of error-free repair. The possibility seems worth considering that it is due to a requirement for a critical amount of alkylation in order to induce an error-prone system. In *Salmonella*, mutagenesis at A:T sites by ENU is promoted by the *mucA,B* operon, which specifies an inducible error-prone system (Zielenska et al. 1988). Could a similar system be functional in spermatogonia? If so, it does present a problem in understanding why. In bacteria, the argument has been made that error-prone systems may be induced when a population is threatened with chemical or other stress in order to generate variability in the population, some members of which may thus become better able to survive. I do not find such an argument attractive for male germ cells, for which a better strategy would seem to be to let damaged DNA "die" if it cannot be repaired in an error-free way.

Clearly, further work is required on this topic. The first step might be to identify the lesion at A:T sites. Experiments with single-stranded viruses in bacteria could be used to decide whether the premutagenic lesion is on the A or the T. Treating ENU-exposed DNA to excess O^6-alkyltransferase could enrich in the A:T lesion and possibly permit the preparation of a specific antibody.

There may, in fact, be lessons from the ENU data of more general relevance to risk estimation. It may be that the greatest risk as far as gene mutations in germ cells are concerned comes from DNA lesions that, perhaps because they cause minimal distortion of the helix, are not seen by error-free repair systems. It may also be that an inducible error-prone system needs to be activated in order for mutations to be formed.

REFERENCES

Sies, H. 1988. A new parameter for sex education. *Nature* **332**: 495.

Zielenska, M., D. Beranek, and J.B. Guttenplan. 1988. Different mutational profiles induced by *N*-nitroso-*N*-ethylurea: Effects of dose and error-prone DNA repair and correlations with DNA adducts. *Environ. Mol. Mutagen.* **11**: 473.

Author Index

Adler, I.-D., 115
Allen, J.W., 133, 155
Andreadis, D.K., 377
Arnheim, N., 363
Aronson, J., 311
Auletta, A.E., 413

Backer, L.C., 133, 155
Balhorn, R., 93
Ball, S.T., 247
Barker, P.E., 377
Barnett, L.B., 237
Beatty, B.R., 377
Beechey, C., 209
Bishop, J.B., 425
Brandriff, B.F., 183
Bridges, B.A., 451

Cattanach, B.M., 209
Chang, A.J., 377
Cimino, M.C., 413
Clegg, J.B., 247
Comings, D.E., 351
Currie, M., 93

Davisson, M.T., 195
Dellarco, V.L., 397
DeLoia, J., 293
de Rooij, D.G., 35

Dove, W.F., 259
Dresser, M.E., 171

Favor, J., 221
Fritz, I.B., 19

Generoso, W.M., 311
Gibson, J.B., 133
Gordon, L.A., 183

Handel, M.A., 51
Hecht, N.B., 67
Hunsicker, P.R., 271

Jones, J., 247
Judd, S.A., 335

Koller, B.H., 321
Kovacs, B.W., 351

Lewis, S.E., 195, 237

McDonald, J.D., 259
McKinney, W.L., 377
Mohrenweiser, H.W., 335
Moses, M.J., 133, 155

Perry, B.A., 335
Peters, J., 247
Poorman-Allen, P., 133, 155

Popp, R.A., 237

Rasberry, C., 209
Rhomberg, L.R., 397
Rinchik, E.M., 271
Russell, L.B., 271
Russell, L.D., 3
Russell, W.L., 271, 385
Rutledge, J.C., 311

Sega, G.A., 79
Selby, P.B., 437
Shahbahrami, B., 351
Shedlovsky, A., 259
Shelby, M.D., 425
Smithies, O., 321
Solter, D., 293
Stanker, L.H., 93
Stilwell, J.L., 93

Tepperberg, J.H., 133

van Buul, P.P.W., 35
van der Meer, Y., 35
van Pelt, A.M.M., 35

Westbrook-Collins, B., 155
Wiley, L.M., 299
Woychik, R.P., 377
Wyrobek, A.J., 93

Subject Index

AA. *See* Acrylamide
2-Acetylaminofluorene, 428
4-Acetylaminofluorene, 428
Acrylamide (AA), 79–84, 86, 89, 115–131, 273–274, 280, 407, 411, 423, 427–430
Actinomycin D, 152
Adriamycin (ADR), 95, 273, 427–428, 430, 434
Alkaline elution, 80, 87, 97
Alkaloids, anti-tubulin, 157. *See also* Colcemid, Colchicine, Vinblastine, and Vincristine
O^4-Alkyladenine, 255
Alkylating agents, 157. *See also* Cyclophosphamide and Mitomycin C
O^6-Alkylguanine, 255
O^6-Alkyltransferase, 453
Amplification refractory mutation system (ARM), 248, 251, 255
Amsacrine (*m*-AMSA), 155, 157–158, 160, 164
Androgen, 26–27
Aneuploidy, 151, 156, 425, 432
  in chimeras, 305
  in humans, 96–97, 183, 185–186
  in mouse, 317–318
  in yeast, 172
Angelman syndrome (AS), 296
Antimitotic agents (AMAs), 133, 143–145, 156. *See also* Colcemid, Colchicine, Vinblastine, and Vincristine
Armenian hamsters, 157, 159–160, 162–164, 167
Aromatase, 26

B-Nerve growth factor (β-NGF), 20
β-Globin, 342–343, 365
β-Hemoglobin, 365
β2-Microglobulin (β2m), 322, 324, 327
β-Thalassemia, 429
Basic fibroblast growth factor (bFGF), 20
Benzo(a)pyrene, 428
Bleomycin, 152–153, 157–160, 164, 169
Bloom's syndrome, 360
Bromodeoxyuridine (BU), 137, 188

Caffeine, 428
Camptothecin, 164, 174
Cancer, 393, 402–403
  chemotherapy, 190, 193
Carbonic anhydrase deficiency syndrome, 429
Cataracts, 390–391, 431, 438, 443–445
Chinese hamsters, 139
Chlorambucil (CHL), 187, 271, 273–277, 280, 427–428

Chloramphenicol acetyltransferase (CAT) gene, 379
Chlormethine, 428
cis-platinum, 14
Colcemid, 143
Colchicine, 133, 143–147, 150–153, 155, 157–158, 160–161, 163
CP. See Cyclophosphamide
Crossing-over, 134, 141, 143, 156, 161, 163–164, 175
Cyclophosphamide (CP) (CPP), 84, 95, 133, 141–144, 155, 157–161, 273–274, 280, 427–428
Cytochalasin, 14
Cytokinesis, 20

DCBP. See Dibromochloropropane
Denaturing gradient gel electrophoresis (DGGE), 335, 337, 341–345
Dibromochloropropane (DBCP), 94, 421–422, 427–428
Diethyl sulfate (DES), 273–274, 280, 312–316, 318, 427–428
Dihydropteridine reductase (DHPR), 260–261
Disomy 1, 185
Drosophila
 chromosomal nondisjunction, 156
 ENU-induced mutations, 232–233, 254
 gene expression, 57–58
 gene loci, 386
 mosaic mutants, 240–241
 sex-linked recessive lethal assay, 415–416
 X-Y pairing, 52
Duchenne muscular dystrophy, 336

Embryonic stem (ES) cells, 321, 323, 325–330
EMS. See Ethyl methanesulfonate
ENU. See Ethylnitrosourea
Escherichia coli (*E. coli*), 229
Ethidium bromide, 368, 371
1-Ethyladenine, 229
Ethylenebisacrylamide, 428
Ethylene dibromide, 427–428

Ethylene oxide (EtO), 79–86, 198, 200, 202, 312–316, 318, 397–407, 411, 427–430
O^6-Ethylguanine, 221, 228–229
N-Ethylmaleimide, 452
Ethyl methanesulfonate (EMS), 79–86, 89–90, 115, 273–274, 280, 312–316, 318, 405–406, 411, 427–429
Ethylnitrosourea (ENU), 95, 127, 197, 199–200, 221–235, 237–245, 247–257, 259–260, 262–263, 266, 273–275, 277–281, 286–289, 312–315, 318, 425, 427–429, 441–442, 445, 448, 453
O^4-Ethylthymine, 229
EtO. See Ethylene oxide
Extracellular matrix (ECM) proteins, 20, 28–29

Follicle stimulating hormone (FSH), 21, 26, 29, 33

Gamma irradiation, 155, 157, 159–160, 163, 169, 196–197, 286, 305, 359
Gap junctions, 33
Genetic imprinting, 293–298
Genetic risk estimation, 430–432
Glucose phosphate isomerase-1 gene (*Gpi-1*), 248–249, 256–257
Golden hamsters, 184–185, 190, 192–193
Growth factors, 20, 26–27
GTP-cyclohydrolase (GTP–CH), 259–260, 263

Hemoglobin α-chain gene (*Hba*), 248, 250, 253–254
Hemoglobin β-chain gene (*Hbb*), 248, 250–254
Hemophilia A, 336
Hexamethylphosphoramide, 427–428
Histones, 67, 81, 85, 98
HOP1 gene, 176, 181
Human chorionic gonadotropin (HCG), 183
Hycanthone, 428
Hydroxyurea, 44, 216

Hyperploidy, 151

Inhibin, 20
Insulin, 13
Insulin-like growth factor, type 1 (IGF-1), 20
Interleukin-1 (IL-I), 20
In vitro fertilization, 184–185
In vitro transcription system, 69
Iodoacetamide, 452
Ionizing radiation, 32, 94, 134, 156, 162, 209, 276, 299–310, 385–386, 388, 441–442, 452
   doubling-dose method, 386–387, 391–392, 437–438
Irradiation, neutron, 35–49, 197–199, 388
Isopropyl methanesulfonate (IPMS), 115

Lipoprotein receptor (LDLr), 366–368, 371
Luteinizing hormone (LH), 4, 20–21, 26

Mannose 6-phosphate, 15
MC. *See* Mitomycin C
Melphalan (MLP), 273–275, 280, 427–428
6-Mercaptopurine (6MP), 273–274, 427–428
Methylene-*bis*-acrylamide, 408, 428–429
Methyl methanesulfonate (MMS), 79–86, 89, 115, 127, 274, 280, 312–316, 427–428
Methylnitrosourea (MNU), 272–275, 277–283, 287, 427–428
Micronucleus test, 116
Mitomen, 84
Mitomycin C (MC) (MMC), 117, 127, 133, 141–144, 157, 164, 169, 278, 428
MMS. *See* Methyl methanesulfonate
MNU. *See* Methylnitrosourea
Myleran, 428

Neisseria gonorrhoeae, 359
Nocodazole, 143, 429

Osteogenesis imperfecta, 336

PCR. *See* Polymerase chain reaction
Phenylalanine (PHE), 259–270
   hyperphenylalaninemia (HPH), 259–265, 268, 270
   phenylalanine hydroxylase (PAH), 259–262, 270
Phenylketonuria (PKU), 259, 263, 266, 268
Phytohemagglutinin (PHA), 301
Plasma membrane proteins, 20
Platinol (PLA), 273, 427–428, 430
Polymerase chain reaction (PCR), 93, 101, 251, 322–324, 342–344, 351, 358, 363–376
Prader-Willi syndrome (PWS), 295–296
Procarbazine (PRC), 127, 273–275, 277–278, 280, 423, 428
Pronuclear transfer, 294, 313, 315
n-Propyl methanesulfonate (PMS), 115
Protamine(s), 79, 81–85, 87, 91, 129–130, 194, 452
   antibodies, 93, 99–105
   arginine, 82, 98
   bull, 101–102, 110
   cysteine, 82, 84, 86, 89, 98, 130, 452
   human protamine 1 (P1), 99–102, 111
   human protamine 2 (P2), 99–102
   mouse protamine 1 (mP1), 67–71, 74, 89, 98–99, 111
   mouse protamine 2 (mP2), 67–74, 89, 98–99
   mouse transition protein 1 (mTP1), 67–69, 71, 74
Proto-oncogenes, 67
Pyrene, 428

RAD50 gene, 175–176
RAD52 gene, 175
Radiomimetic agent, 155, 157–158. *See also* Bleomycin
Restriction enzymes
   *Bam*HI, 337–338, 340
   *Eco*RI, 337–340
Restriction enzyme site (RES), 335, 337–338, 344
Restriction fragment length polymorphisms (RFLPs), 337, 366
Retinoblastoma, 297
Robertsonian translocation, 116, 296, 387

Salmonella, 453
  Salmonella (Ames) assay, 415–416, 423
  Salmonella typhimurium, 229
SC. *See* Synaptonemal complex
Seminiferous growth factor (SGF), 20
Sex body (SB). *See* XY body
Sexual dimorphism, 52–53, 57–60
Sex vesicle. *See* XY body
Short tandem repeats (STRs), 351–360
Sister chromatid exchange (SCE), 360
Skeletal fusions with sterility (*sks*), 51, 53–55, 57–58, 63–64
Skeletal malformations, 438–442
SPO11 gene, 176, 182
Spontaneous abortion, 183–185, 190
Steroidogenesis, 20
Sulfogalactoglycerolipid, 306
Synaptonemal complex (SC), 54–55, 57, 64, 116–118, 133–153, 155–169, 174–178, 181, 451
  axial element, 136–137, 142–144, 150
  central element, 136, 173
  lateral element, 136–137, 156, 160, 162–164, 168, 173, 176
  asynapsis, 157, 160, 163
  foldbacks, 144, 157, 160, 164
  mispairing, 157, 160, 163–164

TEM. *See* Triethylenemelamine
Testis, 19–34
  adluminal compartment, 4, 12–14, 19–20, 25, 32–33, 152, 411
  basal compartment, 4, 19–20, 25, 273, 411
  blood-testis barrier, 451
  epididymis, 400, 408–410
  interstitium, 19, 23
    Leydig cells, 5, 19–20, 23, 25–27, 91
    lymphatic system, 5–6
  rete testis, 23–25
  seminiferous epithelium, 22, 35, 37–43, 94
  seminiferous tubule, 3–17, 19–20, 23–24, 28, 273, 400
    basal lamina, 3–4, 6
    basement membrane, 6, 8, 20, 28
    endothelial cells, 3, 5
    micropinocytotic vesicles, 3, 5–8

  myoid cell, 4, 6–8, 20, 22–23, 25–26, 28
  Sertoli cell, 4, 9–10, 17, 19–20, 22–29, 38, 62–63, 152, 410–411, 451
  Sertoli cell barrier, 9, 11–16, 32–34, 152, 274, 288, 410, 451
  tunica albuginea, 19, 23
  vasculature-testis barrier, 5
Testosterone, 21
Tetrahydrobiopterin (BH_4), 260–262
Tetrahydrocannabinol, 428
Tetramethylammonium chloride, 354
Topoisomerase I inhibitor, 174. *See also* Camptothecin
Topoisomerase II inhibitor, 155, 157. *See also* Amsacrine
Toxic Substances Control Act (TSCA), 413–415, 420, 422
Tracer, testicular, 5, 8, 12–14
Transferrin (Trf), 10, 26, 29, 32–34, 248
Transforming growth factor type alpha (TGF-α), 20
Transforming growth factor type beta (TGF-β), 20, 26
Triethylenemelamine (TEM), 168–169, 175, 198, 205, 272–275, 277–278, 280, 287, 312–315, 318, 405–406, 427–428
Trillium, 135
Trimethyl phosphate, 408
Trisomy, 156, 184–185, 318, 387, 435
Tritiated thymidine ([^{3}H]dThd) (^{3}H-Tdr), 36–37, 40–41, 80–81, 84–85, 137–139, 141–143, 145–146, 161, 300–302
Tubulin, 128, 144–145, 307–308
Tyrosine, 260

Urethane (UR), 273–274, 427–428

Vinblastine (VS), 133, 143, 145, 157
Vincristine, 428
Vitamin A, 35, 37–39, 43

Weibull modeling, 402–406

X-irradiation, 196–199, 302
X-ray(s), 84, 134, 209–219, 241, 286, 301, 312–316, 388, 429, 444
  hydroxyurea-X-ray, 44, 216
XY body, 52–60, 65, 137–138

Yeast
  Saccharomyces cerevisiae, 171–182
  Schizosaccharomyces pombe, 176
Yeast artificial chromosomes (YACs), 178